B. Maidl, L. Schmid, W. Ritz,
M. Herrenknecht

Hardrock Tunnel Boring Machines

Bernhard Maidl, Leonhard Schmid,
Willy Ritz, Martin Herrenknecht

Hardrock Tunnel Boring Machines

In co-operation with
Gerhard Wehrmeyer and Marcus Derbort

o. Prof. Dr.-Ing. Dr. h. c. mult. Bernhard Maidl
Ruhr-Universität Bochum
Fakultät Bauingenieurwesen
Lehrstuhl für Bauverfahrenstechnik, Tunnelbau
und Baubetrieb
44780 Bochum

Cover picture: Gripper TBM for the Gotthard Base Tunnel, Switzerland

Bibliographic information published by the Deutsche Nationalbibliothek
Die Deutsche Nationalbibliothek lists this publication in the Deutsche Nationalbibliografie;
detailed bibliographic data are available in the Internet at <http://dnb.ddb.de>.

ISBN 978-3-433-01676-3

© 2008 Ernst & Sohn Verlag für Architektur und technische Wissenschaften GmbH und Co.KG, Berlin

All rights reserved, especially those of translation into other languages. No part of this book shall be reproduced in any form – i.e. by photocopying, microphotography, or any other process – or be rendered or translated into a language useable by machines, especially data processing machines, without the written permission of the publisher.

Registered names, trademarks, etc. used in this book, even when not specifically marked as such, are not to be considered unprotected by law.

Typesetting: ProSatz Unger, Weinheim
Printing: Strauss GmbH, Mörlenbach
Binding: Litges & Dopf GmbH, Heppenheim

Printed in Germany

Preface

> "... I have discovered methods for tunnels and engulfed secret ways, which are excavated without any noise, in order to reach predetermined locations, even if they have to be dug under ditches or under a river."
>
> "... I have noiseless methods to dig tunnels and winding secrete catacombs in order to reach a pre-planned place, even if they have to be built underneath ditches and rivers."
>
> (Leonardo da Vinci, 1452–1519)

Leonardo da Vinci is considered to be one of the greatest inventors of all times. His revolutionary and futuristic construction plans e.g. in aviation and in the construction of canals and bridges were ridiculed during his time. Many of his discoveries were later 're-invented' at times when they were actually necessary.

A horizontal drilling machine cannot be found among Leonardo's known and published sketches, but he elaborated extensively about their usage and advantages in the construction of shafts. In his writings one can find descriptions of the method of drilling horizontal or vertical shafts, which means that he had already plans for useful applications in this field before Agricula, Brunel and all the following patent holders.

The drilling of tunnels has a long history. First patents were already distributed in the beginning of the 19th Century, but it took almost another century until a similar machine had drilled a longer distance through mountains. There were several reasons for this delay. Among them were ineffective mining tools, which were too soft, but above all it was the lack of sufficient energy sources at the workplace. In those days movable steam engines, even hydraulic devices for blasting rock formations were in use, but those techniques were not effective enough to dig a tunnel through an entire mountain.

Nowadays TBM tunnelling gains more and more significance in hard rock formations even with larger diameters. The number of substantially longer tunnels as well in road and rail traffic, as in the areas of supply and waste are constantly on the rise; not only are those projects worth mentioning like the Alpine tunnels that are in process or in developmental stages, but also those subterranean tunnels underneath straits, which will be realized in the future. Those procedures are still considered as spectacular. Bernard Kellermann wrote about a similar project in his utopian best seller 'The Tunnel' (published in 1921), describing the construction of a railway tunnel connecting Europe with the North American continent by means of four drills.

The varying usage of open tunnel drills with grippers will be extended from that for smaller or medium diameters to larger, which will be effective also under mixed geological circumstances. Those mixed geological circumstances have already led to an enormous push in the development of the shield machines and prefabricated tunnel linings. Open hard rock machinery has no or only a short protective shield and as a result, its usage is rather limited under difficult geological circumstances. Currently there is a development of open hard rock machinery on the way, with the main focus on securing the area behind the drill head and advanced safety. The safety systems, with shotcrete, anchors and plates, which where developed for the traditional construction of tunnels are not of much use for continuous boring with Gripper-TBMs. Still, one cannot foresee if there are modifications of existing procedures possible or if one has to find completely new ways of dealing with these problems. In addition there is further demand for refining the use of tunnel bore machines in high altitude with high pressure.

The goal of the authors, Maidl, Schmid, Ritz and Herrenknecht was to accumulate existing knowledge and experience, to define requirements of use and to present as well as to encourage potential developments. This endeavour requires an interaction of Science and Practice at the highest level.

I am grateful to the entire team of collaborators, primarily, Leonhard Schmid, without whose contribution this book would have never been written, as well as, Martin Herrenknecht and Willy Ritz for their work on specific chapters of this book. Hereby I also acknowledge the support of the associates of my co-authors and the companies for providing the most current information. Furthermore I like to express my gratitude to Gerhard Wehrmeyer and Marcus Derbort for their coordination and detailed analyses, as well as the collaborators from the chair, who were involved in the writing of additional articles, Ahmed Karroum and Volker Stein. Also many thanks to the staff members from the engineering office of Maidl/Maidl, Ulrich Maidl and Matteo Ortu and my technical assistant, Helmut Schmid for his expert draftsman ship, Christian Drescher for his typing and my secretaries, Brigitte Wagner and Ruth Wucherpfennig, for their varied assistance and also to Gerhard Wehrmeyer.

February 2008   *Bernhard Maidl*

Contents

1	**Historical Development and Future Challenges**	1
2	**Basic Principles and Definitions**	15
2.1	Basic Principles and Construction	16
2.1.1	Boring System	16
2.1.2	Thrust and Clamping System	17
2.1.3	Muck Removal System	18
2.1.4	Support System	18
2.2	Definitions and Terms	20
2.2.1	Tunnel Boring Machines with Full-Face Excavation	20
2.2.1.1	Gripper TBM	20
2.2.1.2	Shielded TBM	22
2.2.2	Tunnel Boring Machines for Partial Excavation	23
3	**Boring Operation**	25
3.1	The Boring Process	25
3.2	The Cutter Head	26
3.2.1	Shape of the Cutter Head	27
3.2.2	Clearing the Muck in the Excavation Area	29
3.2.3	Cutter Head Construction and Soil Consolidation	31
3.3	Cutting Tools	32
3.3.1	General	32
3.3.2	Working Method of Cutter Discs	33
3.3.3	Cutter Spacing	36
3.3.4	Penetration	38
3.3.5	Wear	41
3.3.6	Wear and Water	48
3.3.7	Cutter Housing	49
3.4	The Main Drive	50
3.4.1	Types of Main Drive	50
3.4.2	Main Bearing	53
3.5	Advance Rate	53
3.6	Special Types	55
3.6.1	Reamer TBMs	56
3.6.2	Bouygues System	57
3.6.3	Mobile Miner (Robbins)	58
3.6.4	Back-Cutting Technology	58
3.6.4.1	Mini-Fullfacer (Atlas Copco)	58
3.6.4.2	Continuous Miner	60
3.6.5	Shaft Sinking	61
3.6.5.1	Raise Boring	62

3.6.5.2	Blind Drilling	64
3.6.5.3	Combinations	65
4	**Thrust**	67
4.1	General	67
4.2	Advance with Gripper Clamping	67
4.3	Advance with a Shield TBM	72
5	**Material Transport**	75
5.1	Material Transport at the Machine	75
5.2	Material Transport Through the Tunnel	77
5.2.1	Rail Transport	77
5.2.1.1	Diesel or Electric Operation	79
5.2.1.2	Muck Cars	79
5.2.1.3	Loading in the Tunnel	80
5.2.1.4	Train Timetable	80
5.2.1.5	Tunnel Track	81
5.2.2	Trackless Operation	81
5.2.2.1	Transport Vehicles	82
5.2.2.2	Haul Road	82
5.2.2.3	Loading	82
5.2.3	Conveyor Transport	84
5.2.3.1	Conveyor Storage	85
5.2.3.2	Conveyor Belt Extension and Belt Operation	85
5.2.3.3	Advantages of Conveyor Transport and Innovation Potential	87
6	**Backup Equipment**	89
6.1	Backup Concept	89
6.2	Design Specifications	93
7	**Ventilation, Dust Removal, Working Safety, Vibration**	99
7.1	Ventilation	99
7.1.1	Danger	99
7.1.2	Ventilation Schemes, Ventilation Systems	99
7.2	Dust Removal	100
7.3	Occupational Safety and Safety Planning for TBM Operation	103
7.3.1	General	103
7.3.2	International Guidelines and National Regulations	104
7.3.2.1	International Guidelines	104
7.3.2.2	National Regulations	104
7.3.3	Integrated Safety Plan	106
7.3.3.1	The Safety Plan in the Environment of Management Plans	106
7.3.3.2	Safety Aims	107
7.3.3.3	Description of Dangers and Risk Analyses	107

7.3.3.4	Action Plan	109
7.3.4	TBM Details and Specifics Regarding Natural Gas Danger and Rock Support	110
7.3.4.1	Natural Gas Danger	110
7.3.4.2	Rock Support	111
7.4	Vibration	111
8	**Additional Equipment**	**115**
8.1	Investigation and Improvement of the Geological Conditions	115
8.2	Equipment for Rock Support	117
8.2.1	Anchor Drills	117
8.2.2	Steel Ring Equipment	118
8.2.3	Mesh Installation Equipment	118
8.2.4	Innovation Aims	119
8.3	TBM Steering	119
8.3.1	Steering the Gripper TBM with Single Bracing	119
8.3.2	Steering the Gripper TBM with X-Type Clamping	122
8.3.3	Steering a Single Shield TBM	122
8.3.4	Steering a Double Shield TBM	125
8.4	Surveying	126
8.4.1	Surveying the Position of the Tunnel Boring Machine	127
8.4.2	Forward Calculation of the TBM Route	127
9	**Tunnel Support**	**129**
9.1	General	129
9.2	Support Systems and Advance Rates	130
9.3	Systematic Support at the Machine	133
9.3.1	Steel Arch Support	133
9.3.2	Liner Plates	136
9.3.3	Segments	137
9.3.3.1	Invert Segments	137
9.3.3.2	Segmental Lining	137
9.4	Shotcrete Support	142
9.4.1	Shotcrete Support at the Machine	142
9.4.2	Shotcrete Support in the Backup Area	143
9.5	Localised Support	144
9.5.1	Anchors and Mesh	144
9.5.2	Arch Support	145
9.6	Stabilisation Ahead of the Cutter Head	146
10	**Gripper TBM and Shield Machine Combinations**	**149**
10.1	Roof Shields	149
10.2	Roof Shield and Side Steering Shoes and Cutter Head Shields	152
10.3	Walking Blade Gripper TBM	153

10.4	Full-Face Shield Machines	155
10.4.1	Developments	155
10.4.2	Special Characteristics	156
10.4.2.1	Cutter Head and Shield	157
10.4.2.2	Thrust Ring	157
10.5	Double Shields	159
10.5.1	Developments	159
10.5.2	Functional Principle	159
10.5.3	Special Cases	160
10.5.3.1	Shield and Bentonite Lubrication	160
10.5.3.2	Telescopic Shield	161
10.6	Slurry Shield Machines	163
10.6.1	Developments	163
10.6.2	Working Principle	163
10.7	Shields with Screw Conveyors	165
10.7.1	Developments	165
10.7.2	Working Principle	165
10.7.3	Machine Types	167
10.7.3.1	Open Mode (Screw Conveyor – Conveyor Belt)	167
10.7.3.2	Closed Mode (Screw Conveyor – Conveyor Belt)	167
10.7.3.3	Closed Mode (Slurry Circuit)	168
10.7.3.4	EPB Mode (Screw Conveyor – Conveyor Belt or Screw Conveyor/Slurry Pump)	169
10.7.3.5	Open Mode (Conveyor Belt)	170
10.8	Micro Machines for Hard Rock	172
10.8.1	Mini TBM	172
10.8.2	Pipe Jacking	174
10.8.2.1	Press-Boring Pipe Jacking	174
10.8.2.2	Shield Pipe Jacking	174
11	**Special Processes: Combinations of TBM Drives with Shotcrete**	177
11.1	Scope of Application	177
11.2	Construction Options	178
11.2.1	Probe Headings	178
11.2.2	Pilot Headings	180
11.2.3	Enlargement for Stations, Points or Machine Halls	181
11.3	Examples	183
11.3.1	Piora-Mulde Probe Heading	183
11.3.2	Kandertal Probe Heading	184
11.3.3	Uznaberg Pilot Heading	188
11.3.4	Enlargement at the Connecting Structure at Nidelbad Zürich–Thalwil Tunnel	190

12	**Geological Investigations and Influences**	195
12.1	General	195
12.2	Influences on the Boring Process	199
12.3	Influences on the Machine Clamping	202
12.4	Influences on the Rock Support	205
13	**Classification for Excavation and Support**	207
13.1	General and Objectives for Mechanised Tunnelling	207
13.2	Classification Systems	208
3.2.1	Classification According to Rock Properties	208
13.2.1.1	RMR System (Rock Mass Rating System)	208
13.2.1.2	Q System (Quality System)	211
13.2.2	Classification According to Cuttability and Abrasiveness	220
13.2.3	Classification According to Type, Extent and Location of the Support Work Required	222
13.3	Standards, Guidelines and Recommendations for the Classification of Mechanised Tunnelling	223
13.3.1	Classification in Germany	223
13.3.2	Classification in Austria	228
13.3.3	Classification in Switzerland	234
13.4	Classification Suggestion by the Authors	239
14	**Tendering, Award, Contract**	243
14.1	Procedure Examples	243
14.1.1	Procedure in Switzerland	243
14.1.1.1	General	243
14.1.1.2	Tender Evaluation	244
14.1.1.3	Quality Management	245
14.1.1.4	Assignment of Risks in the Contract	246
14.1.1.5	Geologically or Geotechnically Altered Conditions, Altered Orders, Altered Schedules	249
4.1.2	Procedure in the Netherlands	249
14.1.2.1	Tendering and Negotiation Procedure with the Botlek Tunnel as an Example	249
14.1.3	Procedure in Germany	252
14.2	Design and Geotechnical Requirements in the Tendering of a Mechanised Tunnelling as Alternative Proposal	253
14.2.1	Introduction	253
14.2.2	Examples	253
14.2.2.1	Adler Tunnel	253
14.2.2.2	Sieberg Tunnel	254
14.2.2.3	Stuttgart Airport Tunnel	255
14.2.2.4	Rennsteig Tunnel	257
14.2.2.5	Lainz Tunnel	258

14.2.3	Additional Requirements for Mechanised Tunnelling Concept in the Tender Documents	260
14.2.3.1	Geology and Hydrology	261
14.2.3.2	Design and Construction Process	261
14.2.3.3	Specification and Contract	262
14.2.4	Decisions Based on Cost	262
14.2.4.1	Design Phase and Preparation for Tendering	262
14.2.4.2	Tendering Phase	263
14.2.4.3	Construction Phase and Final Payment	263
14.2.5	Forecast	263
15	**Tunnel Lining**	**265**
15.1	General	265
15.2	Design Principles for Tunnel Linings	265
15.2.1	Single-Shell and Double-Shell Construction	265
15.2.2	Watertight and Water-Draining Forms of Construction	267
15.3	Lining with Concrete Segments	269
15.3.1	General	269
15.3.2	Construction Types	271
15.3.2.1	Block Segments with Right-Angled Plan	271
15.3.2.2	Hexagonal or Honeycomb Segments	274
15.3.2.3	Rhomboidal and Trapezoidal Segment Systems	275
15.3.2.4	Expanding Segments	276
15.3.2.5	Yielding Lining Systems	277
15.3.3	Joint Detailing	282
15.3.3.1	Longitudinal Joints	282
15.3.3.2	Ring Joints	286
15.3.4	Steel Fibre Concrete Segments	289
15.3.5	Grouting Annular Gap	290
15.3.5.1	Filling with Gravel	290
15.3.5.2	Mortar Grouting	291
15.3.6	Measures for Waterproofing Tunnels with Segment Linings	291
15.3.6.1	Sealing Bands	292
15.3.6.2	Injecting	294
15.3.7	Segment Production	295
15.3.8	Damages	296
15.3.8.1	Damage During Ring Erection	297
15.3.8.2	Damage During Excavation	297
15.3.8.3	Damage at the Shield Tail Seal	298
15.3.8.4	Damage after Leaving the Shield	298
15.3.8.5	Repair of Damage	299
15.4	Cast in-situ Linings	299
15.4.1	General	299
15.4.2	Construction	300

15.4.3	Manufacture	300
15.5	Shotcrete Layers as the Final Lining	301
15.6	Structural Investigations	302
16	**Examples of Completed Tunnels**	**303**
16.1	Tunnel Excavation with Gripper TBMs	303
16.1.1	Control and Drainage Tunnel, Ennepe Reservoir	303
16.1.2	Manapouri Underwater Tunnel, New Zealand	305
16.2	Tunnelling with Shield TBMs	313
16.2.1	San Pellegrino Tunnel, Italy	313
16.2.2	Zürich–Thalwil Twin-Track Tunnel, Section Brunau–Thalwil	317
16.3	Inclined Shaft Tunnelling with Double Shield	321
16.3.1	Cleuson–Dixence Pressure Shaft	321
References		327
Index		337

1 Historical Development and Future Challenges

Tunnelling developed rapidly during the industrialisation at the start of the 19th century with the building of the railway network. In hard rock, this was by drilling and blasting. The first stage of the developing mechanisation of tunnelling therefore was the development of efficient drills for drilling holes for the explosive [96]. There were also attempts to excavate the rock completely by machine.

The story of the development of the first tunnel boring machines contains, besides the technically successful driving of the Channel Tunnel exploratory tunnels by Beaumont machines, many attempts, which failed due to various problems. Either the technological limits of the available materials were not observed or the rock to be tunnelled was not suitable for a TBM. The early applications were successful where the rock offered the ideal conditions for a TBM.

The first tunnelling machines were not actually TBMs in the true sense. They did not work the entire face with their excavation tools. Rather the intention was to break out a groove around the wall of the tunnel. After this had been cut, the machine was withdrawn and the remaining core loosened with explosives or wedges. This was the basic principle of the machine designed and built in 1846 by the Belgian engineer Henri-Joseph Maus for the Mount Cenis tunnel (Fig. 1-1). The machine worked with hammer drills chiselling deep annular grooves in the stone, dividing the face into four 2.0×0.5 m high stone blocks. Although this machine demonstrated its performance capability for two years in a test tunnel, it was not used for the construction of the Mount Cenis tunnel because of doubts about the drive equipment. The compressed air to power the drills was to be provided by water powered compressors at the portal and fed to the machine through pipes. Considering the 12,290 m length of the tunnel, Maus expected that only about 22 kW of the 75 kW generated would arrive at the machine. It also turned out that the material used at that time could not resist the wear during tunnelling. The result would have been increased wear of the bits. Despite these problems, Maus assumed an average advance rate of 7 m, or considering downtime for cutter change, 5 m per day.

The American Charles Wilson developed and built a tunnel boring machine as early as 1851, which he first patented in 1856 (Fig. 1-2). The machine had all the characteristics of a modern TBM and can thus be classified as the first machine, which worked by boring the tunnel. The entire face was excavated using disc cutters, which Wilson had already developed in 1847 and applied for a patent for. The tools were arranged on a rotating cutter head and the thrust required for cutting was resisted by pressure sideways against the rock. In comparison with modern TBMs, the integration of a rotating mounting for the disc cutters stands out. The mounting plate was arranged with its rotational axis perpendicular to the tunnel centreline in the cutter head holder, which combined with the rotation of the outer cutting head to cut a hemispherical face. Wilson's machine underwent various tests in 1853. After advancing about 3 m in the Hoosac tun-

1 Historical Development and Future Challenges

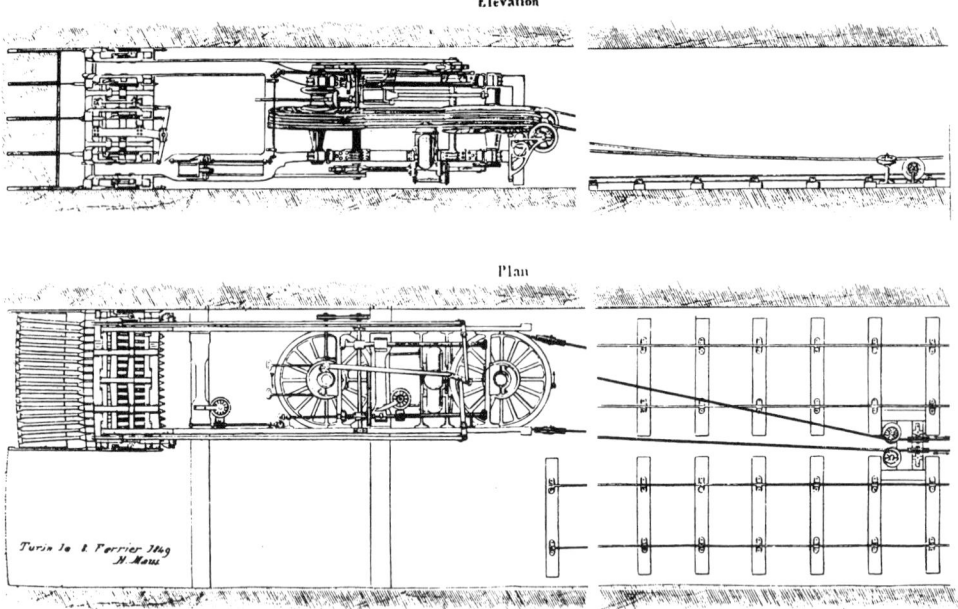

Figure 1-1
Tunnelling machine by H.-J. Maus, Mount Cenis tunnel, 1846 [159]

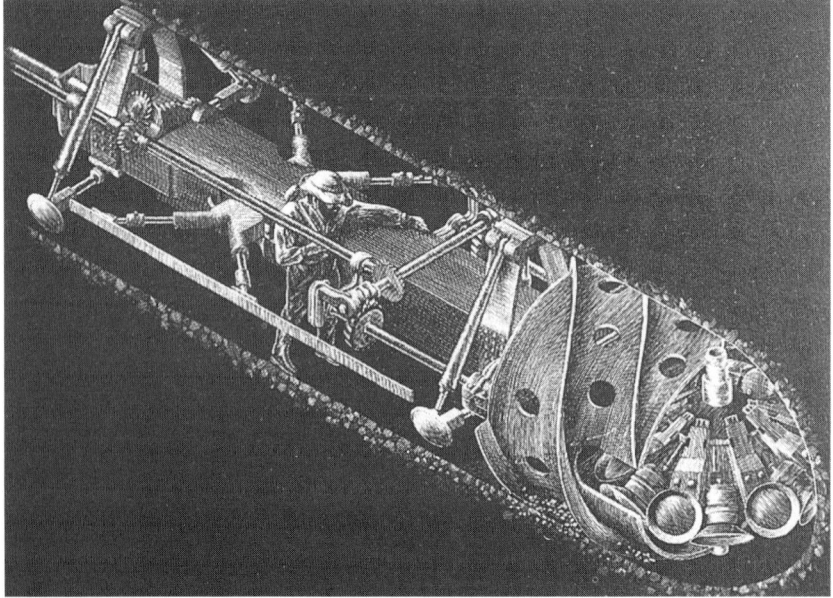

Figure 1-2
First tunnel boring machine by C. Wilson, Hoosac tunnel, 1853 [127]

1 Historical Development and Future Challenges

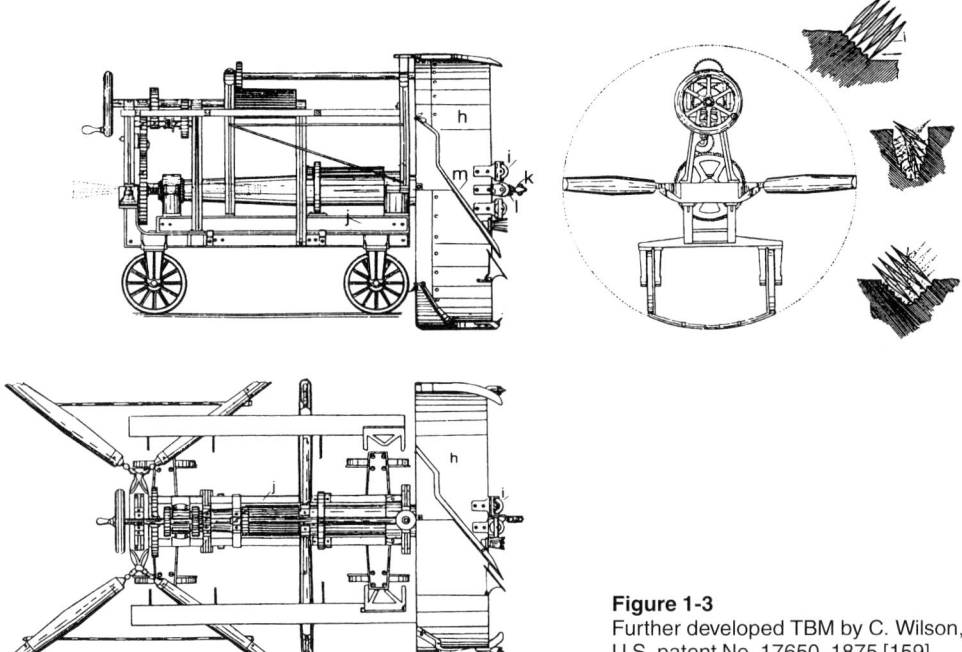

Figure 1-3
Further developed TBM by C. Wilson, U.S. patent No. 17650, 1875 [159]

nel (Boston, USA), the machine proved, because of problems with the disc cutters, unable to compete with the established drill and blast method.

After his experiences with the TBM at the Hoosac tunnel, Wilson applied for a patent in 1875 for an improved version of the machine (Fig 1-3). This was based on a completely new design of cutting head; no longer was the entire face to be excavated with cutting tools, but only an external ring and a central hole. This was to be achieved by mounting disc cutters at the outer rim and the rotational axis of the cutting wheel. After reaching the maximum cut depth, the machine had to be withdrawn to enable the remaining core to be loosened using explosives. The advantage was the precise profile of the excavation. This type of excavation with outer groove and central drilled hole proved to work well and was also used for other early tunnel driving machines like that of Maus, and this type of excavation has also been used from time to time since.

Also in 1853, the same year as Wilson was testing his first machine in the Hoosac tunnel, the American Ebenezer Talbot developed a tunnelling machine, which worked using disc cutters and a rotating cutting wheel. But this construction had the disc cutters arranged in pairs on swinging arms on the cutting wheel (Fig. 1-4). The combination of the rotation of the cutting head and the movement of the cutting arms enabled the excavation of the entire face. Talbot's machine failed in the first tests boring a section of diameter 5.18 m. Looked at with modern eyes, it is possible to recognise in the arrangement of the disc cutters on cutting arms parallels to the System Bouygues (see Fig. 3-36) tunnelling machines used in the 1970s.

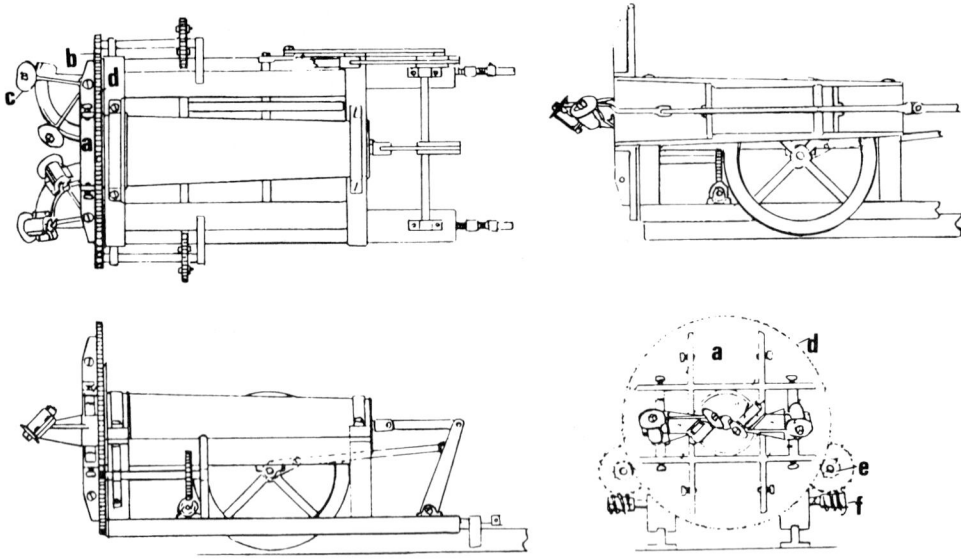

Figure 1-4
Tunnelling machine with drilling head and swinging cutting arms from E. Talbot, U.S. patent No. 9774, 1853 [159]

Cooke and Hunter (Wales) proposed an entirely new system with their patent from 1866 (Fig. 1-5). Instead of a cutting wheel turning about the tunnel centreline, three drums rotated about a horizontal axis transverse to the tunnel. The central drum had the largest diameter and ran ahead of the others, while the outer drums extended the cross section. The excavated section had a box shape with right-angled extensions. The direction of rotation was meant to clear the muck from the face during boring. The machine was never built, but the idea of a rotating extraction drum was found again fifty years later in tunnelling machines like the "Eiserner Bergmann" *(Iron Miner)* (see Fig. 1-8).

After Frederick E. B. Beaumont had already applied for a patent in 1863 for a tunnelling machine equipped with chisels and used this unsuccessfully for the construction of a water tunnel, he applied in 1875 for a patent for a tunnel boring machine with a rotating cutting wheel (Fig. 1-6).

The cutting wheel consisted a number of radial arms mounted on the end of a horizontal shaft. The tapered cutting arms were fitted with steel bits. The tip of the drilling bit formed a large conically ground chisel. The driving force was to be produced by a hydraulic pump driven by compressed air.

This patent was taken up by Colonel T. English and further developed for his own machine, for which he applied for a patent in 1880 [159]. There were cylindrical holes in the cutting arms for the drilling tools, into which chisel bits were screwed. The new idea of this construction was that the bits could be exchanged without having to withdraw the machine from the face. The arrangement of the bits on the two cutting arms was designed to cut concentric rings into the working face, so the remaining rock

1 Historical Development and Future Challenges

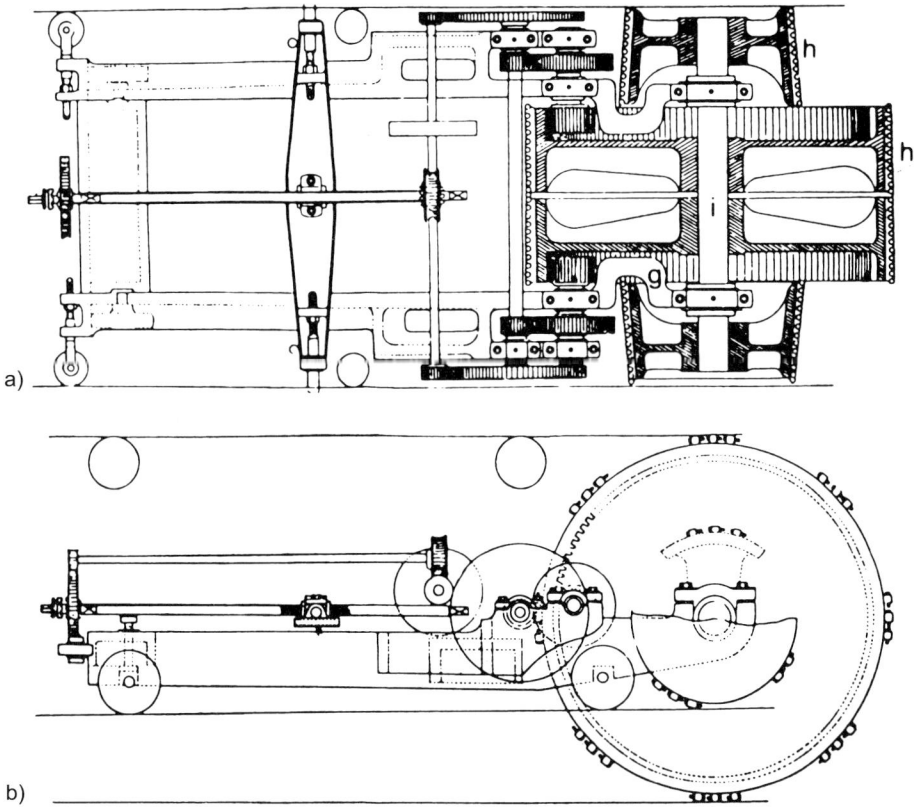

Figure 1-5
Tunnelling machine by Cooke and Hunter, U.K. patent No. 433, 1866 [159]

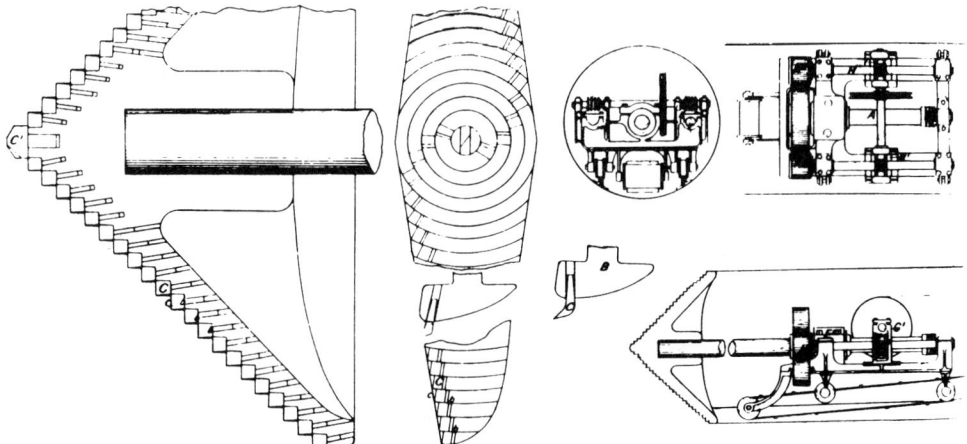

Figure 1-6
Tunnel boring machine by Beaumont, U.K. patent No. 4166, 1863 [159]

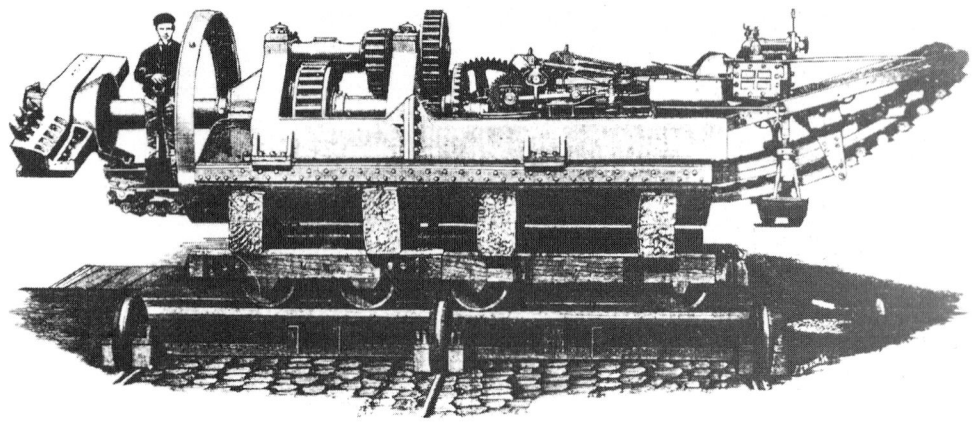

Figure 1-7
Tunnel boring machine by Beaumont/English, ⌀ 2.13 m, Channel Tunnel, 1882 [159]

between the grooves would break off during cutting. A lower frame formed the base frame of the machine with equipment to carry away the muck and the drive for the drilling head. An upper frame held the actual drilling equipment, which was pushed forward by a hydraulic cylinder. So it was possible for the first time to push the cutter head forward without releasing the bracing of the machine to the tunnel walls. This system allowed high blade pressure and is still a principle of modern TBMs.

Beaumont built two machines to the patent of Colonel T. English in 1881 and used them to drive the Channel Tunnel (Fig. 1-7). The machines worked there very successfully from 1882 until 1883, when the work was stopped for political reasons. Altogether 1,840 m were driven on the French side and 1,850 m on the English side. The maximum daily advance rate was 25 m, a considerable achievement for that time [100].

There was no further application of tunnelling machines in the next decades. They were, however, successfully used in mining for cutting relatively soft rock. In the first half of the 20th century, tunnelling machines were used for driving galleries in potash mines. The first version from 1916/1917, called the "Eiserner Bergmann", had a rotating roller fitted with steel cutters as a cutting wheel, which on account of its dimensions produced rectangular sections (Fig. 1-8).

The next generation of gallery cutting machines built by Schmidt, Kranz & Co. from 1931 was more successful. The machine consisted of the main components drill carriage, bracing carriage, cable carriage and loading band (Fig. 1-9). The three-armed cutting wheel was fitted with needles and achieved on average advances of 5 m per shift. Five men were needed to operate the machine. The disadvantages of this machine, which was also used in Hungarian brown coal mining, were considered at the time to be the size, the weight, the poor mobility and the time wasted bringing the machine back. In practice, the machine was used for quickly driving investigation and ventilation headings. The similarity to the TBM built by Whittaker for the Channel Tunnel in the

1 Historical Development and Future Challenges 7

Figure 1-8
Gallery driving machine "Eiserner Bergmann" 1916/17 from Schmidt, Kranz et al. [145]

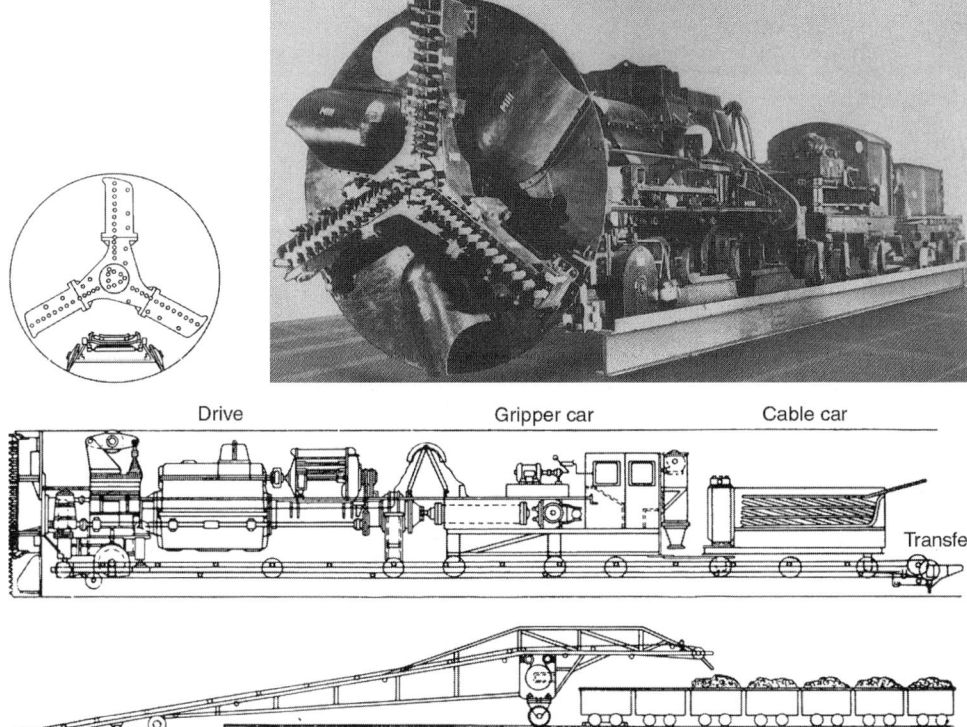

Figure 1-9
Gallery cutting machine from Schmidt, Kranz et al., ⌀ 3 m, 1931 [114, 145]

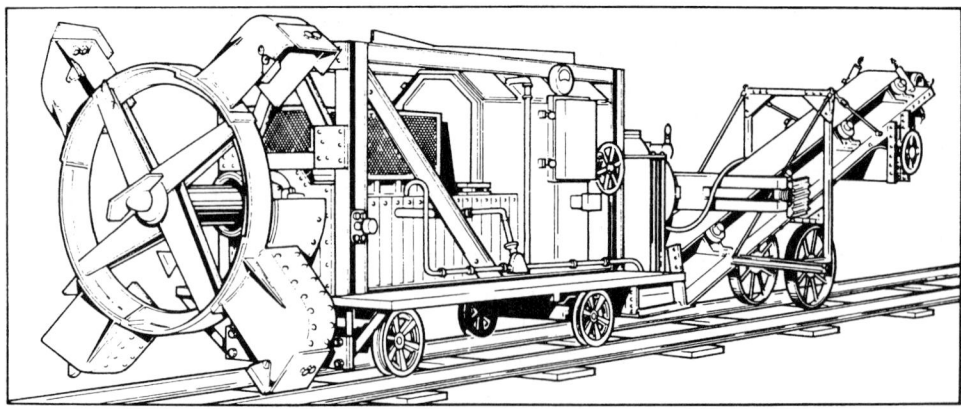

Figure 1-10
Tunnel boring machine by Whitaker, ⌀ 3.6 m, 1922 [72]

1920s is noticeable (Fig. 1-10). This achieved an average advance speed of 2.7 m/h in a test heading in the lower chalk near Folkestone.

The breakthrough to the development of today's TBMs did not occur until the 1950s, when the first open gripper TBM with disc cutters as its only tools was developed by the mining engineer James S. Robbins. Preliminary tests driving the Humber sewer tunnel in Toronto showed that, with only disc cutters and with considerably greater working life, the same advance performance could be achieved as with the intended combination of hard metal cutters and discs of the former TBM. Using this TBM in the Humber sewer tunnel, advances

a)   b)

Figure 1-11
Tunnel boring machines from Robbins [125]
a) First Robbins TBM, model 910-101, Oahe Damm, ⌀ 8.0 m, 1953
b) First modern gripper TBM von J.S. Robbins, model 131-106, Humber River sewer tunnel (Toronto, Canada) ⌀ 3.27m, 1957 (Robbins)

of up to 30 m/d were achieved in sandstone, limestone and clay (Fig. 1-11 a). Mechanical tunnelling at this time was primarily concentrated on stable and relatively soft rock. With the growing success of Robbins, further American manufacturers like Hughes, Alkirk-Lawrence, Jarva and Williams began building tunnel boring machines. Machine types still current today like the main beam TBM or the kelly TBM had their origins at this time.

After a slight delay, the development of tunnelling machines was also taken up in Europe. At first, however, different avenues of development were followed. Based on experience in Austrian brown coal mining with the Czech Bata machine [114], The Austrian engineer Wohlmeyer developed undercutting technology with rotating milling wheels (Fig. 1-12 a). This technology did not catch on, and nor did that used by the Bade company with the cutting head divide into three contra-rotating rings fitted with

Figure 1-12
First European developments of tunnelling machines [170]
a) Wohlmeyer gallery cutting machine SBM 720
 (Österreichisch Alpine Montan-Gesellschaft), ⌀ 3 m, 1958
b) Tunnel boring machine SVM 40 (Bade), operating in coal mining industry, ⌀ 4 m, 1961

toothed roller borers, which were already outdated at the time of the trial (Fig. 1-12b). Both types of machine were unsuccessful in tests in the hard rock of the Ruhr, although other Wohlmeyer machines were used successfully for the Albstollen heading and in the subsidiary headings of the Seikan tunnel [18, 45, 170]. Undercutting technology has been used and further developed over many decades by various manufacturers like Habegger, AtlasCopco, Krupp, IHI and Wirth because of the low thrust force required and the ability to drive non-circular cross-sections. The separation of the Bade TBM into a front section with cutting head and a rear section, which was hydraulically braced by four large pressure plates against the tunnel sides to provide reaction for the boring head carrier, and which is withdrawn after the completed travel of the advance cylinder, is however recognisable in modern double shield machines.

In the 60s, German manufacturers like Demag and Wirth began building tunnel boring machines of North American type. These machines were mainly intended to bore hard rock. Drilling tools from deep boring technology like TCI or toothed bits were mounted on the drilling heads. The developing technology for hardening the disc cutters enabled the use of this type of tool in really hard rock. At the end of the 60s, inclined headings and large tunnel sections were driven for the first time using the reaming method, the development of reamer boring being closely associated with the Murer company (Fig. 1-13).

Progress in the 70s and 80s was directed towards driving in brittle rock and the enlargement of tunnel sections, with the consideration of the stand-up time of the soil/rock becoming particularly important. Encouraged by the successful implementation of a gripper TBM for the Mangla dam project in 1963 with a diameter of 11.17 m, a gripper TBM was also used for the construction of the Heitersberg tunnel ($\varnothing$ 10,65 m) in Switzerland in 1971. The work necessary to secure the rock with steel installation,

a)   b)

Figure 1-13
Special types of Wirth Tunnel boring machines [182]
a) Inclined heading TBM TB II-300 E Emosson pressure tunnel, $\varnothing$ 3 m, 1968 (Wirth)
b) Enlargement TBM TBE 770/1046 H Sonnenberg tunnel, $\varnothing$ 7.70 m/10.46 m, 1969 (Wirth)

1 Historical Development and Future Challenges   11

anchors and mesh-reinforced shotcrete however made the hoped-for advance impossible. The required adaptation to the large cross-section was first achieved in 1980 by the modification of the Robbins gripper machine from the Heitersberg tunnel by the Locher und Prader company to a shielded TBM with segmental lining for the advance of the Gubrist tunnel ($\varnothing$ 11,50 m) (Fig. 1-14a). Robbins and Herrenknecht have made shield machines of this type in diameters from 11–12.5 m.

At the same time, Carlo Grandori developed the concept of the double shield TBM and, in collaboration with Robbins, put it into practice for the building of the Sila pressure tunnel ($\varnothing$ 4,32 m) in Italy (Fig. 1-14b). The main intention of the development of this

a)

b)

Figure 1-14
Tunnel boring machine with shield
a) Single shield TBM, Gubrist tunnel, $\varnothing$ 11.50 m, 1980 (Locher/Prader [144])
b) Double shield TBM 144-151, Sila pressure tunnel, $\varnothing$ 4.32 m, 1972 (Robbins [52])

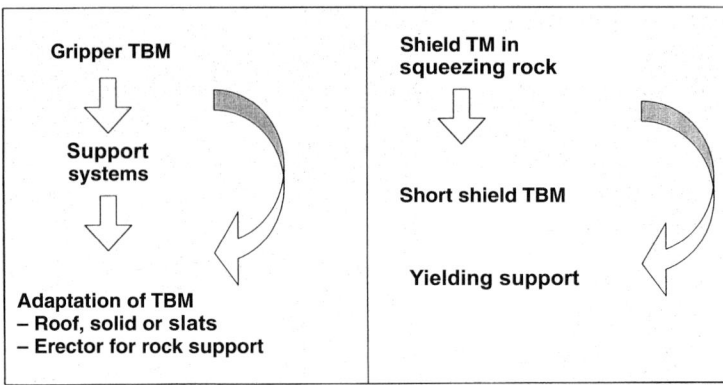

Figure 1-15
Postulated innovation path

machine was to make the gripper TBM, which had then already proved very effective in appropriate geological conditions, more flexible for use in heterogeneous rock conditions. Since their first use in 1972 and the successful modification of this type of machine, double shield TBMs with customised segmental lining designs have achieved high advance rates under favourable rock conditions and have been made by all the well-known manufacturers, mainly in the medium diameter range. The capability of the double shield TBM design was demonstrated impressively at the end of the 80s in the chalk of the Channel Tunnel, which is favourable for tunnelling. [100].

Alongside the development of the TBM with shield, the manufacturers of open gripper TBMs began to investigate possibilities of improving their machines to enable any necessary lining to be installed earlier. Shotcrete around the machine was tested. The state of progress with large diameter TBMs today is the installation of lining elements immediately behind the boring shield or partial areas of the shield and the systematic installation of rock anchors. With smaller tunnel boring machines, the body of the machine obstructs the installation of lining around the machine using mechanical equipment with the result that where lining has to be installed quickly, this has to be done by hand with a corresponding reduction of the advance rate.

The development of gripper TBMs at the moment is to enable the early mechanical installation of the lining around the machine in order to improve the boring performance by reducing the time taken to install the measures to secure the tunnel sides. Further reductions of the boring time would only lead to a marginal increase of advance rates, as today's TBMs already have availability rates of 80–90%.

For future development of tunnel driving with gripper TBMs, it is necessary to adapt the design of linings intended for conventional tunnelling to the special requirements of TBM tunnelling. The fear of a shield TBM jamming fast, which is repeatedly expressed, and the problem of rigid lining also demand innovative developments, although no such case is known for relevant single-shield TBMs.

The route from the ancestors of the modern TBM described here to modern high-technology machines was long, often arduous and even dangerous. To describe the early designs individually in more detail would exceed the space available for this book. Interested readers are recommended the reference book by Barbara Stack [159], which goes into the history of patents in TBM tunnelling in detail. Current developments and innovations are dealt with extensively in the following chapters.

2 Basic Principles and Definitions

The description tunnel boring machine (TBM) refers to a machine for driving tunnels in hard rock with a circular full-cut cutter head, generally equipped with disc cutters. The rock is cut using these excavation tools by the rotation of the cutter head and the blade pressure on the face.

Tunnel boring machines have sometimes also been described as milling machines, but this does not describe their method of operation.

In contrast to drilling and blasting, where it is possible to react flexibly to the interaction of tunnel and rock, either by subdividing the excavated section or by a rapid adaptation of the support to the geological situation, this is not possible driving with TBMs.

Gripper TBMs are suitable for use in hard rock with medium to high stand-up time. The working face must be largely stable, because support by the cutter head is only indirectly possible while driving. When the cutter head is withdrawn from the face for maintenance or to change bits, then there is no more support at the face. Under these conditions support, where necessary, can only be achieved with additional measures. The capability to cut rock up to 300 MPa enables TBMs to be used in most hard rock.

The higher investment cost of TBM driving compared with conventional drilling and blasting can only be compensated by higher advance rates. A greater length of drive is also necessary. If, however, the wear rate of the tools increases too much on account of the rock strength or other negative parameters, frequent cutter changing can lead to high downtime. This reduces the active working time considerably, which is an essential characteristic of the efficiency of the machine. Support measures required in fault zones and limited effectiveness of clamping can also reduce the advance rate significantly.

The reduction in effective working time of the machine can reduce the performance so far that it is no longer economical. This makes the logistical processes of more significance than with conventional methods.

If the effectiveness of the clamping in most of the tunnel cannot be guaranteed, then the use of shield tunnel boring machines is only possible against already installed segmental lining.

A decision to use a TBM also requires better geological investigation than for drilling and blasting and extensive detailed advance planning of the entire driving and supporting process.

Further considerations result from the form of the route. Especially tight radius curves set limitations for shield TBMs with long shields.

In the list below, the essential advantages and disadvantages of a TBM drive in comparison with a conventional drive are shown once more:

Advantages:

- Much higher advance rates possible
- Exact excavation profile
- Automated and continual work process
- Low personnel expenditure
- Better working conditions and safety
- Mechanisation and automation of the drive

Disadvantages:

- Better geological investigations and information are necessary than for drilling and blasting
- High investment resulting in longer tunnel stretches being necessary
- Longer lead time for design and building of the machine
- Circular excavation profile
- Limitations on curve radii and enlargements
- Detailed planning required
- Adaptation to varying rock types and high water inflow is only possible to a limited extent
- Transport of the machine with trailers to the tunnel portal

Although the number of disadvantages exceeds the number of advantages, the technical, safety and economic advantages for longer tunnels in suitable rock conditions are enormous.

2.1 Basic Principles and Construction

The basic elements of a TBM are the cutter head, the cutter head carrier with the cutter head drive motors, the machine frame and the clamping and driving equipment. The necessary control and ancillary functions are connected to this basic construction on one or more trailers.

So there are the following four system groups (Fig. 2-1) [11]:

- Boring system
- Thrust and clamping system
- Muck removal system
- Support system

2.1.1 Boring System

The boring system is the most important and determines the performance of a TBM. It consists essentially of cutter housings with disc cutters, which are mounted on a cutter head.

2.1 Basic Principles and Construction

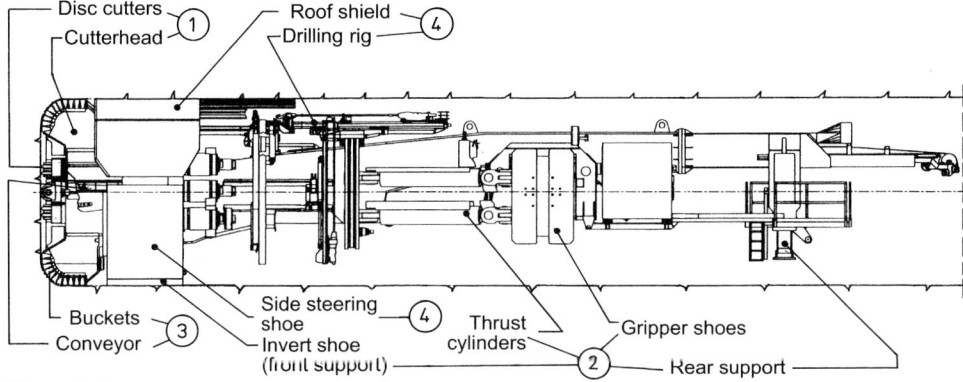

Figure 2-1
System groups of a tunnel boring machine
① Boring system
② Thrust and clamping system
③ Muck removal system
④ Support system

The discs are so arranged that they contact the entire cutting face in concentric tracks when the cutter head turns. The separation of the cutting tracks and the discs are chosen depending on the rock type and the ease of cutting. The size of the broken pieces of rock results from this.

The rotating cutter head presses the discs with high pressure against the face. The discs therefore make a slicing movement across the face. The pressure at the cutting edge of the disc cutters exceeds the compressive strength of the rock and locally grinds it. So the cutting edge of the disc pushes rolling into the rock, until the advance force and the hardness of the rock are in balance. Through this displacement, described as net penetration, the cutter disc creates a high stress locally (splitting tension), which leads to long flat pieces of rock (chips) breaking off.

Because the discs have little effect on the surrounding rock, this method of tunnelling can be seen as gentle on the surrounding rock.

2.1.2 Thrust and Clamping System

The thrust and clamping system is an element, which affects the performance of a TBM. It is responsible for the advance thrust and the boring progress. The cutter head with its drive unit is thrust forward with the required pressure by hydraulic cylinders. The length of the piston of the thrust cylinder restricts the maximum stroke. The TBMs usual today achieve a value of up to 2.0 m.

The thrust system limits the possible thrust and must resist the moments created by the rotation of the cutter head. The limits on the applied clamping forces are not determined by the mechanically produced force, which could be increased, but result from the natural condition of the rock. No greater clamping force can be applied through the gripper than that which can be resisted by the rock of the tunnel walls.

The term gripper describes the curved shoes, which are matched to the excavated section and lie against the tunnel wall in the braced condition. After a bore stroke has been bored, the boring process is interrupted so that the machine can be moved with the help of the clamping system. The gripper TBM is stabilised during this process by the clamping at the back and the shield surfaces around the cutter head, which are pushed radially against the tunnel wall.

During the moving operation, the grippers are loosened by hydraulic cylinders and braced again with the necessary pressure against the tunnel walls in the new machine position. This requires a free tunnel wall, which is only available in stable rock.

For shield TBMs, it is not the rock strength but the segmental lining, which is decisive, because these machines cannot be braced radially against the tunnel walls but axially against the lining. Between these two variants there are combined system solutions.

2.1.3 Muck Removal System

The muck is collected at the face by cutter buckets, which in TBMs are mostly constructed as slots around the perimeter of the cutter head and delivered to the conveyor down transfer chutes. In order to ensure the carrying away of the muck in the entire tunnel, a powerful system should be chosen, which does not interfere with the supply of the TBM and the necessary support measures. Either a rail system or a conveyor system is suitable according to local conditions. The use of large dump trucks is also possible.

Problems can arise, both with the cutter head buckets as with the continuous conveyor, through blockages caused by larger blocks of stone or the accumulation of fine-grained but also cohesive muck. Excessive water inflow at the face can also hinder the muck being carried away, if the overall character of the muck no longer remains and a muddy substance results, which makes sludge pumps or for example a screw conveyor necessary. This would make normal operation with a TBM impossible. In this case, where short fault zones occur, ground improvement, e.g. by injection or even freezing, must be carried out and for longer sections, the entire tunnelling concept will have to be altered to take the problem into account. Constant adaptation is not possible.

2.1.4 Support System

The use of tunnel boring machines in brittle rock sinks with increasing diameter. Support measures for smaller diameters can only be displaced to around the rear carriage behind the TBM. Therefore boring through fault zones with poor geology, where the stand-up time of the rock is shorter than the advance time, is always a problem case. With larger diameters, drilling guides can be used in the area behind the cutter head, which enable the installation of anchors or skewers. The erection of expanding rings is also possible and shotcrete behind the cutter head has been tried. To secure the fault zones, advance support methods like bolts, piles, injection or even freezing could be used over or in front of the cutter head, which stabilise the rock sufficiently to enable further driving with the TBM.

The advance rate can be considerably reduced by the effort required for the support measures required or the drive can even come to a standstill. The use of shotcrete around the machine directly behind the cutter head has still not been satisfactorily solved until today, because rebound especially leads to problems. Further development work is needed here, not only in shotcrete technology but also on the mechanical side. One exception is the enlargement TBM, where the entire tunnel section directly behind the cutter head is available for support measures and this means that shotcrete is already in successful use today.

If the rock is only lightly fractured, i.e. only occasional locations with smaller rock falls are to be expected, it can be sufficient to support the top with a roof shield. This roof shield protects the crew immediately behind the cutter head from and rock falls.

The gripper TBM has a short cutter head shield and this ends behind the cutter head, so that the support measures discussed above can be carried out more or less directly behind; behind comes the rear carriage.

In modern tunnel driving with a gripper TBM, a single invert segment is used in the invert. This segment is the temporary and the permanent support and can be equipped with mountings for the rails of the rear carriage and the rail conveyor, so that the tracks can be installed promptly. Drainage pipes can also be integrated into the invert segment. The final support of the remaining tunnel section is then performed in in-situ concrete using formwork carriages.

With the shield TBM, in contrast to this, the shield ensures the temporary support of the rock around the shield. The shield casing begins directly behind the circumferential discs and also encloses the area where the support elements are installed. Reinforced concrete segments are mostly used for the support. The segments are installed singly by the erector and form an immediate support. If high pressures are to be expected from flowing soil, then in extreme cases invert segments with a thickness of up to 90 cm can be installed. A shield TBM can be equipped with compressed air, hydraulic or earth pressure support and can then be used under the water table. However, in this case particular problems have to be solved in hard rock, because experience has been mostly in fractured rock.

The rearmost part of the shield, the shield tail, overlaps the last segment and supports the rock until the annular gap has been grouted. The shield tail of hard rock shield machines is also called the tail shield. Grouting the annular gap by blowing or filling avoids any possible loosening of the rock and a connection is created between rock and lining. In order not to hinder the advance or interrupt it for a longer period, the rear carriage must contain all the equipment necessary for a rapid installation of the support.

2.2 Definitions and Terms

Various types of machine are in use today for the mechanised tunnelling in hard rock. Figure 2-2 shows according to the DAUB system [33].

2.2.1 Tunnel Boring Machines with Full-Face Excavation

Figure 2-3 shows the various machine systems in full-face excavation, which are briefly described below:

2.2.1.1 Gripper TBM

The gripper TBM, often also widely described as open TBM, is the classic form of tunnel boring machine. The area of application is mostly in hard rock with medium to high stand-up time. It can be most economically used if the rock does not need constant support with rock anchors, steel arches or even shotcrete.

In order to be able to produce the thrust behind the cutter head, the machine is braced radially against the tunnel wall by hydraulically moved clamping shoes, the so-called grippers. As time went on, two different clamping systems were developed, single clamping (Robbins) and double clamping (Wirth and Jarva).

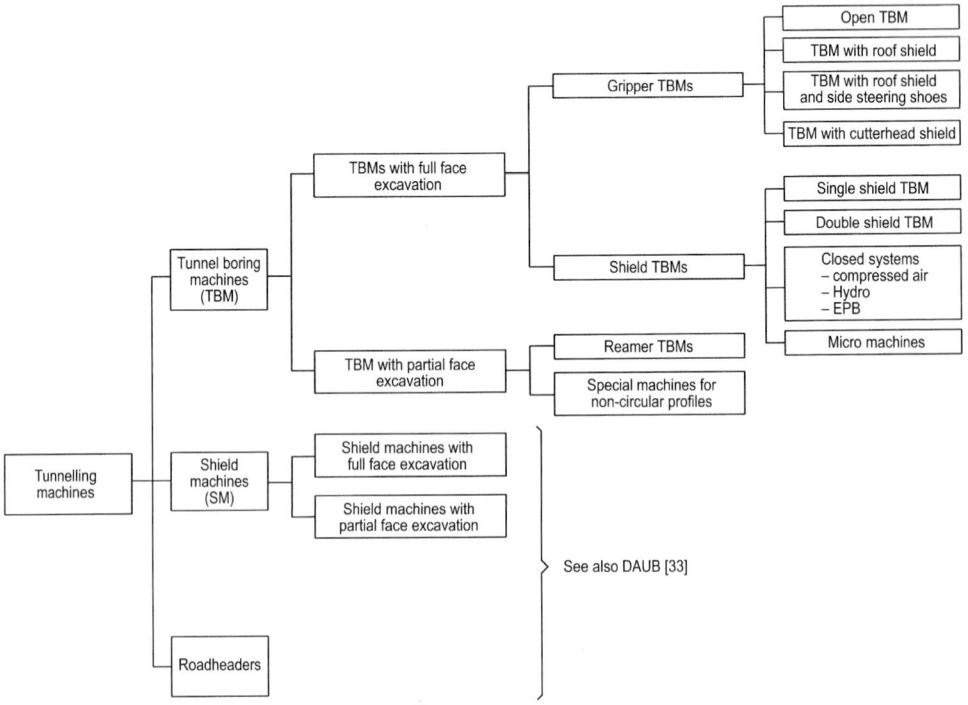

Figure 2-2
Overview of tunnel driving machines (according to the DAUB [33])

2.2 Definitions and Terms

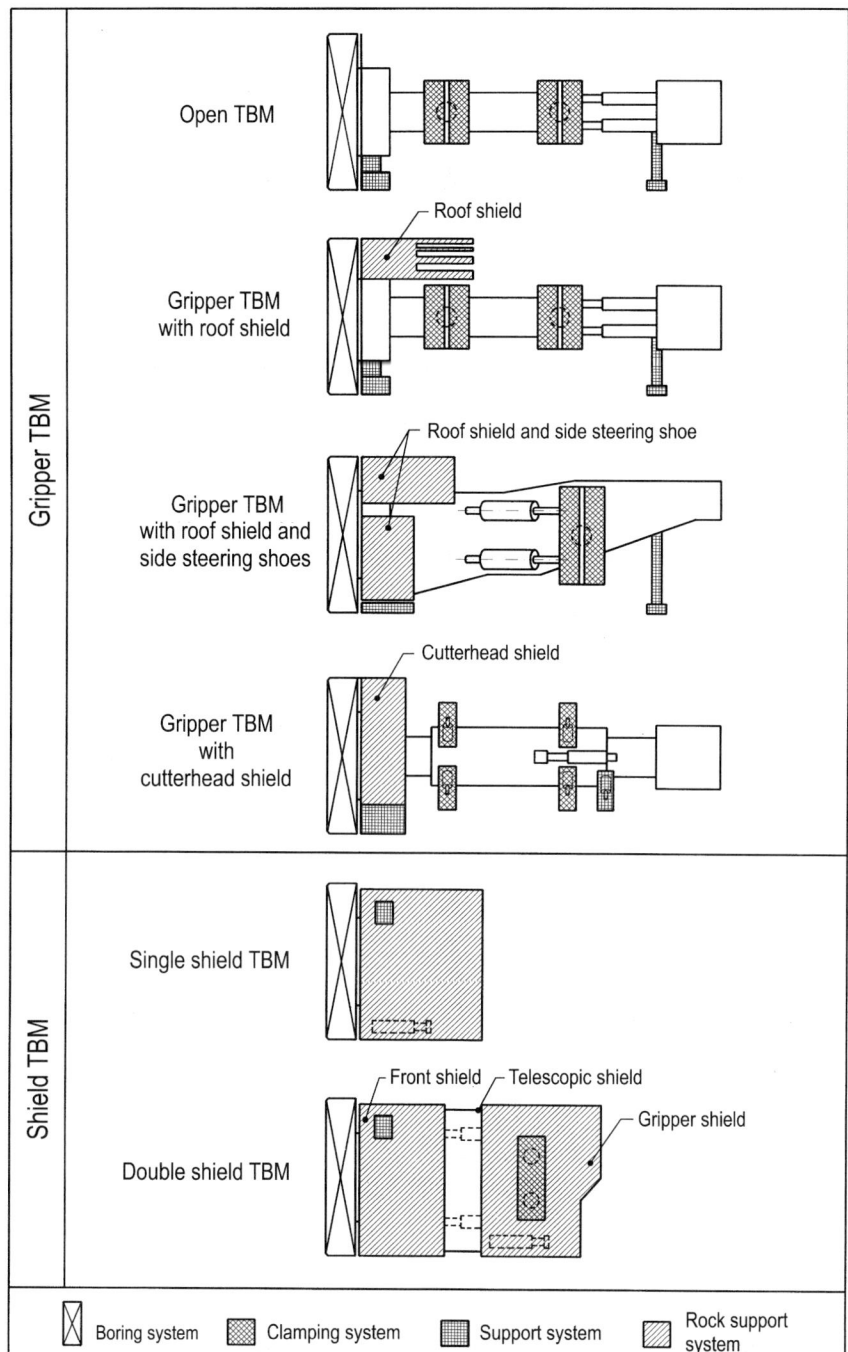

Figure 2-3
Overview of the various machine systems of tunnel boring machines with full-face excavation

Gripper TBMs are further categorised into open TBMs, TBMs with roofs, with partial shield and with cutter head shield:

Open TBM. The description open TBM is limited to TBMs without static protection units behind the cutter head. Machines of this type are today only found in smaller diameters.

TBM with roof shield. The construction of the TMB with roof shield corresponds to that of the open TBM. If, however, isolated rockfalls are to be reckoned with during excavation, then this type of machine has static protection roofs, so-called roof shields, installed behind the cutter head to protect the crew.

TBM with roof shield and side steering shoes. The side steering shoes have, in addition to the protection function, the purpose of support at the front when moving the machine and steering during boring. The side surfaces can be driven radially against the tunnel walls.

TBM with cutter head shield. The cutter head shield serves in this type of machine to protect the crew in the area of the cutter head. When moving the machine, the short shield liner forms the forward support.

2.2.1.2 Shielded TBM

Single shield TBM. Single shield TBMs are primarily for use in hard rock with short stand-up time and in fractured rock. The cutter head is not essentially different from that of a gripper TBM in relation to excavation tools and muck transport. To support the tunnel temporarily and to protect the machine and the crew, this type of TBM is equipped with a shield. The shield extends from the cutter head over the entire machine. The tunnel lining is installed under the protection of the shield tail. Support with reinforced concrete segments has become the most commonly used system nowadays. According to the geology and the application of the tunnel, the segments are either installed directly as final lining (single shell construction) or as temporary lining with the later addition of an in-situ concrete inner skin (double shell lining).

In contrast to the gripper TBM, the machine is thrust forwards with thrust jacks directly against the existing tunnel support.

Double shield or telescopic shield TBM. The double shield or telescopic shield TBM is a variant of the shield TBM. It enables, like also the single shield TBM, driving in fractured rock with low stand-up time, but has the following differences from the single shield TBM:

The double shield TBM consists of two main components, the front shield and the gripper or main shield. Both shield parts are connected with each other with telescopic jacks. The machine can either adequately clamp itself radially in the tunnel using the clamping units of the gripper shield; or where the geology is bad, can push off the existing lining in the direction of the drive. The front shield can thus be thrust forward without influencing the gripper shield, so that in general continuous operation is possible, nearly independent of the installation of the lining.

The double shield TBM has, however, essential disadvantages compared to the single shield TBM. When used in fractured rock with high strength, the rear shield can block due to the material getting into the telescopic joint. This is falsely described as the shield jamming. Blocking and jamming are however caused differently and should therefore be clearly differentiated.

The apparent advantages of the rapid advance of a double shield TBM only apply with a single shell segmental lining, which requires installation time per ring of about 30–40 minutes. With a double shell lining with installation time per ring of about 10–15 minutes, the higher purchase price and the greater need for repairs are no longer economical.

Closed systems. Closed TBM systems with shield are combined system solutions for use under the water table, with water inflow being prevented by compressed air or by supporting the cutting face according to the slurry or EPB principle. These systems are used in hard rock and also in fractured rock [95].

Micromachines. These are nowadays also equipped for use in hard rock. For excavation, the same basic principles apply for the design of the cutter head as for larger machines. Micromachines are equipped with a shield.

2.2.2 Tunnel Boring Machines for Partial Excavation

Enlargement machines. The enlargement machine [57] is a special case of gripper TBM. It is used for tunnels with a diameter of over 8.00 m. In a drive with an enlargement machine, the section is enlarged to the required diameter, starting with a continuous pilot heading, which is driven completely in the centre of the tunnel before beginning enlargement boring. The enlargement can be carried out in one or more stages. The clamping is in the pilot heading in front of the cutter head of the enlargement stage.

Special machines for non-circular sections. All types of machines, which excavate the face in a partial process and thus enable sections varying from circular, are categorised as special machines for non-circular sections. Examples for special machines are the various developments by the different manufacturies like the Mobile Miner (Robbins), the Continuous Miner (Wirth), the Mini-Fullfacer (AtlasCopco) and Japanese developments [87].

3 Boring Operation

Modern TBMs have a closed cutter head. The disc cutters are simple discs and sit in the cutter head with a fixed spacing between tracks. The rotational speed of the cutter head depends directly on the radial speed of the gauge cutter. This varies today between 140 and 160 m per minute.

Disc cutters are simple discs, less often multiple discs or TCI bits, and roll against the face with the rotation of the cutter head (Fig. 3-1). The rock is crushed by the rolling disc cutter tools because of the thrust on the cutter head and, because the shear or tension strength is exceeded, spalled off between the roller tracks.

Figure 3-1
Tracks of the cutters on the face

3.1 The Boring Process

The performance of the boring process depends on many factors, which can be collected into three groups.

The factors determining the performance alter from site to site and can vary considerably even on one site. The successful use of a TBM is therefore always dependant on the selection and adaptation of the overall system to the particular conditions of the planned project.

The basic requirement is the best possible realistic description of the rock to be bored through. The project engineer above all is responsible for this. He must establish, in intense discussions with the geologists, what the risks are. Finally the risks have to be evaluated for the corresponding method of construction and made available to the con-

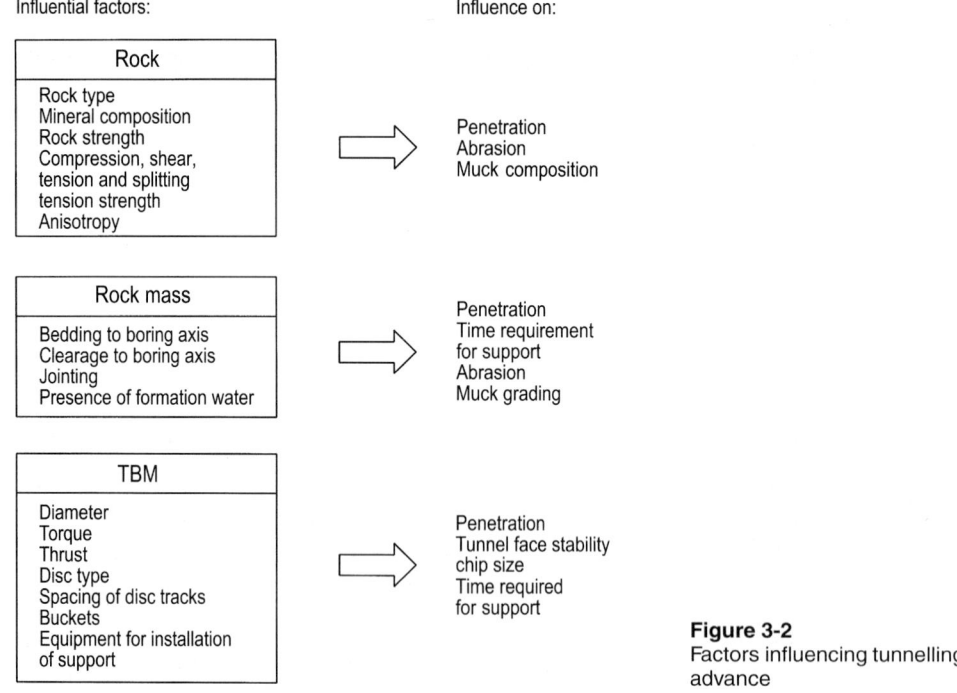

Figure 3-2
Factors influencing tunnelling advance

tractor in a suitable form. First then can a decision be made whether the use of a TBM is suitable and which type.

In addition to the usual site investigation (this is dealt with in Chapter 12), there are new possibilities available today as additional methods like:

- The seismic reflection method, which is capable of displaying the rock surface and the course of the strata together in high resolution.
- The borehole photography, which very rapidly makes a complete display of the rock bored through with joints and stratification, even without withdrawing cores.

3.2 The Cutter Head

The cutter head has many functions:

- It is the carrier for the excavation tools.
- The excavated muck is collected by the muck bucket with the rotation of the cutter head, picked up and deposited on the conveyor.
- It supports the face during interruptions of normal operation.
- It supports the face, in case of a cave-in, until the fractured rock has been secured by suitable measures.
- In special cases it carries overcutters.

3.2.1 Shape of the Cutter Head

Lightly domed shaped cutter heads have become established for tunnel boring machines with diameters up to 4–5 m. For larger TBMs, flat cutter heads are suitable. They lead to a favourable stress distribution in the rock and thus to the least possible breaking out of the face.

Various shapes of cutter head have been developed in the course of time (Fig. 3-3). The small TBM with applied cutters and practically fully open cutter head in front (Fig. 3-3 a) is quite suitable in hard rock. To use it in fractured rock or in rock liable to bursting would lead to serious problems regarding safety and the advance rate would then fall off rapidly.

Figure 3-3
Shapes of cutter heads
a) Open cutter head with attached tools
b) Closed, domed cutter head
c) Closed cutter head of a shield TBM with external buckets
d) Closed cutter head of a shield TBM with external and internal buckets

The inclined shaft TBM as double shield (Cleuson-Dixence) in Fig. 3-3 b has a closed, domed cutter head. This type of cutter head also makes it suitable for use in fractured rock. A shield TBM cutter head is shown in Fig. 3-3 c (TBM Gubrist, Zürichberg, Sachseln, Arrissoules) with 16 external buckets to cope with the excavated muck.

The cutter head of the Murgenthal shield TBM shown in Fig. 3-3 d with eight muck buckets. It also has front buckets with adjustable lips in front of the cutter head.

In fractured rock, grill bars between the peripheral buckets have a favourable effect on the stability of the bored cavity (Fig. 3-4).

Figure 3-4
Cutter heads with rims
a) Shield TBM, San Pellegrino (Herrenknecht)
b) Gripper TBM, Lesotho Highlands Water Project (Robbins)

a)

b)

3.2 The Cutter Head

A cutter head as short as possible, in order that the rotating parts have as little contact as possible with the tunnel walls, also has a good effect on the advance rate. Roofs or cutters of partial and full shields should be kept as far forward as possible. The cutter head itself should be kept as closed as possible.

3.2.2 Clearing the Muck in the Excavation Area

The disc cutters sitting on the cutter head excavate the rock continuously. The buckets scooping up the muck transport away the loosened rock in pulses or intermittently. The muck is also tipped onto the conveyor intermittently. This is an unfavourable transport flow.

With smaller diameter machines, this intermittent material flow causes no problems. With larger and large diameter machines, however, this material flow can lead to reduced performance through reduction of the penetration if the cutter head is not laid out correspondingly. Figure 3-5 shows the limit range of the penetration for a TBM ∅ 12 m, which is determined by the performance of the conveyor belt. It is clear that, due to the pulsating transport of material, higher penetration can be produced by a greater number of buckets on the cutter head without reaching the limiting capacity of the conveyor belt.

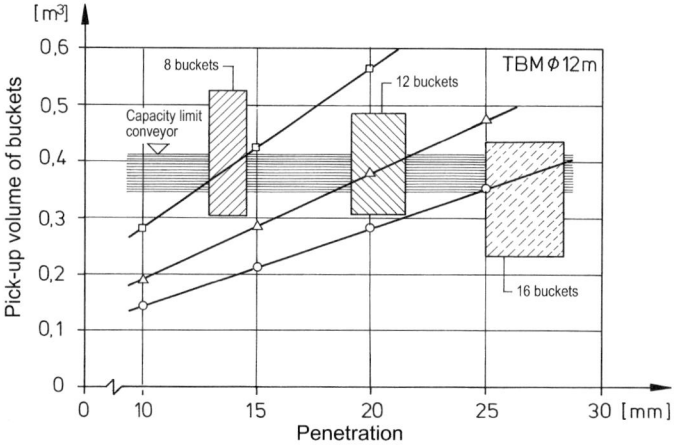

Figure 3-5
Limiting range of penetration, determined by the conveyor belt capacity of the TBM [143]

The contrasting material flows of the continual excavation and the discontinuous material removal leads, with a low number of buckets, to a jam of material between cutter head and face. With a large number of buckets, this process is relatively insignificant. This material jam causes a loss of torque of the TBM. In Fig. 3-6, this loss of torque is shown against the number of buckets and the overall characteristics for a TBM ∅ 12 m. The torque loss increases much more rapidly with increasing penetration and a low number of buckets than with a larger number of buckets because of the excavated rock between face and cutter head.

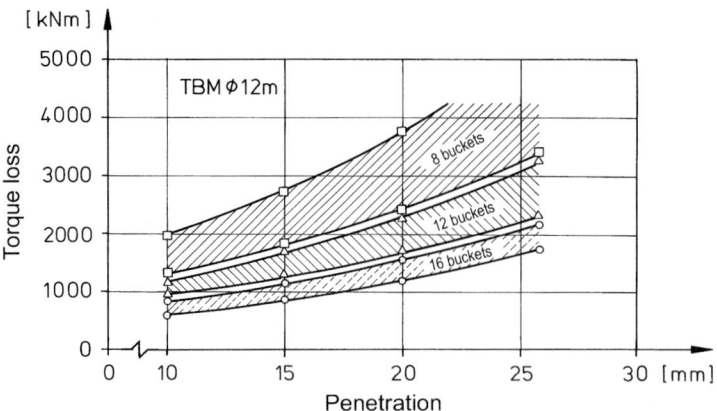

Figure 3-6
Torque loss caused by loosened rock between face and cutter head,
taking into account number of bucket openings and rock characteristics

When there is a jam of material, there is also the danger of the face tearing, resulting in fractures or cave-ins. So the cutter head should be equipped with an adequate number of buckets.

Tunnel boring machines are rarely designed for boring a hard class of material, like granite or amphibolite. If the rock allows a high penetration, then the contractor will want to exploit this, therefore cutter heads and their buckets should always be designed correspondingly.

Front buckets, which have been in use for about 10 years and partly extend into the centre of the cutter head, certainly have their origin in the scraper head used for tunnelling in loose rock. The effectiveness of such front buckets is very debatable; the edge of the bucket must in any case be so far recessed against the cutting face that it never scrapes the face, even at high penetration and with worn disc cutters, otherwise tearing and possible cave-ins can be expected.

These front buckets could also purely geometrically take over 20–25% of the muck, even with complete contact with the face, which cannot be allowed to happen. Therefore the effective capacity of such front buckets is under 10–15%. This statement is also demonstrated by the experience in the Murgenthal tunnel (Fig. 3-3 d). Front buckets were used there for the first time. In an extremely abrasive Rinnen sandstone, the front buckets showed scarcely any wear, while the outer buckets had to be renewed many times.

Moreover, the slots for front buckets weaken the cutter head considerably. In consideration of the not very convincing advantages of such front buckets, their suitability should be investigated before each new construction of a TBM.

The outer buckets are not in the position to take all the muck. The rest of the excavated material remains in the invert and builds up, according to type of TBM, in front of the cutting shoe on the stator or, in case of a shield TBM, in front of the shield. Trailing

3.2 The Cutter Head

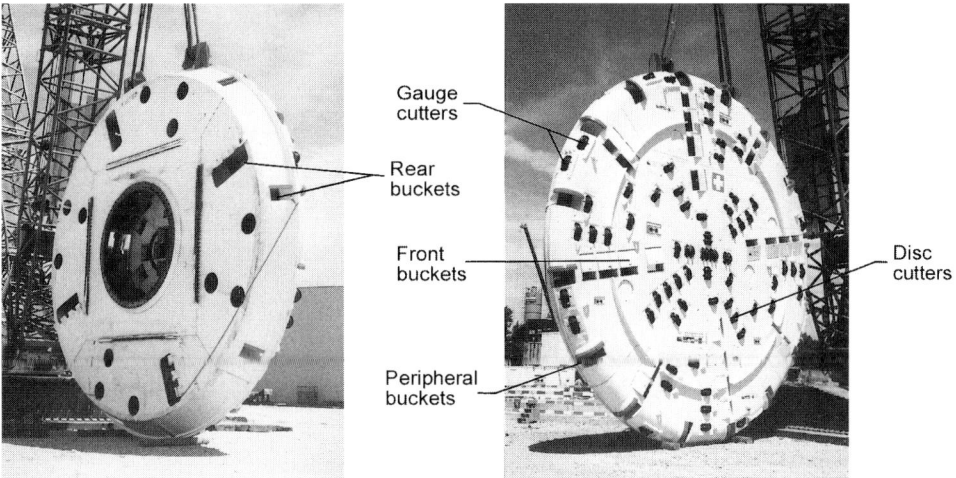

Figure 3-7
Cutter head of the Zürich–Thalwil TBM with 16 gauge buckets, 24 front buckets, 24 front buckets with adjustable flaps and 8 trailing buckets

buckets provided on the cutter head pick up this built up muck and transfer it with the rotation of the cutter head to the conveyor belt. Such trailing buckets should best not have any connection with the gauge buckets or the front buckets. In case of a collapse of the face and continuous bucket systems, the space between stator and cutter head fills up with fallen material. The clearing of this space is in such a case very tedious.

Transport chutes inside the cutter head can simply be part of the steel construction or they can be designed in order that the muck slides as well as possible. Too often the first is the case, which results in unnecessary blockages of cohesive muck and leads to considerable trouble to clear the blocked material. To work well, such chutes should have no edges and corners, especially no narrowing. Rounded corners and in extreme case coatings should be considered in some cases.

3.2.3 Cutter Head Construction and Soil Consolidation

Even when the geological investigations have been very thorough, a collapse over or in front of the cutter head can never be ruled out. The mechanical equipment of TBMs, especially around the cutter head, should enable such collapses to be consolidated by injection.

Injection openings should be provided. With larger machines, it should also be possible to drill injection lances through the cutter head and undertake the consolidation with suitable material (see also Chapter 9).

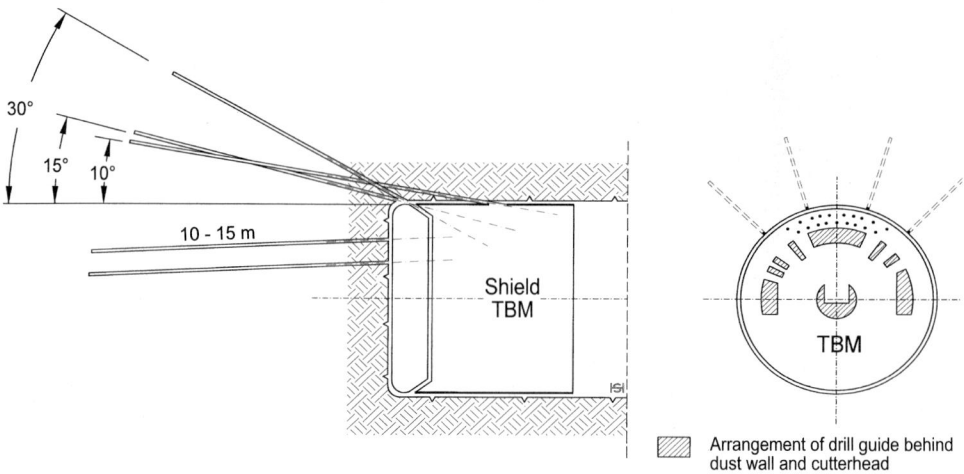

Figure 3-8
Arrangement of the drilling and injection channels through the shield casing and cutter head on the Adler tunnel shield TBM, ⌀ 12,53 m (Herrenknecht) [94]

3.3 Cutting Tools

3.3.1 General

Tunnel boring machines were equipped very early with disc, toothed and Tungsten carbide insert bits (TCI) cutters. This choice of tools is therefore different from tunnel milling according to the Wohlmeier principle, which works by detaching chips.

The first cutting tools came from oil well drilling technology. Soon the TBM manufacturers began with the development of their own tools. Cutter development began with a diameter of 10″, later replaced with 12″, which allowed average cutter loadings of 100–120 kN. In cuttable stone like molasse sandstone and marl, or also relatively soft Jurassic limestone, reasonable advances were reached. Contractors, convinced by the method of mechanical tunnelling, looked to use TBMs in hard rock too. The machine manufacturers took up the challenge and produced ever stronger cutters with diameters of 15½″, 17″ to 19″. These larger cutters not only made a considerable increase of average cutter loading possible, but also led to a meaningful increase of the cutter life. This

Figure 3-9
Cutting tools: disc cutter, TCI bits cutter (Palmieri)

3.3 Cutting Tools

Figure 3-10
Roller assembly with cutter disc and shaft for fitting to the roller housing

higher cutter loading first made reasonable penetration possible in rock, which is hard to bore. Cutters in common use today are shown in Fig. 3-9.

The disc cutters as a consumable item are mounted on a specially sealed roller assembly with tapered roller bearings. Figure 3-10 shows such a roller assembly with installed disc cutter.

The disc cutters must be secured in a particular position on or in the cutter head, in order to rule out swinging motion under high loading. But it must also be possible to change them easily. Above a cutter head diameter of 4m, the disc cutters can usually be changed from behind.

Housings vary according to manufacturer in the method of securing the rollers. Figure 3-11 shows two typical roller housings, one face-mounted corresponding to a TBM with open cutter head (Fig. 3-3 a), and one set into the cutter head and intended for changing the discs from behind, for TBMs with closed cutter head (Fig. 3-3 b–d).

3.3.2 Working Method of Cutter Discs

The rotation of the cutter head causes the discs to roll in concentric tracks on the cutting face. Singly mounted discs roll in their own tracks. The average disc loading of 100–250 kN, according to the geological conditions, thrusts the discs into the face. The advance per rotation is called the penetration.

There are two different theories about the process of the rock spalling off between neighbouring roll tracks. The former assumption, probably inspired by the wedge section of the discs, presumes a shearing process initiated by the flanks of the disc. Working on this basis, the cutting angle has led various authors to different performance assumptions. The typical wear of the discs by material displaced sideways puts not only the shearing theory but also the essential influence of the cutting angle into question (Fig. 3-12).

Newer assumptions, influenced especially by disc cutters of almost the same thickness in the working area, assume rather a splitting tension loading, presuming the penetration reaches values of 4–15 mm, in softer rock up to 20 mm or more.

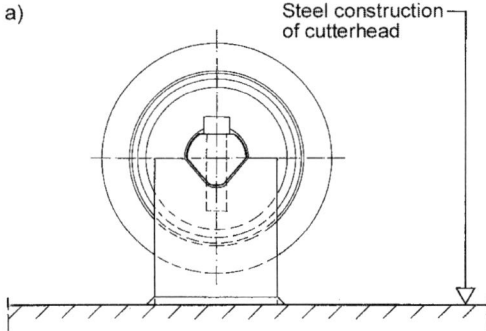

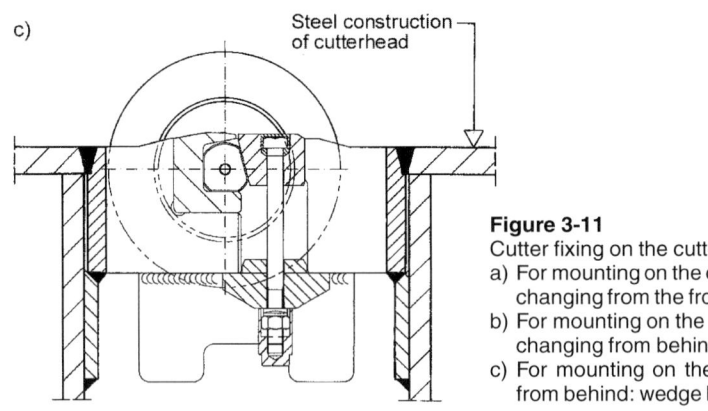

Figure 3-11
Cutter fixing on the cutter head
a) For mounting on the cutter head and changing from the front
b) For mounting on the cutter head and changing from behind: bayonet system (Wirth)
c) For mounting on the cutter head and changing from behind: wedge lock system (Robbins)

The splitting process described by Büchi, which is also comparable with the work of other authors, is plausible, especially with the shape of newer discs (Fig. 3-13). The deep effect of the formation of radial cracks under the disc is largely unknown. It can be investigated by using thin sections obtained from many bore cores arranged over the rolling track.

3.3 Cutting Tools

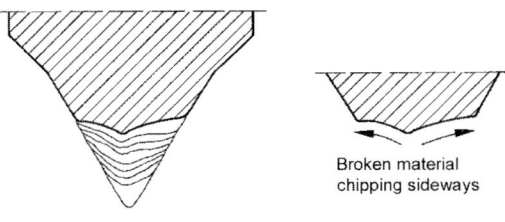
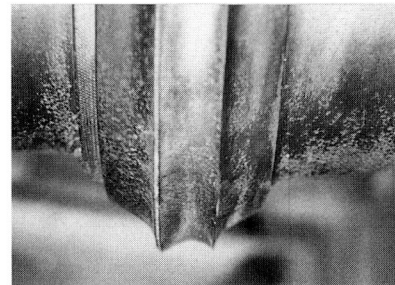

Figure 3-12
Cutter disc in worn state, blade of older type [21]

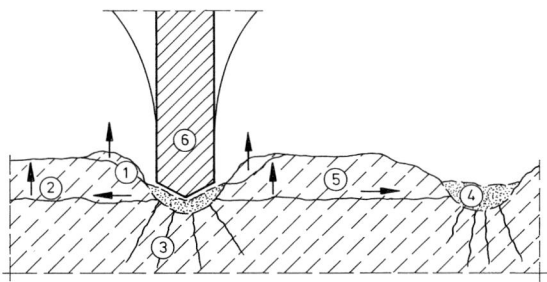

Figure 3-13
Schematic diagram of breaking out process, based on [21], with a cutter disc of constant width
① Spalling caused by tension cracks
② Shear failure or tension crack failure
③ Formation of radial cracks under the disc
④ Flowing of material out of the furrow of the disc
⑤ Typical shape of a larger chip
⑥ Disc cutter with almost constant thickness in cutting area

The average disc loading during the rolling process is often considerably exceeded. The breaking out of the rock by the rolling process never leads to clean, regular screw curves. Unevenness caused by anisotropy or discontinuities leads to strong peaks in the disc loading. These can well amount to twice the average disc loading; various measurements on site show such peaks.

The disc cutter bites into the face, depending on penetration, along 60–90 mm of its perimeter. It can easily be seen that with the very low loading phase of 0,025 s for the profile disc, also called the gauge cutter and 0,3 s for discs near the centre, extreme peak loads are applied to rock and disc.

These dynamic loadings with their high peaks stresses are what enable a penetration in rock, which is hard to bore, like fine-grained granite or amphybolite.

It has been tried out, in rock of very high compression strength, to cut a groove between the disc roller tracks with high-pressure water jets. This would result in a weakened rock in front of the rolling disc and would have the advantage of not only reducing the required thrust behind the discs but also the dust would be considerably reduced. With the advance of cutter development and especially the high loadings now possible on the cutters, these attempts were fully uneconomical on account of the additional energy needed, 4,000 kW for a cutter head of 50 cutters [76].

3.3.3 Cutter Spacing

Until now the greatest possible advance rate was the only aim of the separation of the cutter tracks or cutting track spacing. The determination of the spacing of the roll tracks was done empirically. These spacings were, according to manufacturer, between 65 and 95 mm. With the use of larger cutters and the resulting greatly increased disc loading, spacings of 80 to 95 mm are now used.

The scarceness of resources of gravel and the increasing difficulties of finding suitable landfill sites for tunnel muck material make it necessary for measures to be adopted in the future leading to better reuse of this material.

Trials by Atlas Copco-Robbins in Äspö with larger spacing seem to offer the possibility of creating coarser chips in the muck (Fig. 3-14), even in rock, which is hard to bore through. The grading curves of the tunnel muck are shown in Fig. 3-15.

Figure 3-14
Chip-shaped muck from a TBM

The spacing of the cutters of the TBM used in the trials with a diameter of 5 m was 86 mm. These spacing was increased 150% to 129 mm and 200% to 176 mm. In granite with compressive strength in the region of 200 to 250 Mpa, the sand component in the muck was reduced in favour of larger components. The increased spacing led to thicker and still more to longer chips. In Fig. 3-16, axis a) is the longitudinal axis, axis b) is the width and axis c) is the thickness of the chips.

An increase in the coarseness of the tunnel muck is possible even in rock, which is hard to bore. The precondition is a TBM, which can apply corresponding disc loadings to the rock.

3.3 Cutting Tools

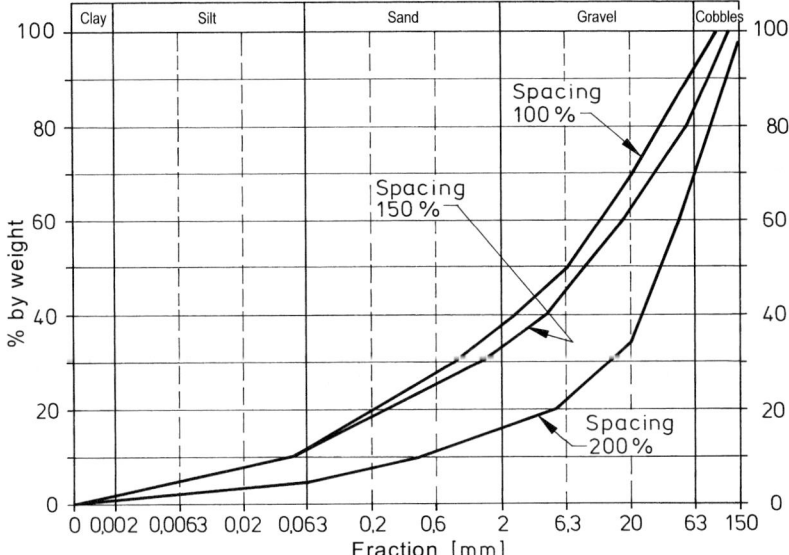

Figure 3-15
Grading curves from the Äspö tunnel test [23]

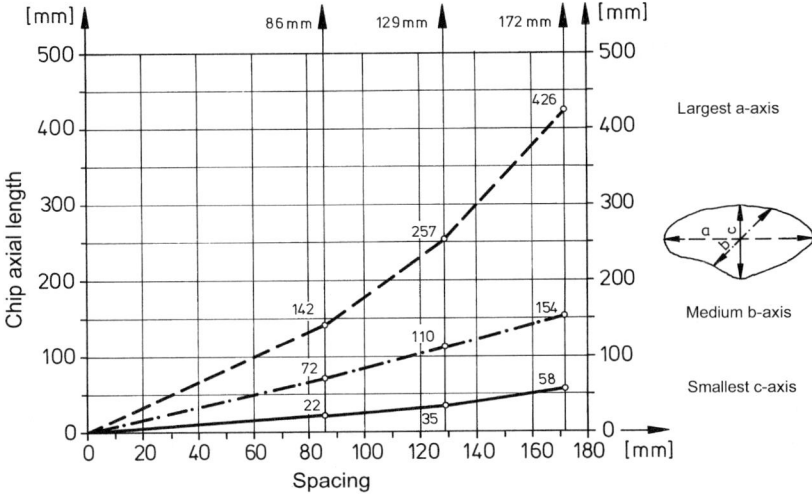

Figure 3-16
Chip formation from the Äspö tunnel test [23]

3.3.4 Penetration

As already shown in Section 3.1 "The boring process", the penetration depends on three main factors:

- Geological conditions
- Rock
- TBM

These influence factors cannot be defined by a simple mathematical formula. Robbins, at the South African tunnelling conference in 1970, showed the qualitative relationship geological conditions/cutter loading/penetration for disc cutters (Fig. 3-17).

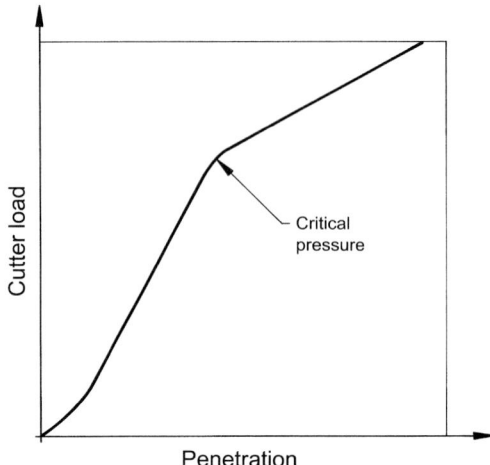

Figure 3-17
Qualitative diagram of the relationship between penetration and cutter loading [122]

Below this critical cutter loading, the TBM only works like a stone mill. If a rock is encountered with excessive strength, with which the critical cutter loading cannot be exceeded, then the penetration with normal disc cutters decreases so strongly that a drive would only be sensible technically with TCI bits; the use of a TBM will then no longer be economic.

Toothed cutters and TCI bits produce a higher loading under the hard metal tooth, so that small rock chips spall off, making an application in such extreme conditions possible.

The curves for various rock types shown by Robbins in 1970 have a remarkable agreement with today's experience. It was, however, to take about 20 years until high-strength rock could be bored.

The values shown in Fig. 3-18 entered additionally into the qualitative Robbins curves give a good overview of the possible penetrations. The area of the uniaxial compressive strength, with the penetration resulting from the relevant cutter loading, represents a good approximation to the relationships penetration/compression strength published by Gehring in 1995 [48].

3.3 Cutting Tools

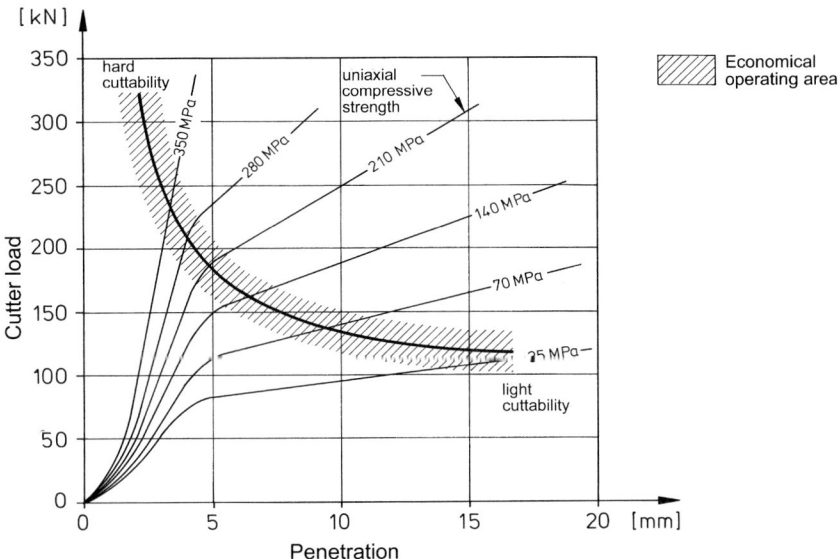

Figure 3-18
Robbins curves from 1970, with cutter pressure and penetration superimposed as a function of rock strength

Many papers [83, 132] have attempted to establish the relationships between disc cutters, contact pressure, rock strength and penetration in simple formulae. Such works mostly include the uniaxial compressive strength as a parameter. Although this is a meaningful, but by no means the most important parameter in the excavation interactions, it is the best-known parameter and the simplest to determine.

With the justified assumption that the splitting tension process is predominant, the uniaxial compressive strength cannot be the most important factor in determining the penetration. Tough rocks, exactly the ones with high compression strength, are hard to bore. The penetration then reduces slightly, only however when the discontinuities are unimportant. This difficulty only has an effect in rock, which is hard to bore with low penetration.

The penetration curves (Fig. 3-19) represent without exception the curves of the bored rock against the cutter loadings. The relatively large spread of values in example 7, Amsteg/CH Aare granite, results from the changes of anisotropy and the varying joint densities.

Real penetration trials must include all possible influences from the geology, the rock and the TBM.

It is often argued that partial collapses of the face lead to a transfer of loading from those cutters, which are no longer in contact, to concentrate loading on their neighbours, and so the entire thrust is distributed over a few cutters. Tunnel boring machines are, however, not controlled by power but geometrically. There is therefore no additional loading of the

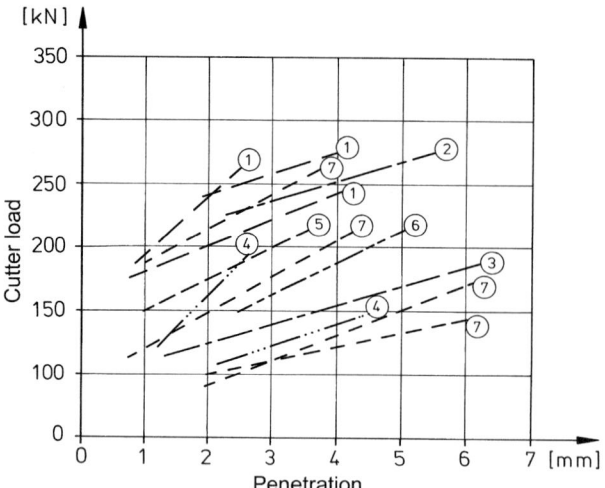

Figure 3-19
Penetrations in hard cuttable rock [3, 116, 128]
① Granodiorite, Paute c (Ecuador)
② Calaveras, California (USA)
③ Quartz diorite
④ Grimsel granite
⑤ Äspö granite (Sweden)
⑥ Bergen granite (Norway)
⑦ Aare granite, Amsteg (Switzerland)

neighbouring cutters, but extra loading results from the elastic deformation of the cutter head if this is not stiff enough because of its size or construction.

Wanner and Aeberli [179, 180] referred to this dependence of the advance rate in the function of the anisotropy based on their investigations of the mechanical tunnelling of the sloping headings of the Gotthard road tunnel. The results obtained then have since been confirmed many times.

Büchi [21] showed in his investigations considerable penetration increases in distinctively slaty rock (Fig. 3-20).

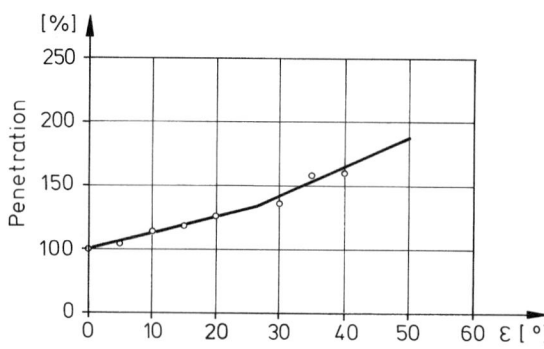

Figure 3-20
Penetration rate as a function of the angle ε between tunnel axis and main joint orientation [21]

3.3 Cutting Tools

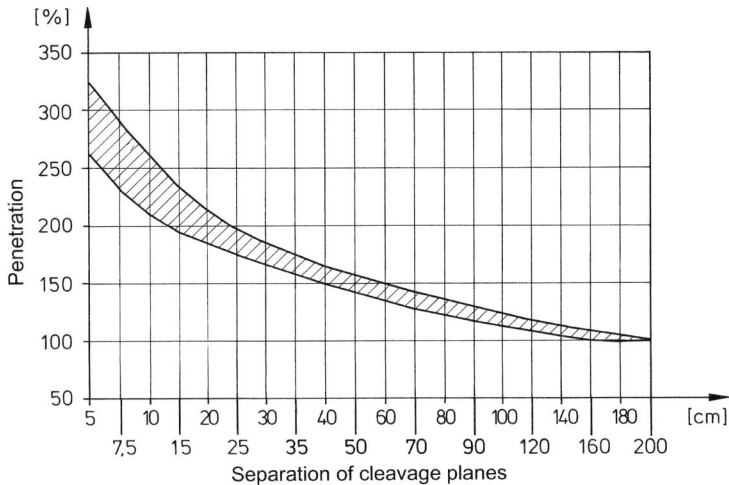

Figure 3-21
Influence of joint spacing on the excavation performance of a TBM [47]

According to investigations at the Technical University of Trondheim [172], the maximum influence occurs at an angle of 60° between the joint orientation plane of the slate and the tunnel axis.

The influence of joint was recognised by Simons and Beckmann and has a stronger effect on the penetration, the closer the spacing of the joints [157].

Gehring and Büchi established the same factors in-situ (Fig. 3-21).

The increased penetration does not, however, mean an increased advance rate. This reduces sharply due to increased safety measures when the spacing of cleavage planes falls under a critical length. For TBMs of smaller diameter this critical spacing is about 10–20 cm and with larger gripper TBM about 50–60 cm. Shield TBMs are less sensitive in this respect because collapses above the shield remain limited in extent and in no case are the crew or the equipment endangered.

3.3.5 Wear

The effectively of a TBM application and the economic limit are essentially determined by the wear, primarily to the cutting tools.

For the life of a disc cutter, measured in kilometres travelled, the essential factors are

- penetration and
- abrasiveness of the rock

Small penetrations lead, related to the distance driven, to a longer rolling distance and therefore to increased wear. In order to increase the life of the disc cutter, the diameters of the discs have been increased constantly as time went on; toughened and hardened steels have also been developed for very hard and abrasive rock.

Starting with 11″ discs, through to 12″, 13″, 15 ½″, 16¼″, the standard equipment of a TBM today is 17″, in extreme cases even 19″ disc diameter. Drilling tools with large diameters are heavy and unwieldy and without mechanical aids it would be impossible to change them.

The life of such discs varies widely, corresponding very strongly to the geological and geotechnical conditions.

Quartzite, Norway	100 km
Granite, Amsteg	280 km
Altkristallin, Amsteg	650 km
Alpine, Limes	900 km
Molasse, upper sweet water	8.000 – 10.000 km
Molasse, lower sweet water	1,000 km
Jura limestone and marl	11,000 km
Basalt-Lava, Lesotho	16,000 km
Marl chalk, Cyprus	22,000 km

The cutters sitting on the cutter head travel very different distances with the rotation of the cutter head. This means that cutters in outer positions travel much further and have to be replaced sooner. The frequency of changing the cutters for a TBM of 5 m diameter is shown in Fig. 3-22. There is a large increase in the number of changes after position 19, approx. 1.5 m from the centre of the cutter head. The cutters 35–38 experience significantly less wear, because many of these travel along the same track.

The wear leads to more difficult boring and increasing abrasiveness leads to considerable costs for disc cutters and the changing of drilling tools. It is therefore a good idea, in addition to the penetration that determines the distance rolled, to also find a measure for the abrasiveness, in order to be able to calculate these costs for drilling tools.

Many wear tests have been thought out, tested and compared with each other, from simple tests of metal hardness, through Brinel or Rockwell tests, impact measurements like Schmitt hammer and on to genuine wear tests like the Cerchar Abrasiveness Index test (CAI test).

The consideration of mineral hardness alone does not lead to usable results. These can, however, be used if the grain binding is extremely poor and the abrasion comes not from the loosening of the rock but from excavating it. Methods similar to Brinel have been successful with the undercutting technology of the AC-Habegger TBM. A very simple version of the CAI test has established itself for TBM use, which can be carried out on hand samples.

In the CAI Test, a conical metal point is drawn for a length of 1 m across the broken surface of the rock with a given contact pressure. The wear on the point is measured and compared with an index value. This process is repeated six times. Each time a new metal pin is used and always in different directions to the texture. The diameter of the stump of the conical metal point, worn down by rubbing on the test body, is measured. One index point corresponds to a conical stump diameter of 0.1 mm.

3.3 Cutting Tools

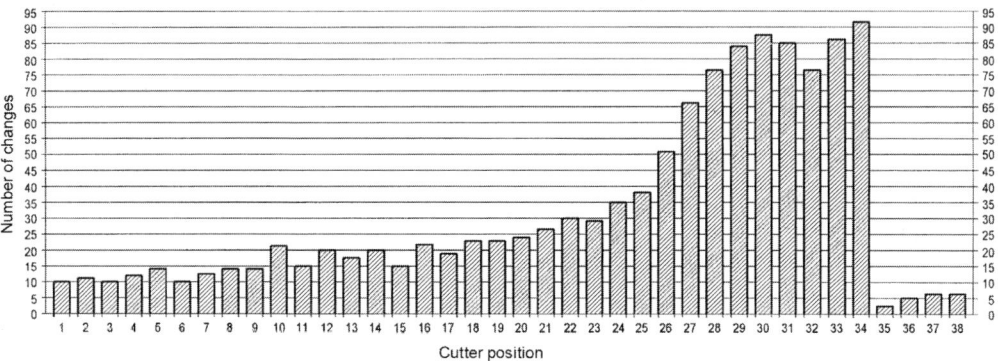

Figure 3-22
Frequency of cutter disc changes for a TBM ⌀ 5 m (Amsteg pressure tunnel) [3]

Büchi, in 1984, collected together many average values of measurements of the CAI dependant on the rock type (Fig. 3-23).

Gehring made use of project experience to plot the relationship between the CAI and the weight loss due to wear on a disc cutter in mg steel per metre rolled (Fig. 3-24). This diagram also shows high a wear rate for today's steel alloys. Instead of the correction factor of 0.74, a value of 0.65 seems to fit the reality better.

The Laboratoire Central des Ponts et Chaussées (LCPC), Paris, has developed a further, but not very practical test of abrasiveness (ABR-Index) [22]. The test is carried out on 500 g ± 2 g of test fraction 4/6,3 mm. The material to be tested is filled into a cylindri-

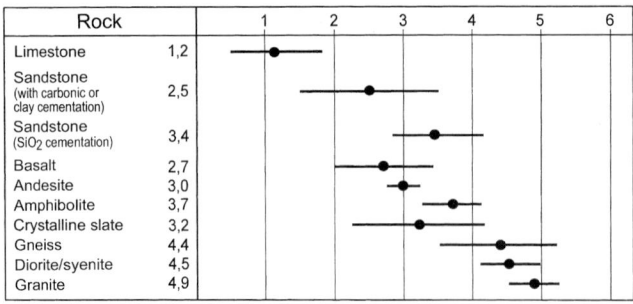

Figure 3-23
Rock-dependant CAI values with average and standard deviation [21]

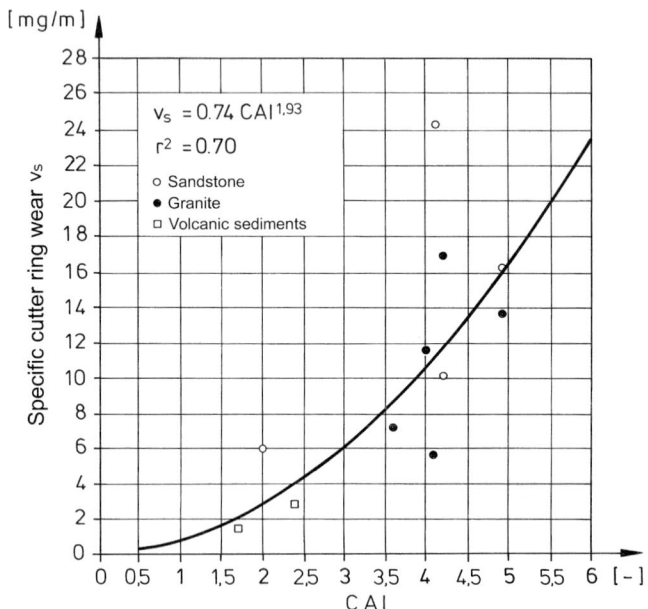

Figure 3-24
Cutter disc weight loss as specific cutter disc wear in mg/m rolling distance as function of the CAI [48]

cal container. In this container, a small metal plate (5 × 25 × 50 mm; hardness: Rockwell B 60–75 HRB) rotates at 4,500 rpm. The metal plate has a sandblasted surface and weighs between 46 and 48 g.

An advantage of the ABR index, which should not be underestimated, is the ability to test the wear rate of rock samples with different water contents. The wear rate is known to increase with increasing water content. The ABR index of the rock can be tested with the known water content in the mountain.

3.3 Cutting Tools

The abrasiveness coefficient ABR is defined as the weight loss of the metal plate per Tonne of sample material:

$$ABR = \frac{M_0 - M_1}{G_0} \left[\frac{g}{t}\right] \tag{3-1}$$

M_0 weight of the metal plate before the test
M_1 weight of the metal plate after the test
G_0 weight of the samples

LCPC gives the following classification scheme for the ABR:

Table 3-1
Classification scheme for the abrasiveness coefficient [22]

ABR (g/t):	0	500	1000	1500	2000
Abrasiveness:	very weak	weak	medium	strong	very strong

This LCPC is primarily suitable for stone processing machines, like crushers etc. It does not take the texture or the cohesion of the grains into account at all, which is the case with the CAI test. The LCPC test can, however, be useful for TBM tunnelling in cases where the CAI cannot give results, like loose sandstone with low cementation and a high content of sharp abrasive minerals.

Various properties of the cutter roll track, like discontinuities or differing rock hardness, can lead to breaking out of the cutter surface. Figure 3-25 shows a blocked disc

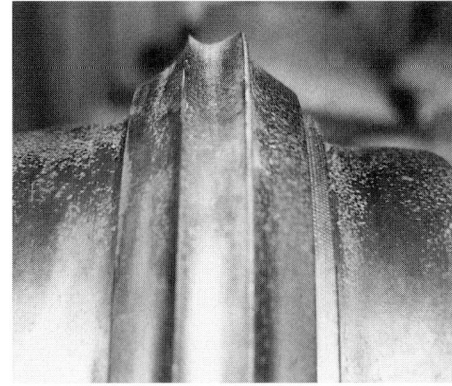

a) b)

Figure 3-25
Wear patterns of cutter discs
a) Blocked cutter disc
b) Cutter disc with normal wear

cutter (Fig. 3-25 a) compared with normal wear Fig. 3-25 b). Experience shows that blocked cutters and cutters worn on only one side due to seized rollers will amount to about 20–30 % of the discs having to be replaced.

With the use of TBMs being made possible in increasingly difficult hard rock through larger disc cutters and increased contact pressure, steel qualities are required, which have enabled better cutter life.

Figure 3-26 shows seven various disc cutters of 17″, all seven with new sections, but in different steel qualities.

Figure 3-26
17″ cutter discs of varying steel quality
a) Robbins normal ³⁄₄″ wide, welded bearing surface
b) Palmieri normal, welded bearing surface
c) Palmieri HP
d) Rock-Tech Robbins medium-duty ³⁄₄″ extra
e) Rock-Tech Robbins 1″
f) Robbins ³⁄₄″ normal
g) Robbins ½″ normal

Of primary interest to the contractor is the relationship between rock, CAI, compression strength and average disc cutter life measured in m³ rock excavated per disc cutter. This is shown in Fig. 3-27 for a 17″ cutter from a multitude of results on site. The lack of precision of such a diagram should be pointed out. The scattering of the individual factors is too large to allow precise predictions to be achieved.

Ewendt [43] developed a prognosis formula based on the wedge angle theory for the tool wear of disc cutters. The basis for this formula is the advance formula according to Sanio [132]:

$$FN_m = \frac{3}{25} \cdot \sqrt{IS_{50} \cdot d \cdot s \cdot p} \cdot \tan \frac{\varepsilon}{2} \tag{3-2}$$

3.3 Cutting Tools

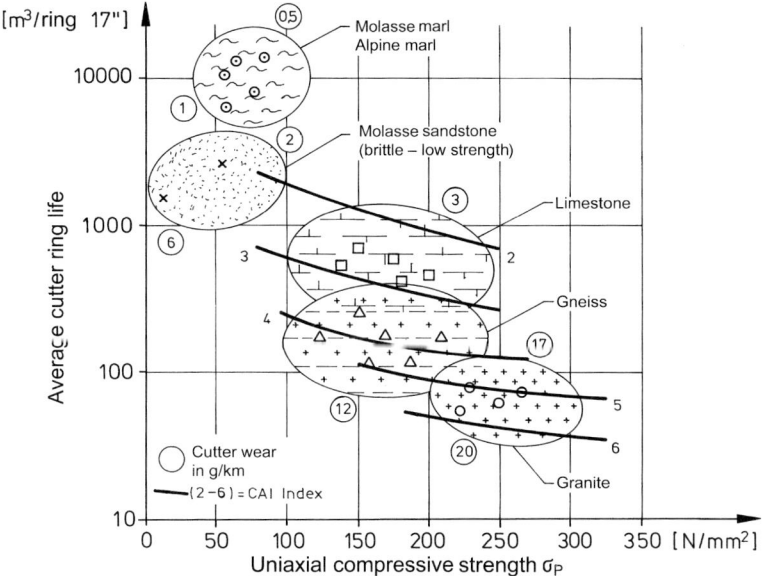

Figure 3-27
Cutter disc wear as a function of uniaxial compressive strength and the CAI

where
FN_m average axial force on the cutter [kN]
IS_{50} point load index for the rock parallel to the excavation surface [N/mm²]
d disc diameter [mm]
s disc track spacing [mm]
p penetration (= advance) [mm]
ε wedge angle of the cutter [8]

According to this advance formula, there is a quadratic relationship between the thrust and the blade pressure required to reach it. Ewendt shows that, for a machine with defined cutter spacing, the wear rate w is also proportional to the square of the thrust. In the general formula, this relationship between force and wear is:

$$w = C2 \frac{FN_m^2}{\sqrt{s \cdot d}} \tag{3-3}$$

To determine the dimensioned proportionality factor *C2*, Ewendt presents two procedures. Proportionality factor *C2* can be determined for each rock from cutting tests. Table 3-2 shows values for the proportionality factor in relation to the rock in each case.

As an alternative to these cutting tests, Ewendt modified the wear value *F* [134] of a rock developed by Schimazek and Knatz for cutter heads of partial excavation machines and developed from this a relation to the proportionality factor *C2*. The wear

Table 3-2
Proportionality factor $C2$ according to [43]

Rock	$C2$ $\left[\dfrac{mg \cdot mm^{1,5}}{m \cdot kN^2}\right]$
Basalt	1.74 ± 0.63
Gabro	0.66 ± 0.32
Gneiss 0/90	1.30 ± 0.38
Gneiss 90/90	1.99 ± 0.70
Granite	3.89 ± 1.30
Quartzite	4.27 ± 1.14
Sandstone	1.01 ± 0.25

value F of a rock is a product of quartz content, the average grain size of the quartz and its strength. The modification was chosen to restrict the influence of the grain size. The formula for the modified F value is then:

$$F_{\text{mod}} = Qz \cdot IS_{50} \cdot \sqrt{dQz > 1\,\text{mm}} \tag{3-4}$$

where
F_{mod} modified F value [N/mm^{1,5}]
Qz quartz content of the rock [%]
IS_{50} point load index [N/mm^2]
dQz quartz grain size of the rock [mm]

According to Ewendt, the proportionality factor $C2$ can also be determined using the modified wear value F:

$$C2 = 0.45 \cdot F_{\text{mod}} = 0.45 \cdot Qz \cdot IS_{50} \cdot \sqrt{dQz > 1\,\text{mm}} \tag{3-5}$$

The complete wear prognosid formula now gives for the wear rate w [mg/m]:

$$w = 0.45 \cdot Qz \cdot IS_{50} \cdot \sqrt{dQz > 1\,\text{mm}} \; \frac{FN_m^2}{\sqrt{s \cdot d}} \tag{3-6}$$

The general opinion today is that Ewendt's work is no longer valid, because it is based on the former assumption of the wedge angle theory for the cutting process.

3.3.6 Wear and Water

The wear on disc cutters, roller housings, roller assemblies, bucket edges and the rim of the cutter head is considerably increased by just a relatively minor ingress of water [43]. Observations in the Murgenthal tunnel have confirmed this observation from other sites.

3.3 Cutting Tools

Apparently, running water increases the wear only slightly. With a small ingress of water, however, the surface tension agglomerates the water film around the grains and makes an abrasive paste with the ground-off muck, which develops an enormous abrasive effect. This is secondary abrasion from the already excavated rock. Figure 12-4 shows such wear to the rim of the cutter head of the Murgenthal tunnel TBM.

3.3.7 Cutter Housing

The cutter housing is the mounting of the disc cutter to the steel construction of the cutter head. As a result of the high loading on the disc cutters, cracks have been discovered in the housings, especially of larger TBMs, which have led to interruption of the drive for the necessary welding work. Such cracks have been seen on the attachment plate under the housing and also in the corners of the upper part of the housing.

It can be assumed that the occurrence of cracks on the surface of the cutter head is caused by the overall deformation of the cutter head structure (Fig. 3-28), while the cracks on the mounting strips are caused by the transfer of the cutting force into the steel construction of the cutter head.

To test these assumptions, Herrenknecht carried out trials with various types of housings, in order to achieve in improvement of the long-term strength of the entire system housing – cutter head.

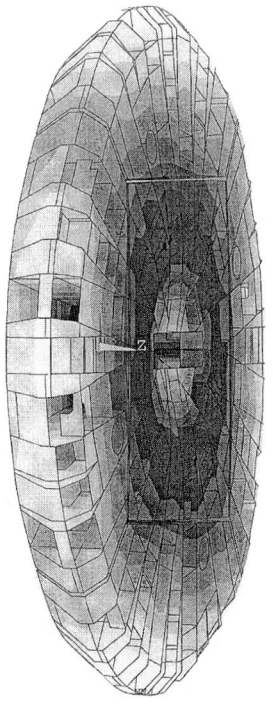

Figure 3-28
Overall deformation of the cutter head structure of a TBM [61]

Various types of housings were investigated in a test rig. First, an example of damage was produced on a reference housing through a pulsating loading similar to a real TBM drive with force transfer through the disc cutter, which corresponded to the cracks to the mounting plate of the cutter head. After this, the reference housing was subjected to a bending load to simulate the overall deformation of the cutter head and to reproduce the resulting pattern of cracks in the test piece. It was demonstrated that the overall deformation of the cutter head determines the occurrence of damage requiring repair.

The test was repeated with altered bearing housings, in order to evaluate the alterations in comparison to the reference housing. Many detailed improvements led finally to a new bearing housing, which was considerable optimised regarding crack pattern and crack extension under standing load. Uncontrolled crack extension in the steel construction was avoided. The total number of loading cycles, after which it was necessary to repair the bearing housing, was increased by a factor of 2.5.

The new bearing housing (Fig. 3-29) has, for example, large rounded corners to reduce stress peaks and a wider housing. The mounting plate and the connection to the surrounding cutter head steel construction were also modified.

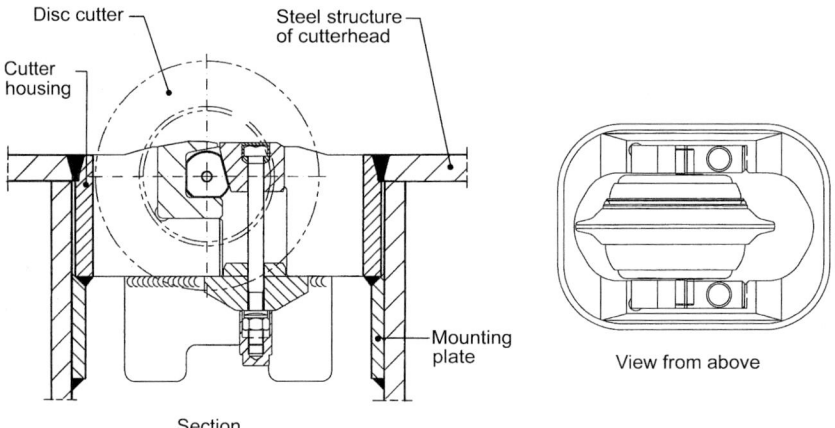

Figure 3-29
Modified cutter housing [61]

3.4 The Main Drive

3.4.1 Types of Main Drive

Parallel to the increasing size of disc cutters and the associated increase in contact pressures, the installed torque of TBMs has also increased since the 1970s.

Figure 3-30 shows this development through a representative selection of tunnel boring machines from various manufacturers in the range of diameters between 3.5 m and 12.5 m. The figures are for the continuous torque at normal revolutions.

3.4 The Main Drive

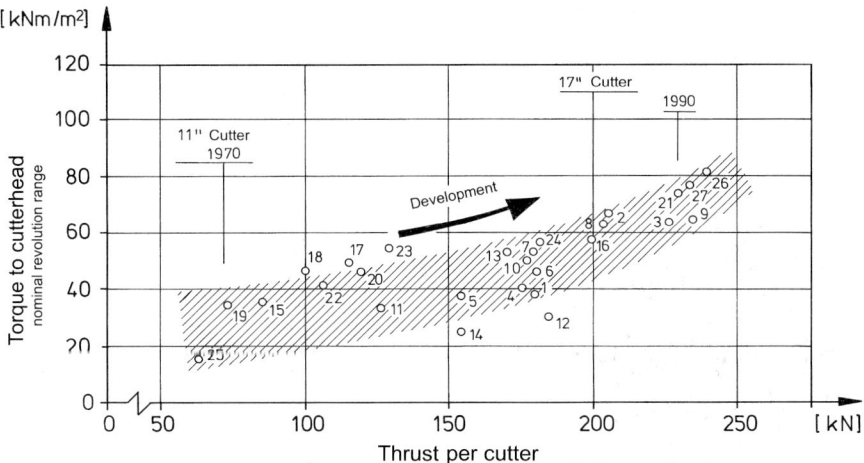

Figure 3-30
Development of cutter discs and torque for TBMs [143]

TBMs are driven by electric or hydraulic motors through clutch and gearbox with teeth on a large gearwheel mostly connected to the main bearing of the cutter head. The different types are:

- Electric drive with totally enclosed water-cooled motors, frequently dual-speed with two revolution speeds.
- Electric drive with frequency control and a continuously adjustable control of revolutions in a wide range.
- Electro-hydraulic drive. Electrically driven hydraulic pumps supply the hydraulic motors of the main drive with the required amount of oil and appropriate pumping pressure. The control of revolution speed is continuously adjustable over the entire operating range of the motors.

The high revolution speed of the electric and most hydraulic motors necessitates the use of a reduction gear between motor and drive pinion.

To lower the extreme starting current and to ensure a high starting torque for the cutter head, clutches are installed between electric motors and gear units.

This enables on the one hand the motors to be started in neutral in pairs – for a TBM of about 12 m diameter it could well be 10–14 motors – and finally all can be connected to the cutter head by using the clutch. This, however, requires friction clutches to be used. In hydraulic clutches, the oil would rapidly become too hot, which would cause the fuse to cut out. Machines with hydraulic clutches should therefore only be used in geotechnically favourable conditions, where there is no risk of overbreaking or collapses. The advantage of hydraulic clutches in being kind to the machine is clear. Tunnel boring machines should, however, be built for rough drives and primarily to bore tunnels.

Electric motors with constant revolutions show a very high breakdown torque. Breakout torque normally increases to three times the standard torque. For hydraulic drives, the breakdown torque is much lower. The high stalling moments help to keep the cutter head in movement after small collapses in the face. This is very favourable, especially for larger TBMs in lightly fractured or fractured rock.

Newer TBMs are equipped with variable frequency motors. These have a considerable number of advantages compared to motors with a fixed revolution speed:

- Low starting current, so that lower demands are made on the primary side circuit.
- The motors can mostly be loaded at 50 Hz to a torque of 150% of the standard torque for a short time.
- The starting torque may be up to about 180% of the nominal torque, even at a current requirement of only about 40–50%.
- The revolution speed can be adapted to the geological requirements.

Figure 3-31 shows a typical torque curve for a variable frequency motor.

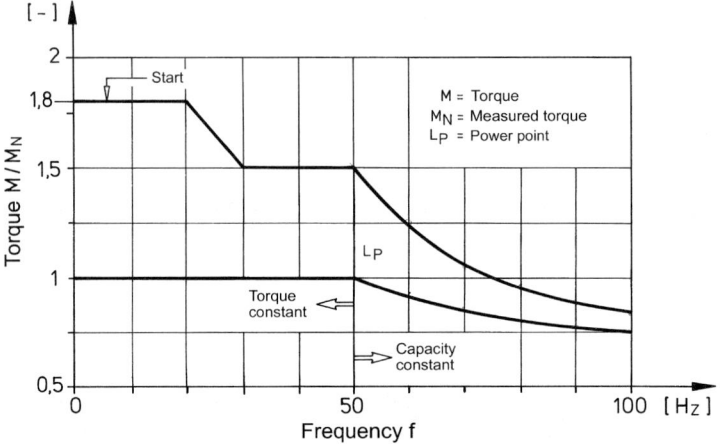

Figure 3-31
Torque curve of a variable-frequency electric motor according to ABB [4]

Up to the power point L_P, a constant torque of about 150% of the nominal torque is available, starting off even 180%. Above the power point L_P, the torque sinks but the power remains constant.

The TBM operates very economically above the power point L_P with a relatively low torque in normal operation. When the geological conditions alter, requiring an increase of torque, the revolution speed can be reduced to the power point, which causes the torque to increase considerably. To the left of the power point, the penetration would have to be reduced in the constant torque range. The curve to the right of the power point is limited by the revolution speed of the cutter head, which is limited by the limiting speed of the disc cutters. A precondition of such usage is adequate dimensioning of the frequency converter. Variable frequency motors at low revolution speeds also require well-organised supervision of the lubrication and water-cooling.

Hydraulic motors had advantages over electric motors above all in the fixed revolution range. The very advanced technology of variable frequency electric motors today, however, makes them more suitable. Hydraulic drives show much lower overall efficiency, require more effort to install and de-install and produce a lot of waste heat on account of the low efficiency, which is undesirable in a TBM drive and has to be dealt with corresponding expense.

3.4.2 Main Bearing

The main bearing does not just have to take the reaction forces of the cutter loadings. Loading from eccentricities from ruptures and from collapses at the face also has to be supported without damage. Main bearings are, however, rarely damaged by loading alone. Their life is mainly determined by the quality of the sealing system. It is therefore important to make sure that the seals never run dry, because they would then rapidly lose their function and a bearing failure would be predictable. Sealing lips must therefore always run in a grease film and greasing points in the sealing system should not be more than 70–80 cm apart.

Main bearings with moderate loading are still frequently constructed as crossed roller bearings or as ball bearings. Heavy bearings have roller tracks in three axes (Fig. 3-32).

3.5 Advance Rate

The penetration alone does not determine the actual boring advance over a particular period. The actual advance rate is also affected by other factors like rock support, changing cutters, repairs and not least the efficiency of the back-up logistics for removing the muck from excavation and delivering construction material and supplies.

The diagram in Fig. 3-33 shows two fundamentally different working processes with the percentage time for the individual activities; on the left, advance characteristics for a gripper TBM with 5 m diameter and on the right, the situation for a shield machine of 12 m diameter in the lower sweet-water molasse as intercalation of sandstone and marl. With the gripper TBM, all three situations caused considerable time to be spent changing cutters, 54–59% of the boring time required, as can be read from the site logs. Rock support leads to varying, mostly however very long interruptions of boring. In Fig. 3-33a, in geology with rock burst occurrences, rock support requires 121% of the actual boring time; in geological conditions with ruptures at the working area and around the machine (Fig. 3-33b) 20% of the boring time was required for rock support. Repairs to the cutter head increased strongly, taking up as much time as the boring process itself. In geology with ruptures and rock bursts (Fig. 3-33c), the time for dealing with the rock support rises to 270% of the actual boring time.

It is therefore no wonder that the average advance for gripper TBMs is often only 35% of the peak advance. Increasing the boring rate achieves little in such a case, it is much

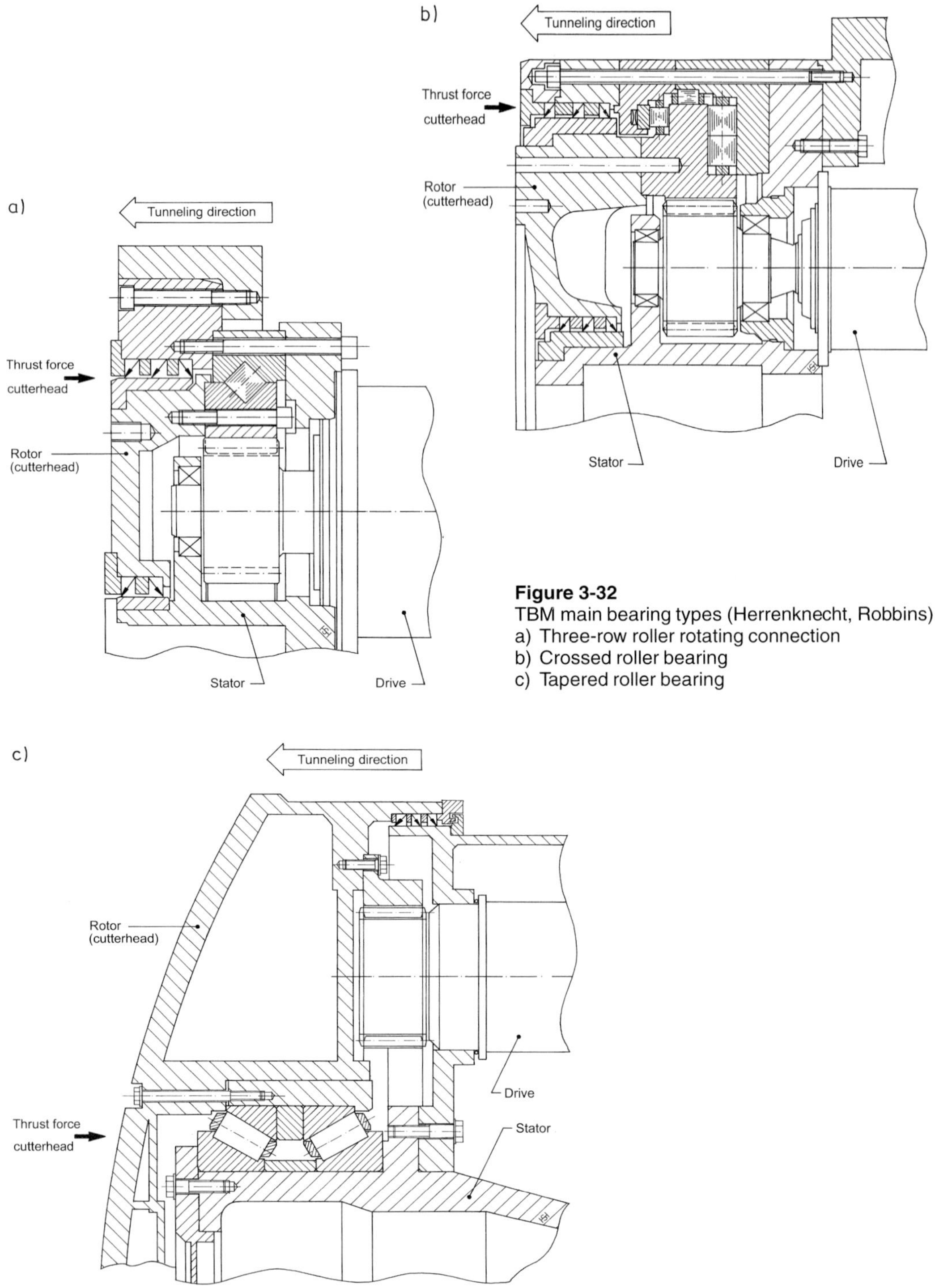

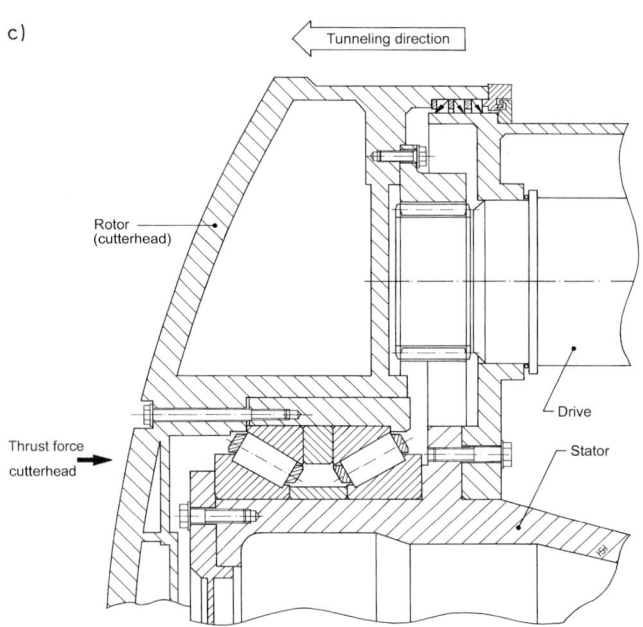

Figure 3-32
TBM main bearing types (Herrenknecht, Robbins)
a) Three-row roller rotating connection
b) Crossed roller bearing
c) Tapered roller bearing

3.6 Special Types   55

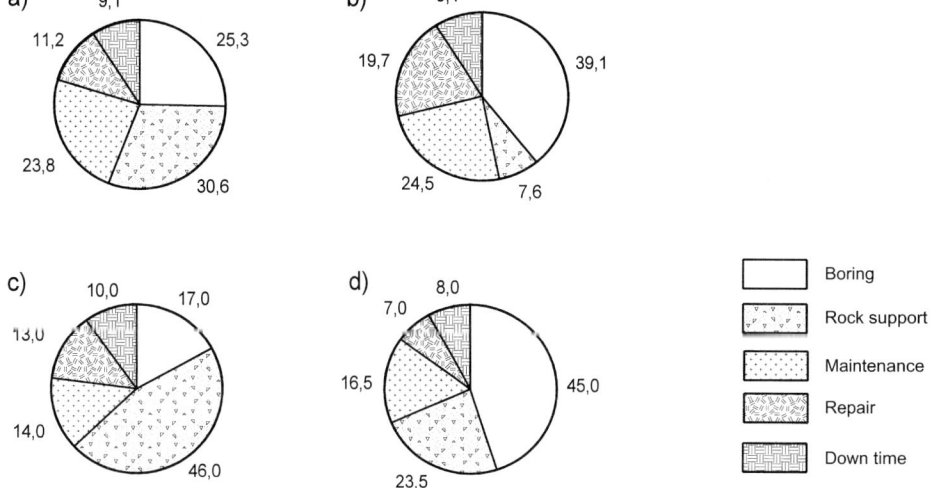

Figure 3-33
Distribution of activities during TBM excavation [%]
a) Gripper TBM ⌀ 5 m with rock bursts
b) Gripper TBM ⌀ 5 m with ruptures at face and around machine
c) Combination of a) and b)
d) Shield TBM ⌀ 12 m

more effective to improve the cuter changing system and improve the system of rock support against rupture.

The shield TBM produces the same advance as the gripper TBM in almost all rock types. The advance can be improved with longer segments. The time taken to install the rings is practically the same for short or for long rings. This means that, for an advance cycle of excavation and ring construction, time is gained for excavation because the overall advance can be increased through the reduction in time to install the rings. In past decades, the length of segments has been increased constantly, starting with a 1.0 m length of segment in 1970 to 2.0 m in 1998. This might turn out to be the upper limit, because on the one hand, larger segments are too heavy to handle, and on the other hand the topology of tunnels determines the length of the segments in curves.

3.6   Special Types

Machine manufacturers have developed special types of tunnel boring machines under pressure, on their initiative or in cooperation with contractors. Finally it has been cost considerations with the intention of the investment as far as possible, or geometrical requirements, because the relevant cavity had an excessively large section, too little length or had a cross-section other than circular. With the exception of reamers, all special types were initially developed as alternatives to drilling and blasting, where this was ruled out by environmental problems, or in rock types, which could not be economically worked by the mining technique of cutting a slot and collapsing the rock above and where, however, hard rock machinery was required for mechanical excavation.

3.6.1 Reamer TBMs

Tunnel boring machines once represented a very high investment in comparison to the wage costs at the time. The idea of constructing reamers was therefore sensible. The total costs of a smaller pilot machine and an enlargement stage were about 150% of two TBMs of the diameter of the pilot and the enlargement stages. The use of reamers in sloping headings has proved itself, on the one hand because boring larger diameters often led to instability in the face and on the other hand because the enlargement from downwards is less dangerous work, as long as the rock stands up.

A single stage reamer TBM is shown in Fig. 3-34a. The use of these reamer TBMs led in some cases to technical, but never to economic success. Comparison with shield TBMs shows the technical disadvantages of the reamers drastically (see also Chapter 9.2).

The enlargement of tunnels often led to precarious situations. It was hard to control the air to prevent impermissible levels of dust, and on the other hand instability often occurs during the enlargement at the edges of the pilot bore, no longer allowing safe working (Fig. 3-34b). In the Walensee tunnel, a two-stage reamer was used and along considerable stretches an additional support had to be installed with rapid-hardening

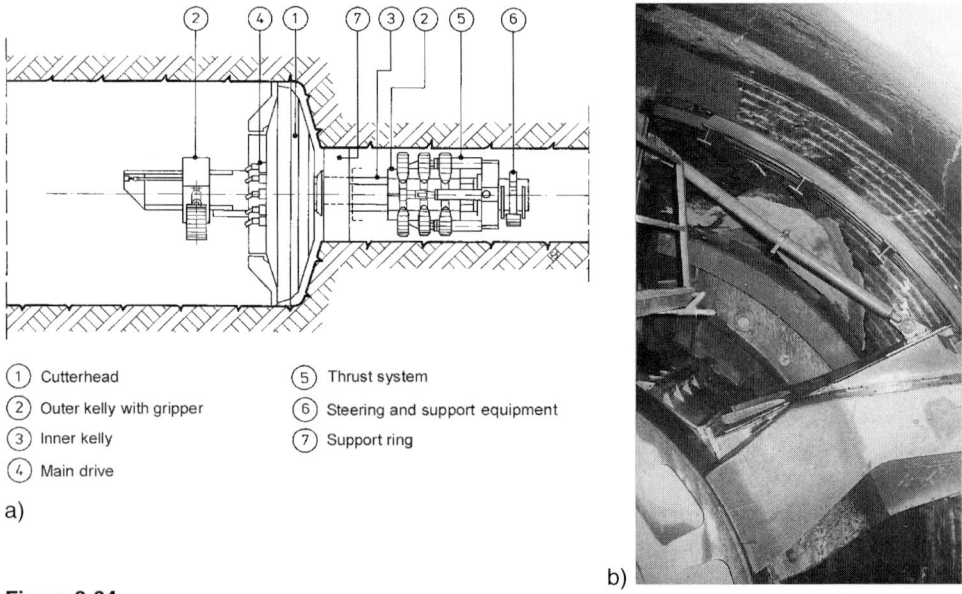

① Cutterhead
② Outer kelly with gripper
③ Inner kelly
④ Main drive
⑤ Thrust system
⑥ Steering and support equipment
⑦ Support ring

a)

b)

Figure 3-34
Reamer TBM
a) Diagram of a single-stage reamer TBE IV 450/1080, Locarno (Wirth) [107]
b) Ruptures at the intersection to the second enlargement stage of a 2-stage reamer, Kerenz tunnel

3.6 Special Types

mortar in the changeover from stage 2 to stage 3 and also over the spoked cutting wheel of stage 3 in order to be able to make any progress at all.

Figure 3-35 makes clear that only limited support measures are possible at all behind the cutter head. It is therefore understandable that the working procedure pilot heading – enlargement and corresponding support measures considered overall can only lead to a disappointing overall advance rate.

Figure 3-35
Reamer TBM with drilling guide for rock anchoring

Respectable advances for the single activity of enlargement have only been achieved under very favourable geological conditions like in gneiss for the Locarno bypass or in Jurassic limestone with a good stand-up time in the Sauges tunnel at Neuenburgersee lake.

3.6.2   Bouygues System

With TBMs, direct access and sight of the face through the drilling head is not generally possible. The type of machine used in the 70s under the name "Bouygues tunnel machine" was intended to get round this problem. The face was excavated with disc cutters, but there were only a few discs provided for excavation. The cutters were each mounted on the end of a swinging arm. The swinging arms were arranged on a carrier construction, which rotated like an ordinary cutter head but much faster. The resulting oscillating movement enabled the cutters to cover the entire face. Slewable buckets removed the muck from the arched face (Fig. 3-36).

This type of machine could not establish itself in competition and disappeared from the market after a few deployments.

Figure 3-36
Cutter head of a Bouygues tunnelling machine TB 500, ⌀ 5 m, Arenberg mine, France

3.6.3 Mobile Miner (Robbins)

In 1970, tenders were requested for a three-lane motorway tunnel in Zurich. The environmental restrictions, especially the foundations of very dense building very near to the tunnel, made drilling and blasting impossible. The contractor Schafir & Mugglin AG therefore cooperated with Robbins to develop the concept of equipment for the drive, which would later lead to the Mobile Miner. An excavation drum equipped with disc cutters, which could be rotated and axially driven, was mounted in a shield of 14 m diameter (Fig. 3-37). This drum was intended to rotate and so excavate the rock similarly to the then familiar processes. The development was never put into practice because of project delays and variations.

The idea of the rotating drum led to the Mobile Miner (Fig. 3-38) and to a machine for sinking shafts (see Fig. 3-45).

The Mobile Miner has cutter head drums of 3.5–4 m diameter, equipped with 15″ or 17″ Disc cutters. The installed power of 300–500 kW enables remarkable excavation performance of 75–80 m^3/h in hard rock. Such Mobile Miners have been used in Japan in 1996 and in Australia in 1992.

3.6.4 Back-Cutting Technology

3.6.4.1 Mini-Fullfacer (Atlas Copco)

The further development of the Wohlmeyer machine (see Fig. 1-12a), firstly by the company Habegger AG, later by Atlas Copco with at that time four rotating heads, was at the end of the 1960s the only machine capable of producing respectable performance

3.6 Special Types

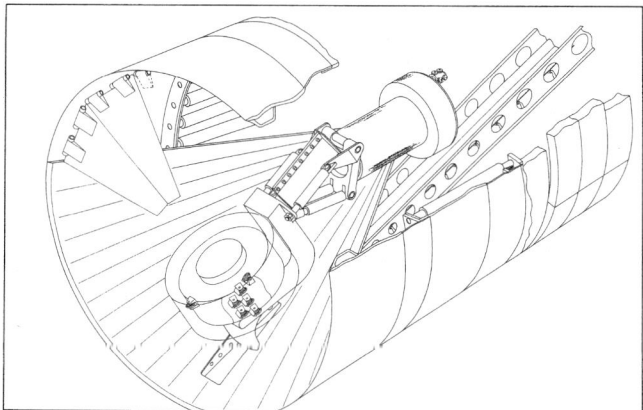

Figure 3-37
Planned excavation system for Milchbuck tunnel, 1970, with cutter heads equipped with disc cutters (not built)

Figure 3-38
Mobile Miner, Model 130 (Robbins)

in rock, which was hard to bore. In connection with sewerage, there were problems with short spur headers with mostly very small dimensions to excavate. A machine was developed analogous to the Wohlmeyer machine with four milling heads except with only a single milling head slewing about a horizontal axis. The resulting cross-section was man-high and about 1.2 m wide (Fig. 3-39 b).

This machine could, however, only be used in really solid rock. The few times it was used in Switzerland under such conditions were successful. It was not successful when used in fractured rock making necessary support directly behind the face, as was the case with a site in Bochum [162]. It was not possible to install support early next to the machine because of lack of space, so the cutter head of the machine was constantly buried in muck.

Some machines are still being used in Australia.

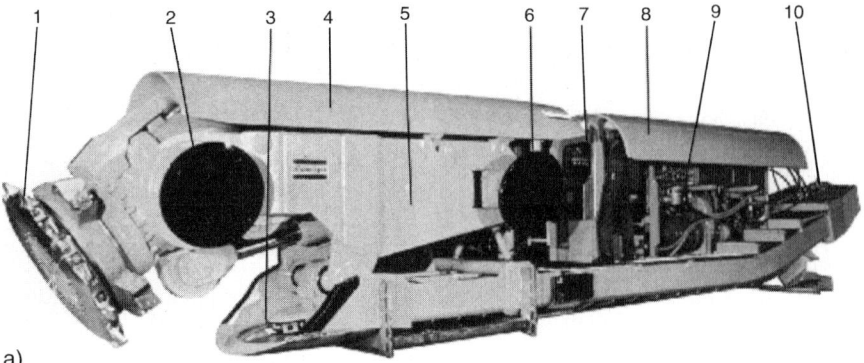

(1) Cutter head
(2) Front gripper
(3) Chain conveyor
(4) Protection roof
(5) Machine body
(6) Rear gripper unit
(7) Operator position
(8) Drive trailer
(9) Hydraulic unit
(10) Drive for chain conveyor

Figure 3-39
Mini Fullfacer (Atlas Copco)
a) Whole machine [5]
b) Section bored by a Mini Fullfacer

3.6.4.2 Continuous Miner

Deutsche-Montan-Technologie (DMT) developed the Continuous Miner in collaboration with Wirth, a under-cutting machine, which can offer advantages in particular conditions. Disc cutters are mounted on a number of arms (Fig. 3-40 b), which are not however aligned at right angles to the face like the cutters of a full-face machine. The special features of the under-cutting process (Fig. 3-40 a) lead to course muck and low thrust forces.

Such a TBM can scarcely compete economically with a TBM with fully equipped cutter head for main drives. The achievable advances are not sufficient to meet deadlines by the client.

3.6 Special Types

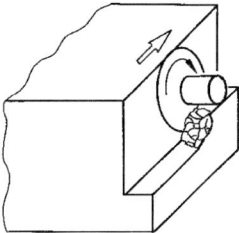

Undercut technology

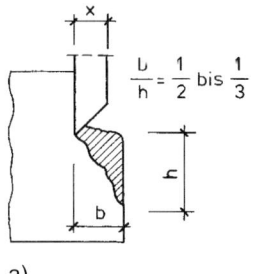

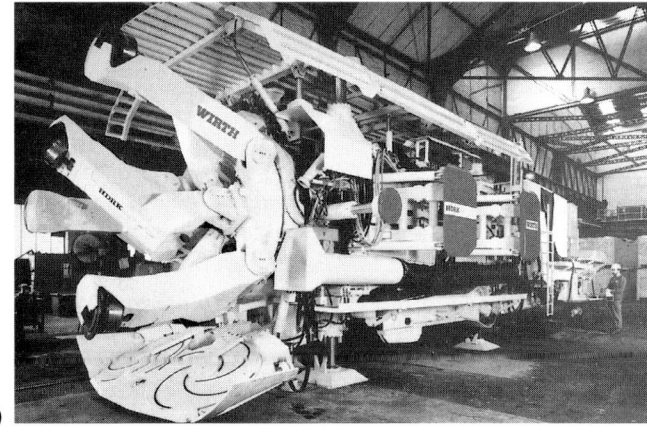

Figure 3-40
Undercutting method
a) Use of discs in under-cutting method
b) Continuous Miner (Wirth)

It can indeed be suitable for driving side passages, dead end headings or very short tunnels, if the geotechnical conditions rule out a partial excavation machine, but mechanical excavation is prescribed for environmental reasons.

The advantage of such a TBM is the range of sections possible. Circular sections are not essential, arched sections are also possible.

The use of this back-cutting technique with the one-sided tapered cutters shown in Fig. 3-40 can come into consideration for rock, which is easy to bore. If the rock is too hard, the cutters will wear so rapidly that back-cutting is impracticable.

3.6.5 Shaft Sinking

One of the most tedious operations in mining is sinking shafts by drilling and blasting. After the first successes of tunnel boring machines in smaller and medium sections, contractors looked for possibilities to mechanically excavate the many shafts required for mines, hydro power stations, or the ventilation of tunnels.

Boring of shafts cannot only fully mechanise access shafts and blind shafts, but also the in-situ rock and loose material can be continuously removed. The shaft boring processes used here, as shown in Fig. 3-41, can be categorised into [59]:

(A) With removal of muck downwards to a surface: Shaft boring with pre-drilling.
   1. Raise boring: Shaft boring from the bottom upward by a machine at ground level connected through the pilot hole.
   2. Raise boring combined with downward boring: boring the shaft on the pilot hole with a downward-boring machine.

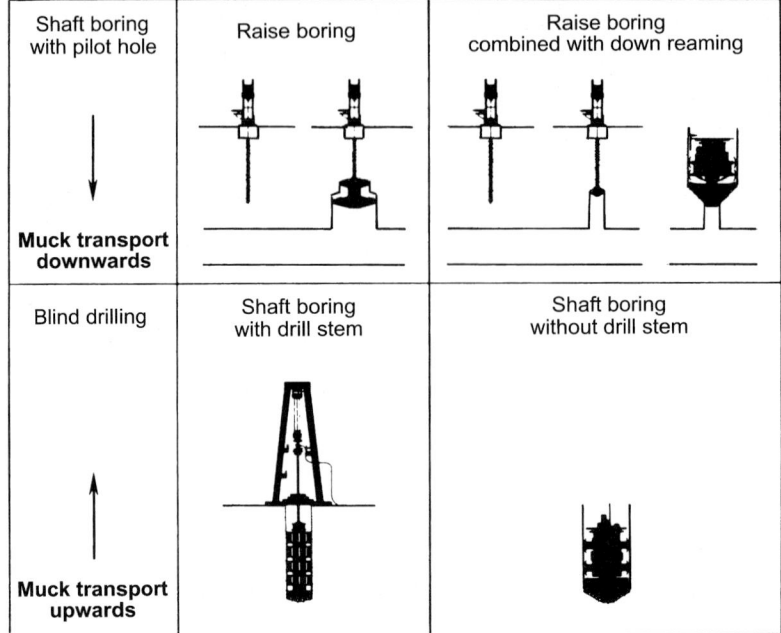

Figure 3-41
Categorisation of shaft sinking processes [59]

(B) With muck removal upwards to the mouth of the shaft or to ground level: Shaft boring without pilot hole (Blind drilling).
3. Shaft boring out of the full hole with drill stem.
4. Shaft boring out of the full hole without drill stem.

3.6.5.1 Raise Boring

Raise boring, which originates from North American mining, is the most common shaft sinking process. The boring equipment developed from borehole technology and called a Christmas tree, is scarcely suitable in middle to hard rock. Wirth was already building large hole enlargement boring equipment in the 1960s, later categorised under the term raise boring. These machines were used successfully in mines (Fig. 3-42). The devices are capable of boring only modest diameters at 50–80 cm, in exceptional cases up to 1.2 m.

Only after the cutter heads of tunnel boring machines had been adapted for such shaft raising equipment was the boring of large diameter shafts of diameter up to 5 m made possible, at depths of a few hundred metres with a tension force of 10.000 kN. Directional drilling rigs can pre-drill with sufficient accuracy.

The first step is to use a boring rig to drill an aiming, pilot or pre-drill hole of max 380 mm diameter to the target cavity (chamber, tunnel, heading). In the target cavity,

3.6 Special Types

a)

b)

Figure 3-42
Raise boring: Shaft boring on pilot hole with drill stem
a) Sandvik cutter head
b) Raising equipment

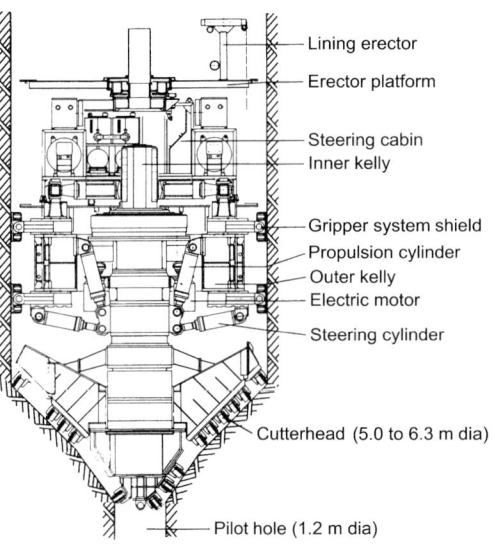

Figure 3-43
Construction of a shaft-boring machine without drill stem (Wirth) [59]

the cutter head is attached to the drill stem (Fig. 3-42b) and raised up rotating. Raise boring shafts can also be bored on a slope, up to the limiting slope of the material.

The most important manufacturers of shaft raising equipment are Wirth, Robbins and Rhino. Sandvik belongs to the leading manufacturers of cutter heads (Fig. 3-42a).

Another process of boring shafts with muck removal downwards to a bottom is the combination of raise boring equipment, which drills a pilot hole of 1.5–2.0 m diameter, with a sinking bore machine without drill stem, which enlarges the hole to the final diameter. This principle is similar to a tunnel boring machine standing on its head and has all the advantages of mechanical excavation over the process with drill stem. Fig. 3-43 shows the construction of a shaft-boring machine without drill stem.

The maximum depth, which can be bored by this method, is only limited by the requirement of keeping the pilot hole vertical and the requirement that the tunnel at the bottom must be constructed first.

3.6.5.2 Blind Drilling

With shaft boring without pilot hole, where the muck has to be transported upwards in the shaft itself, there are processes guided by a drill stem and processes without a drill stem.

The oldest process is shaft boring guided by a drill stem, with the muck being transported out of the shaft to ground level. About 40 shafts were bored with the Honigmann shaft sinking process, first used in 1896, until it lost out in competition to the shaft freezing process. A large diameter shaft was bored in one or more stages, with walls stabilised by a clay suspension. The removal of the muck was already done then by blowing [59]. The former disadvantage of not being able to correct deviations of position of the bored hole have been compensated in modern drill-stem-guided shaft sinking by the use of steerable boring machines. This shaft-boring machine resembles a TBM standing on its head and is steered by a gripper. The drill stem only serves to transport the muck.

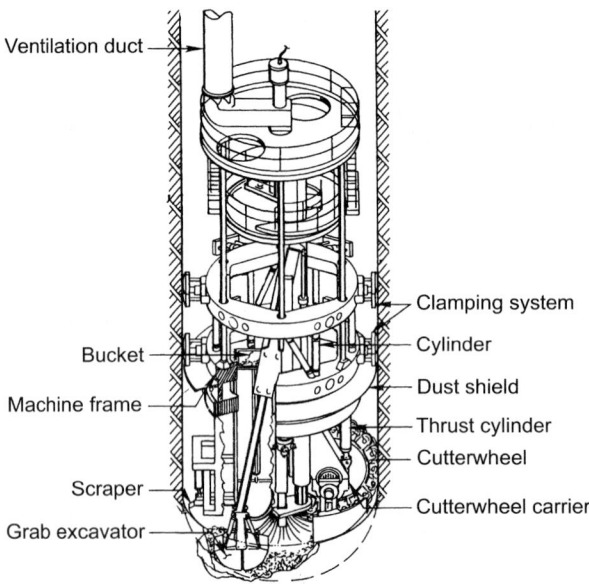

Figure 3-44
Shaft boring machine by Robbins (Model SBM), with mechanised preliminary and main muck transport (Robbins)

3.6 Special Types 65

Shaft boring without a drill stem is still in development. The particular problem is how to lift the muck from the bottom (preliminary transport) and how to transport it up the shaft (main transport). Versions with mechanical (Fig. 3-44) and hydraulic preliminary and main transport have been tried, although the main transport should be carried out mechanically out for economic reasons [169].

Manufacturers of shaft boring machines are the companies Zeni, Drilling, Robbins, Hughes, Wirth and Turmag.

3.6.5.3 Combinations

The machine shown in Fig. 3-45 represents a combination of blind drilling and raise boring. Robbins has developed this shaft-boring machine under the name BorPak, which can economically produce blind bores of 1 to 2 m diameter over 100 m between 30° and 90° to the horizontal. It bores from bottom to top and the muck is carried down to the starting level.

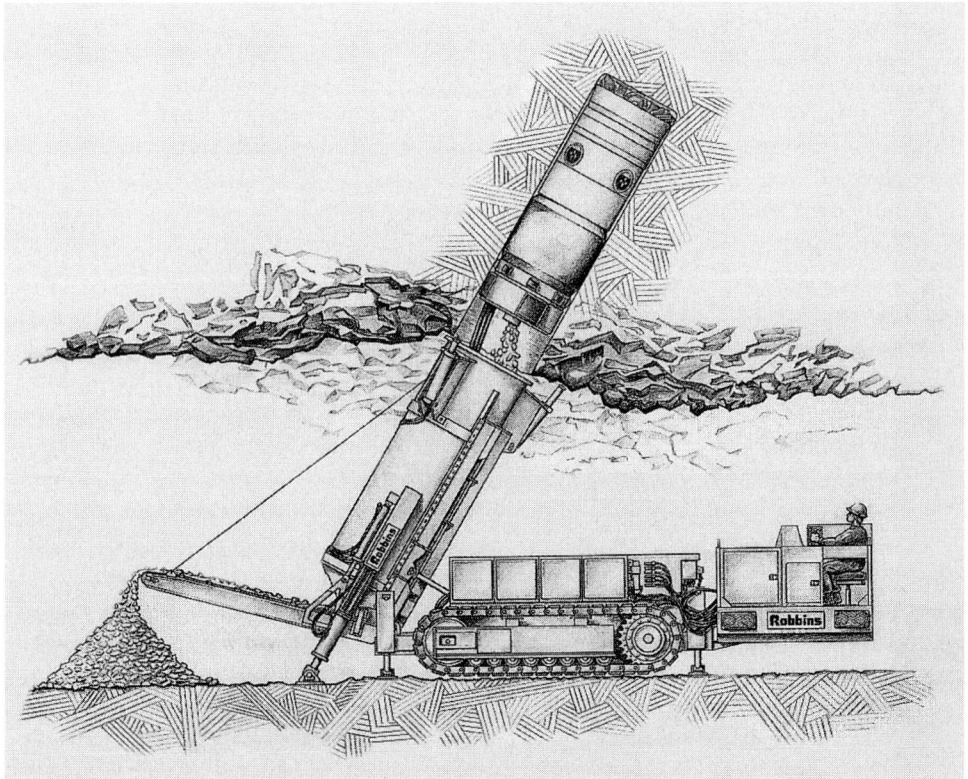

Figure 3-45
BorPac, Shaft boring machine by Robbins

4 Thrust

4.1 General

The required penetration requires the appropriate cutter loading. The sum of the cutter loadings plus the friction forces caused by the body of the machine, which is dependant on type of TBM, gives in each case the total thrust, which has to be provided for the rock to be bored through.

A TBM is normally built for a particular type of rock, i.e. for one class of rock according to difficulty of boring. Tunnel boring machines are often designed for hard or very hard-to-bore rock and then have considerable reserves in softer rock. If longer stretches of softer rock are expected, it makes little sense to lay out the entire back-up for the larger capacity or to load the tunnel walls and thus the surrounding rock with excessive forces with a gripper TBM. Excessive thrust forces could also overload the main bearing of a middle class TBM. Long-term overloading could lead to erosion of the bearing tracks, which in an extreme case could require a change of the bearing in the tunnel. This is of course possible, but would cause delay and cost.

4.2 Advance with Gripper Clamping

The total thrust force, which has to be exerted axially, is transferred by the clamping mechanism into the rock. Among the various manufacturers of TBMs, two different clamping systems and type of thrust have been developed in the course of time.

Robbins and Herrenknecht prefer a simple clamping system with the machine body sliding on one sliding shoe, often formed as a partial shield (Fig. 4-1).

The stator of the TBM with the entire drive unit lies in a protection unit consisting of one or more parts resembling a shield.

The machine slides on an invert shoe and side steering shoes protect the sides. Over the unit there is a roof shield, often with trailing fingers for protection against falling stones.

The actual clamping unit braces itself with powerful hydraulic rams against the tunnel walls. Within the clamping unit, the main beam can be oriented according to the topology to be excavated. The clamping rams additionally serve to steer the machine.

Jarva and Wirth prefer X-type gripping (Fig. 4-2). The front and rear clamping grippers form a unit with the main body (Outer Kelly), in which the main beam of the TBM, a rectangular hollow section (Inner Kelly) is born with sliding bearings. The cutter head sits on this rectangular hollow section. Inside the rectangular hollow section, the main drive shaft connects the drive and gear unit at the back with the bearing housing on the cutter head.

The TBM is aligned with the clamping unit before each stroke. In contrast to the system with single clamping, the TBM is held in position for each stroke. The steering of the

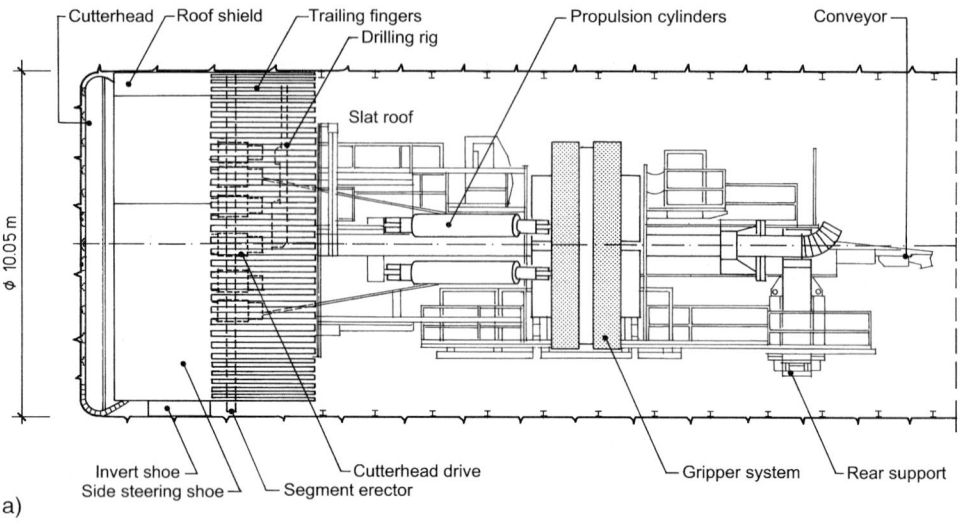

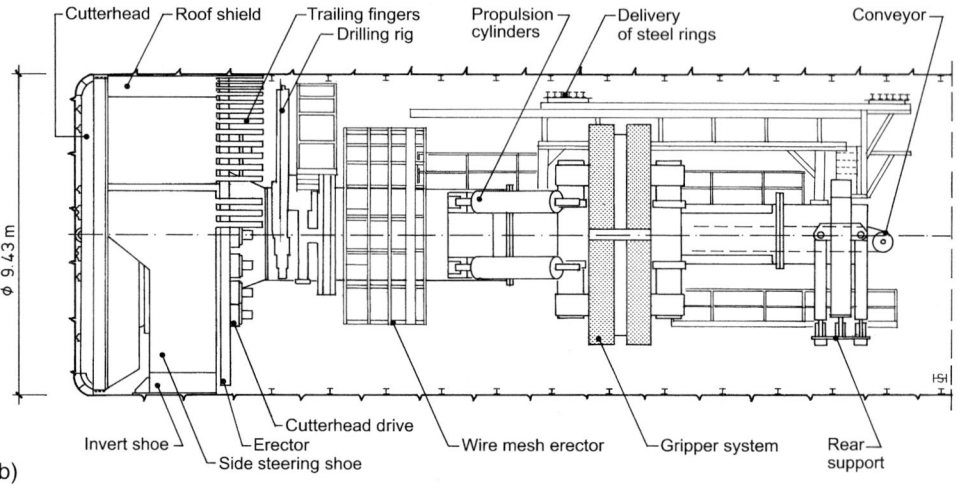

Figure 4-1
TBM with single gripper clamping system
a) Main Beam TBM MB 323–288 (Robbins) Manapouri, ⌀ 10,05 m, 1997 [128]
b) Main Beam TBM S-167 (Herrenknecht) Lötschberg base tunnel, Steg section, ⌀ 9,43 m, 2000 [61]

4.2 Advance with Gripper Clamping

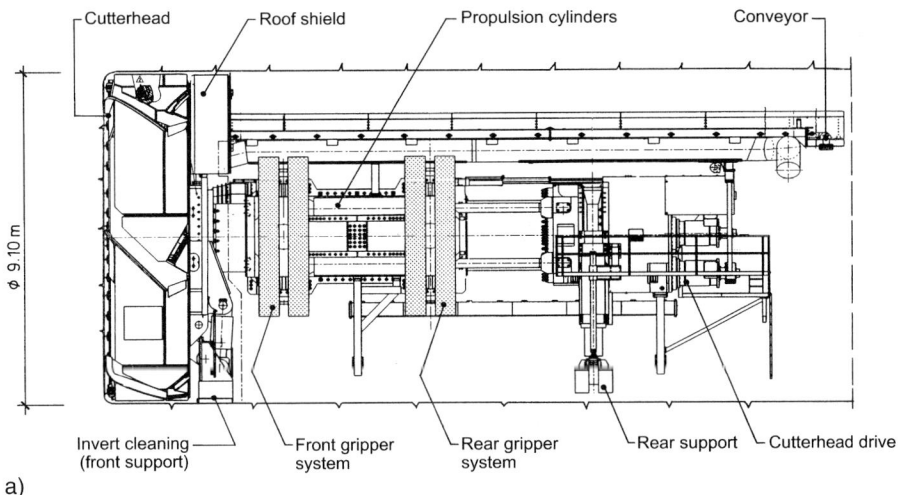

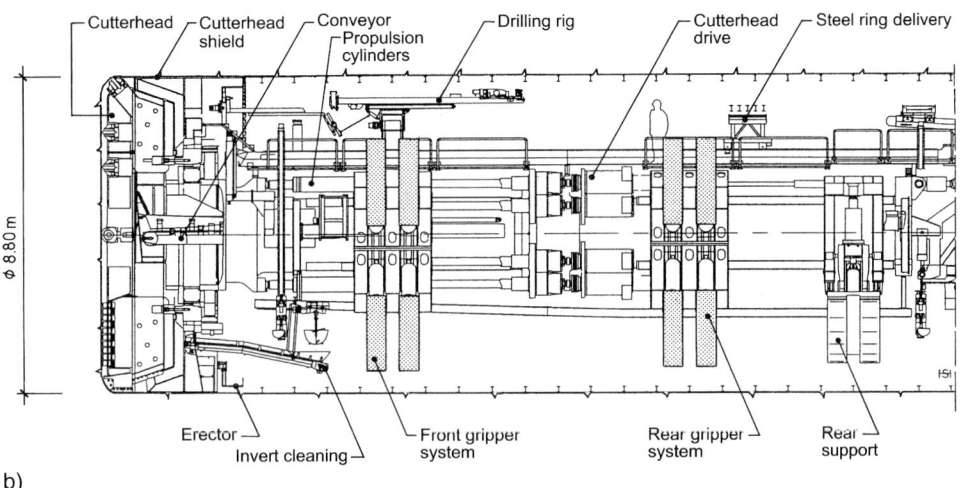

Figure 4-2
TBM with X-type gripping
a) Kelly TBM MK 27 (Atlas Copco) Hållandsas, ⌀ 9,10 m, 1994 [5]
b) Kelly TBM 880 E (Wirth) Qinling, ⌀ 8,80 m, 1997 [182]

machine always necessitates a displacement and new setting of the clamping unit. For machines over 6m diameter, cross clamping is normally used. Both TBM systems have their supporters.

The single gripper system with side grippers (Fig. 4-3) seems to be more robust. This system has advantages above all in geology requiring considerable support against ruptures, because more space is available around the machine.

Figure 4-3
Gripper with propulsion jacks, Gripper TBM S-167 (Herrenknecht) Lötschberg base tunnel, Steg section

The clamping forces of these gripper-braced tunnel boring machines amount to about twice the thrust. The assumption that it is always possible to clamp the machine without damage against the rock and without sliding is indeed obvious, but definitely cannot always be confirmed.

Soft ground gives way under the clamping units. Excessive clamping pressure causes unfavourable stress conditions in the surrounding ground (Figs. 4-5 and 4-6). These known clamping pressures under the gripper shoes of 2–4 MPa apply for softer ground.

Really hard rocks with a cutter load of 250–270 kN lead to a considerable increase, because the gripper shoes cannot be enlarged without limit.

Gripper clamping leads in any case to an exceptional loading of the surrounding ground, above all because the alternating loading rapidly loads and unloads, which is constantly repeated while driving.

Figures 4-5 and 4-6 are practical examples showing the stress distribution in the surrounding rock for a tunnel ($\varnothing$ 6 m) in granite with a cover of 400 m and a side pressure coefficient K = 0.9. The tunnel excavation causes a nearly circular stress distribution around the cavity, calculated on an elastic plastic basis.

Assuming a massive granite requiring a cutter loading of 270 kN and a cutter head fitted with 37 cutters, the resulting thrust force will be 10,000 kN with a clamping force of 20,000 kN. The TBM would be clamped sideways with 20,000 kN, and each clamping point of a X-type gripping would be loaded to 10,000 kN in order to achieve the same total clamping.

The stress increases slightly under the gripper shoes with both clamping systems. Overall, however, the stress distribution is quite different. With side clamping, the stress at

4.2 Advance with Gripper Clamping

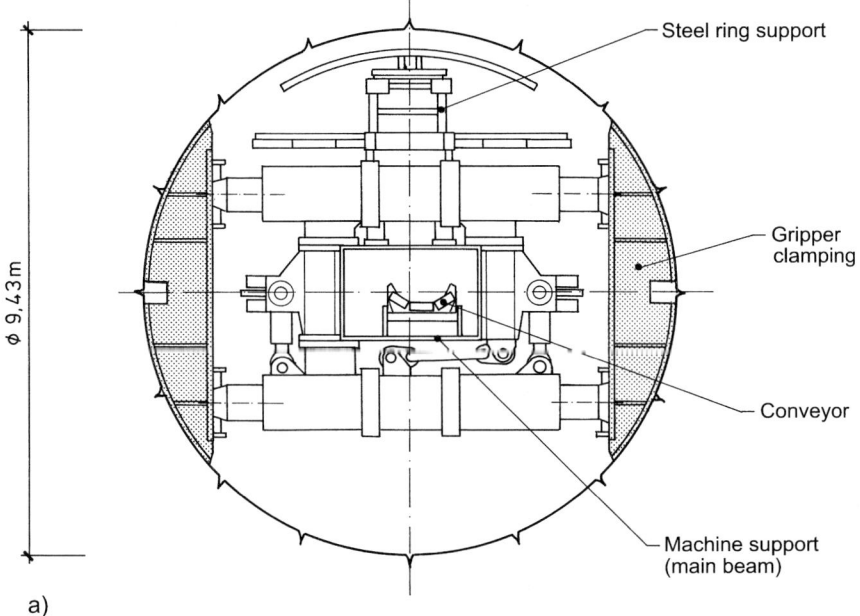

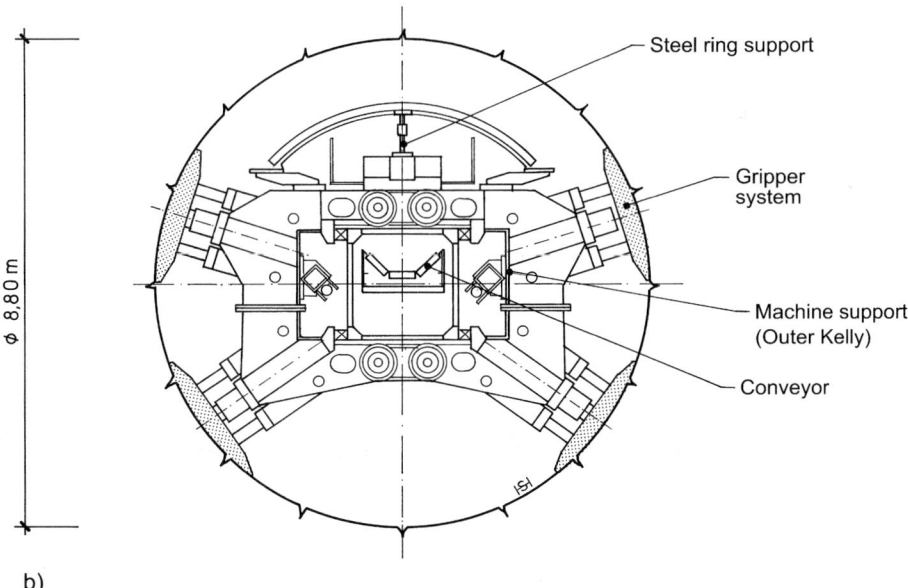

Figure 4-4
Gripper clamping
a) Side clamping
b) X-type clamping

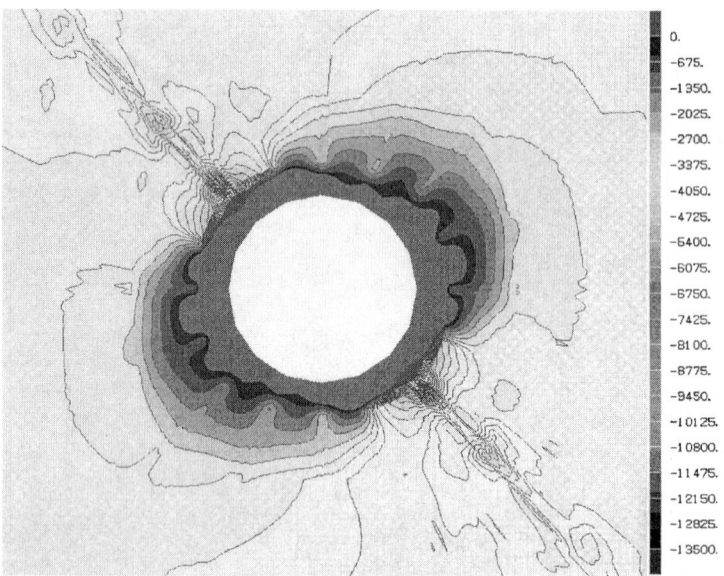

Figure 4-5
Stress plot without loading of the surrounding rock by clamping forces in massive granite with ring stress as the main stress of about 27 MPa

crown and invert sinks to about one quarter of the values. The deformation shows slight deflections of approx. 2 mm into the rock under the grippers, but in the crown an additional 4mm into the cavity.

With the X-type clamping, the lower loadings from the four shoes, which actually correspond to two staggered clamping units, have a slightly more favourable effect. The deformations remain a bit. The stresses at the quarter points 3, 6, 9, and 12 o'clock are rather less compared to the side clamping.

The large and constantly repeating stress and deformation redistributions cannot remain without effect. It is no wonder that this can cause ruptures and that the deformation can cause collapses in fractured rock, causing increased support work. The reduction of the drive advance rate, whether caused by the increased support work or by increased damage to cutter head and discs, is an inevitable result of the system.

Analogous calculations in chalk, with the associated lower clamping forces because the cutter loadings are lower, show similar problems.

4.3 Advance with a Shield TBM

The entire drive unit supported in the shield by hydraulic jacks with cardan joints (torque box drive). Cylinders between shield and the TBM stator, controlled according to travel, provide the thrust required for the penetration. These have a safety override in

4.3 Driving with a Shield TBM

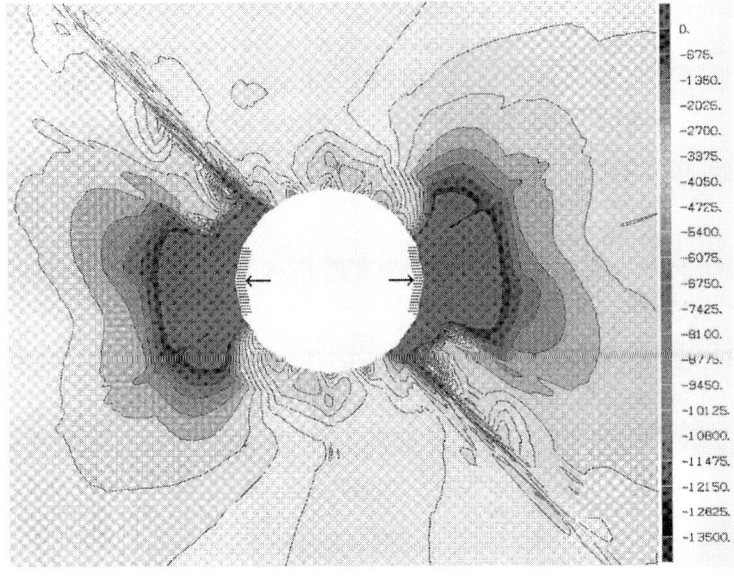

a)

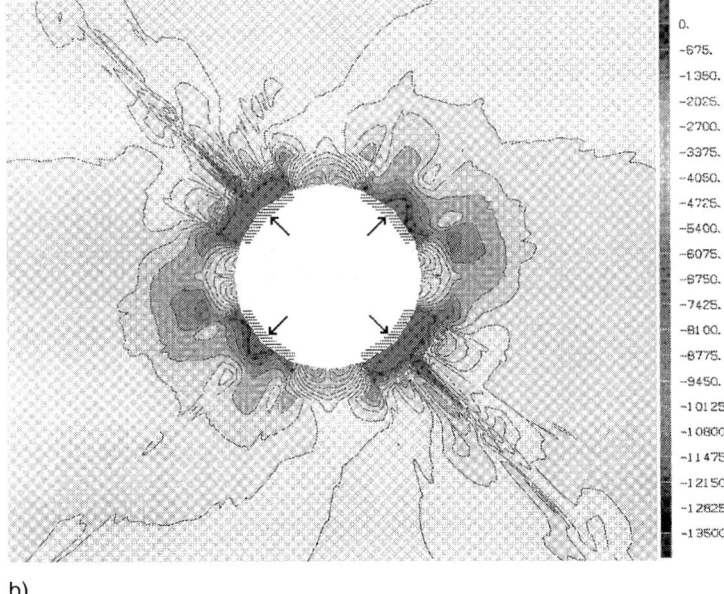

b)

Figure 4-6
Stress plot overlaid according to Fig. 4-5 with gripper clamping
a) Side clamping
b) X-type clamping

case of an overload of the main bearing. The forces required for the advance of the shield unit are far larger than the actual cutter head advance force. The larger forces to thrust the shield result from thrust forces and the shield friction.

The shield friction of a large shield of around 11 to 12 m diameter, without considering the broken-off rock around the shield, about 12,000 to 15,000 kN. This type of shield friction, however, only affects the invert area from 4 to 8 o'clock.

A shield TBM with single shield does not create continuous stress redistribution like a gripper TBM (Fig. 10-7). A double shield TBM with gripper clamping is a special case. This is described in chapter 10. With gripper machines, the clamping unit resists this contra-rotation. With shield machines, a distortion of the thrust jacks out of the shield axis is sufficient. The contra-rotation that this causes is enough to resist the rotation of the shield caused by the TBM. The friction between the supports of the jacks or the pressure ring and the segment ring as well as within the segment ring is sufficient, given the existence of the axial thrust load for driving, even without bolting the segments.

A detailed diagram of the calculation of the thrust forces can be found in [95].

5 Material Transport

5.1 Material Transport at the Machine

The disc cutters excavate the rock continuously. As described in Chapter 3.2, the muck mostly falls down to the tunnel invert. The buckets mounted on the cutter head pick up the excavated material like a scoop. The muck slides down chutes to the conveyor belt with the rotation of the cutter head, at about 2 o'clock for counter-clockwise rotation and at about 10 o'clock for clockwise rotation.

The continuity of the excavation gives way to an intermittent material flow. Fig. 3-5 shows the basic features of the cutter head design with the number of buckets and the conveyor belt for the excavated material.

Tunnel boring machines of large diameter, with the corresponding bearings, permit a layout with central conveyor belt. Fig. 5-1 shows a gripper TBM with the central conveyor belt and the muck ring.

Figure 5-1
Conveyor belt in centre of cutter head with muck ring; Gripper TBM S-155 (Herrenknecht), Tscharner, ⌀ 9,53 m, 1999 [61]

Machines of smaller diameter, with their cramped main bearings, do not permit a central mechanical conveyor belt. The conveyor belt runs above the main bearing into the bucket wheel of the TBM (Fig. 5-2).

The intermittent transport of material through the cutter head requires good matching with the capacity of the conveyor belt and essentially also with the muck ring. The width of this conveyor belt at the machine results from the maximum practical penetration of the cutters, the number of buckets and the speed of the belt. From experience, this speed should not exceed 1.3 m/s. Higher speeds lead to a shortened life of the belts.

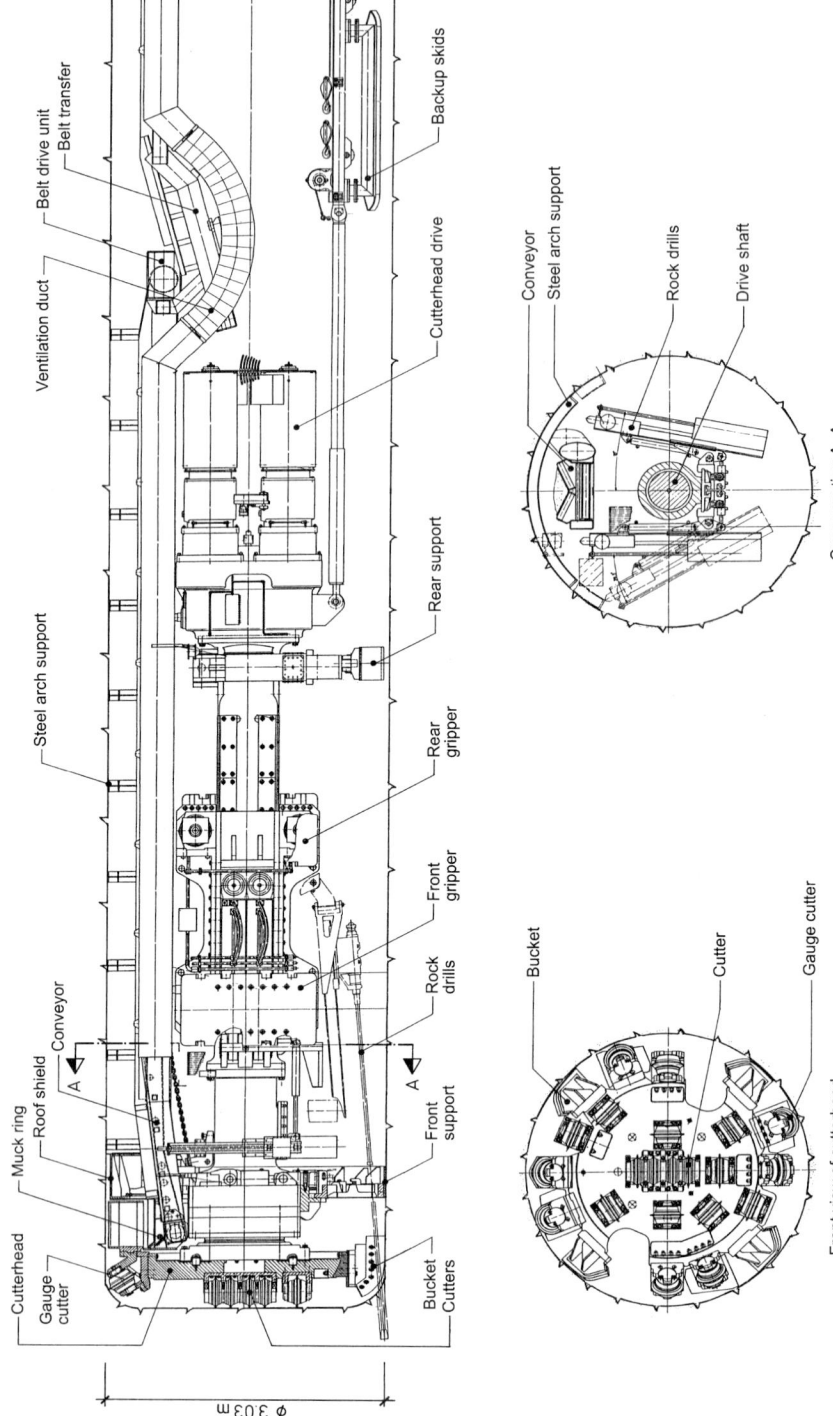

Figure 5-2
Gripper TBM S-96 (Herrenknecht) with high-level conveyor belt Kompakt TBM 3000, ⌀ 3,03 m [61]

Normally, the conveyor belt is started without load. The conveyor drive must, however, be capable of starting the conveyor with a full load, even under awful conditions like a water inflow. Special devices for cleaning the inside of the belt belong to the standard equipment of a conveyor belt.

Material transfers, whether the tipping of the material out of the cutter head onto the conveyor belt or a belt transfers, nearly always cause a considerable amount of dust, causing an unacceptable worsening of the air in the tunnel.

The reduction of dust with water nozzles only works sufficiently if the water is atomised. The fine water mist then surrounds the dust particles, which ball together due to the surface tension of the water until heavy enough to sink to the ground. What work much more efficiently are encapsulated transfers with dust extraction and a dry dust reducer installed for a number of transfers.

The length of the conveyor belt from the cutter head and the material transfer to the tunnel transport is arranged from case to case as depending on the backup concept and any safety work, which has to be done at the same time as the advance.

5.2 Material Transport Through the Tunnel

In the course of time, various systems have been developed for transport through the tunnel. Each has its own area of application; the contractor can investigate which one is appropriate from case to case.

The pure boring time of a TBM is so limited by cutter changing, repairs, re-gripping and rock support (see Chapter 3.5) that a further operating restriction would be unbearable. The removal of the muck should therefore not be allowed lead to any further disruption of the advance. The usually very considerable bulking should also be taken into account. The transport volume is typically 170–200% of the undisturbed volume.

5.2.1 Rail Transport

Rail transport, formerly the only possibility of transporting spoil, is still useful in smaller cross-sections and above all over longer distances. Its limitations are technical; a maximum gradient of 3%, even if all wagons in the relevant train are equipped with air pressure brakes. This limitation of gradient can be overstepped a bit in special cases, but the performance then reduces sharply.

So, for example, the Sörenberg heading, with only 4 m diameter, was equipped with a belt conveyor because of the steep gradient of 5%.

Rail transport can be categorised into the following types of operation:

- Track with sight rules control
- Railway with signal control

a)

b)

Figure 5-3
Track operation, Sörenberg tunnel
a) Segment wagons as supply train
b) Diesel locomotive and wagons for conveyance

The normal case is the track controlled with sight control – the engine driver ensures that the track is clear. This is, however, less suitable for long distances.

In the Channel Tunnel, the entire transport system failed to achieve the necessary performance while operating under sight rules, and this slowed the advance rates down. Not until the operation was changed over to a railway system with signals was the necessary performance achieved.

5.2.1.1 Diesel or Electric Operation

Only diesel and electric locomotive have been used for underground construction work. The advantages of the battery locomotive with its low ventilation requirements are losing importance because the diesel motors used are becoming ever cleaner, above all with the fine particle filters, which are nowadays demanded by insurance companies. Diesel locomotives are much easier to operate and maintain.

Figure 5-3 shows a train with diesel locomotive and wagons for conveyance (Fig. 5-3b) as well as segment wagons as a supply train (Fig. 5-3a). The muck in this case was cleared by conveyor belt.

5.2.1.2 Muck Cars

Wagons with a capacity of 2,5–20 m^3 are used, adapted for the excavated cross-section. Most of these are, as ever, one-side self-tippers (Fig. 5-4a), with the name of the supplier Mühlhäuser often being used as a name for them.

Simple trucks are sometimes used instead of expensive self-tippers. They require corresponding stationary tipping equipment (Fig. 5-4b) in a simple version or even a rotating tipper, with which the wagon is rotated fully upside down. Equipment of this sort is suitable for cohesive muck.

a)   b)

Figure 5-4
Material transfer above ground [116]
a) Tipping point with self-tipper
b) Wagon tipper with feed and tip belts

Figure 5-5
Wagon shunting equipment with sliding stage, LEP Ring, CERN, 1987 [66]

5.2.1.3 Loading in the Tunnel

In smaller tunnels, the loading mostly takes place directly at the discharge from the conveyor belt because of lack of space. The train has to be monitored and pushed under the discharge point according to the amount of muck. This longitudinal shunting is done by mechanical or also hydraulic equipment (Fig. 5-5). The locomotive of the train is probably unsuitable for such manoeuvring. The TBM driver is not in a position to shunt the train along according to how full the wagons are. Wagons are often overfilled. The muck material, which then falls to the tunnel floor, has to be tediously removed by hand.

Larger tunnels have enough room for smaller or larger material holding bunkers for the loading the muck material onto the wagons. These bunkers or their compartments are ideally matched to the size of the wagons.

5.2.1.4 Train Timetable

One train in the tunnel is seldom sufficient. A train timetable dependant on the one hand on the transport volume and the resulting capacity of the trains and on the other hand on the distance to be travelled is a precondition for an efficient tunnelling. Figure 5-6 shows such a timetable for a travel distance – travel distance is not always the same as the excavation length – of 5,000 m, in the example for a distance of 2,500 m.

Such a timetable makes clear where the trains in the tunnel meet. Passing loops will need to be created for a one-track system. As the excavation proceeds, these passing loops often do not remain at the same location as determined by the timetable. Moveable passing loops, so-called California switchs (Fig. 5-7), enable such flexible passing loops in one-track system.

5.2 Material Transport Through the Tunnel   81

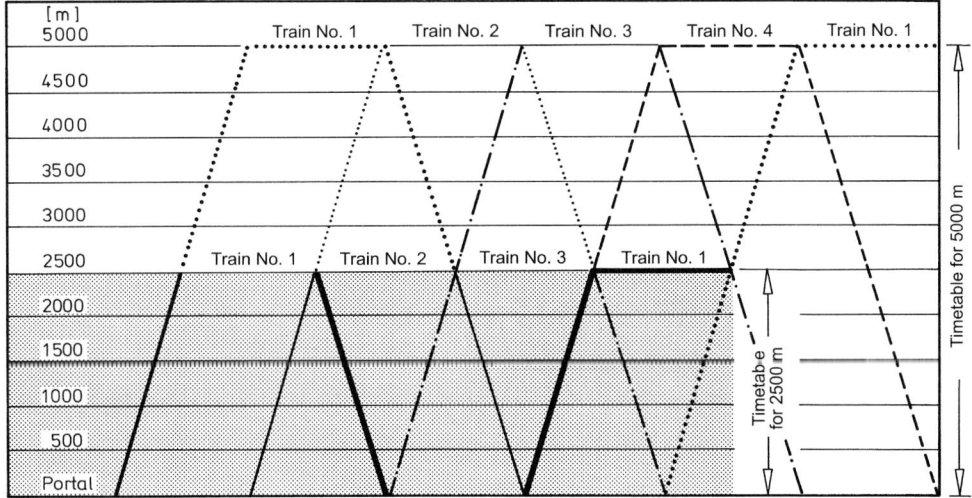

Figure 5-6
Train timetable, basics: loading time 20 min/train unit, travel speed 15 km/h

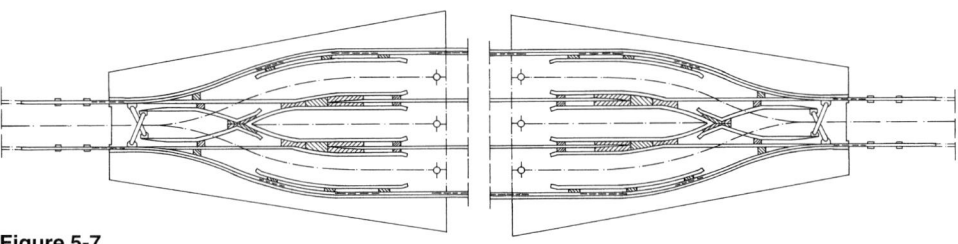

Figure 5-7
California switch

5.2.1.5 Tunnel Track

If the excavation is to be advanced without delays due to supply and disposal, the track in the tunnel should be well constructed. The large volumes to transport require corresponding track, constructed to similar standards as real railway tracks. Trains should be able to circulate with the planned speed and with the corresponding safety.

5.2.2 Trackless Operation

Trackless operation simplifies many operational processes. It should not, however, be a reason for negligence. The daily preparations for track operations require increased attention.

5.2.2.1 Transport Vehicles

Practically all earthmoving dump trucks are suitable as transport vehicles. Vehicles constructed specially for tunnel building, like the Kiruna dumper, are especially suitable for muck transport because of their lowered construction. They also have a suitable motor performance with positive effects on the ventilation of the site.

Figure 5-8
Wheeled transport operation with Kiruna dumpers for muck transport and segment supply, Zürichberg tunnel

5.2.2.2 Haul Road

The requirement formerly prevalent in earthmoving, that a haul road should be constructed so that a car can drive on it without problems at 50–60 km/h, remains the standard to be aimed for in tunnelling. On the one hand, the repair costs are reduced, and on the other hand the driving time of the dumpers and so the high ventilation requirement. Haul roads should therefore ideally be constructed with an asphalt surfacing.

5.2.2.3 Loading

Loading the dumpers directly form the conveyor belt of the TBM would interrupt the drive advance frequently. To ensure continuous advance, sufficient stacking silos must be installed for temporary stacking of the muck in the backup of the TBM. These temporary stores, which have to be arranged along the tunnel because of the limited space, should match the working load of the dumper, either completely or per compartment. The filling of the temporary store requires extendible distributor belts (Fig. 5-9).

Where there are a number of stacker silo compartments, a suitable signal shows the driver which position to drive to (Fig. 5-10).

5.2 Material Transport Through the Tunnel

Figure 5-9
Material stacker with distribution belt on backup III

Figure 5-10
Material transfer from the temporary stacker in the backup to dumper, Bözberg tunnel

When the gate of the silo is opened, the entire load falls into the dumper in a short time. With dry material this produces an impermissible level of dust, which spreads into the tunnel. The loading equipment should therefore be equipped with effective equipment for dealing with the dust (fans and dust filters), which extract the dust cloud and remove the dust.

Even in double-track railway tunnels or in road tunnels, the width is not sufficient to use dumpers without considerable tyre wear. Contractors have therefore developed turntables (Fig. 5-11), which are always stationed relatively near to the reloading point in

Figure 5-11
Turntable

the backup. They can be moved along easily, as periodically required with the advance. The safety in the tunnel is improved by the reduced necessity of reversing and at the same time the job of the dumper driver is made easier.

5.2.3 Conveyor Transport

Conveyors were quickly taken up in mines. In tunnelling proper, the first known use of conveyor transport for tunnel transport was in the construction of the Great St. Bernard tunnel (1959–1964), the road connection from Wallis to the Aosta valley. It is not clear whether this was economical due to the drilling and blasting method then used.

Not until 1985–1990, with the construction of the town motorway tunnel in Neuchâtel, was the muck from TBM excavation removed by conveyor. The gallery conveyor did not yet have temporary storage. A relatively long transfer belt connected to the conveyor belt of the reamer TBM carried the muck to the tunnel belt. The extension of the conveyor construction took place under the transfer belt. The belts had to be periodically lengthened after reaching the connecting stretch of the transfer belt.

The first use of conveyor belt stacking, still as simple equipment with only one short conveyor stacker, was in the Grauholz tunnel. The conveyor equipment in Murgenthal, with an Eickhoff conveyor and Zürich-Thalwil with a conveyor stacker from H+E Logistik GmbH, enabled considerably more storage with their multiple belt deflection. Today, conveyor belts have become established in large tunnels, even up to 5 km long. Even with small tunnels with a diameter of 4.4 m, but with a gradient of 5%, which would be unsuitable for rail transport, the contractor of the Sörenberg tunnel is using conveyor transport. Because of the restricted working conditions, the conveyor belt outside the tunnel is arranged vertically as a stacker tower for protection from wind and weather (Fig. 5-12).

5.2 Material Transport Through the Tunnel

Figure 5-12
Conveyor storage in stacker tower, Sörenberg tunnel

Fundamentally, a relatively low belt speed of < 2,2 m/s should be aimed for. The width of the belt for larger machines should not be less than 800 mm because of the safety consequences of large chunks of rock. 650 mm should be the minimum for smaller TBMs.

5.2.3.1 Conveyor Storage

Modern horizontal conveyor cassettes enable storage of up to 600 m of spare belt or 300 m of installed belt (Fig. 5-13). The roller sets in the conveyor stacker, over which the belts run, are tensioned by a braked cable winch. With the advance of the TBM and the resulting conveyor extension of the belt, the spacing of the sets of rollers reduces to the minimum permissible spacing. For the extension of the belt, the conveyor belt is cut at a particular point and the additional piece is vulcanised in.

5.2.3.2 Conveyor Belt Extension and Belt Operation

Modern installations permit the extension of the conveyor construction while the belt remains running, even while material is still being transported. Figure 5-14 shows such an extension point; the lower run of the belt is enclosed in flat duct constructed so that a fitter can install the extension in the safe working area. The tunnel belt can be suspended or also propped from the carriageway. The suspension of the tunnel belt allows more unimpeded traffic, even under the belt (Fig 5-15).

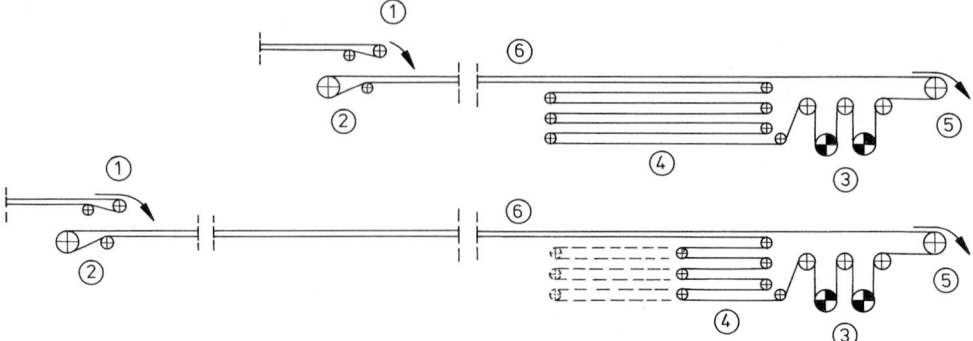

Figure 5-13
Conveyor belt storage with mobile deflection station (H+E Logistics)
① Backup belt dropping point
② Travelling deflection station of the gallery belt
③ Drive station
④ Belt storage
⑤ Tunnel belt, muck dropping point
⑥ Tunnel belt

Figure 5-14
Extension point of the gallery belt of a backup

A conveyor belt of a few kilometres length cannot be stopped immediately, whether empty of full. In order to keep the necessary stopping distance within limits after an emergency stop, a braking mechanism should be provided. With a long tunnel conveyor, the loaded belt runs on considerably despite the braking. At belt transfers, chutes or silos, therefore, enough room must be planned to take the muck when the belt runs on.

5.2 Material Transport Through the Tunnel 87

Figure 5-15
Conveyor belt attachment, Zürich–Thalwil tunnel, lot 3.01

5.2.3.3 Advantages of Conveyor Transport and Innovation Potential

Conveyor transport, with its continuous operation, has led to remarkable performance increase compared with a comparable use of dumpers with their intermittent transport. The Bözberg and Murgenthal tunnels were driven with the same machine. The difference was wider segments in the Murgenthal tunnel with a 1.5 m length of ring compared to 1.25 m for the Bözberg tunnel and in the conveyor transport as against the use of trucks in the Bözberg tunnel.

The advance rates for the Bözberg tunnel over 3,500 m are shown in Fig. 5-16, with an average advance rate of 54 m per week. Figure 5-17 shows the advances in the Murgenthal tunnel over 4,250 m with an average weekly advance of 100 m.

The investment for a conveyor system is higher than for a pure dumper operation. Against this, the overall wage costs are less with a conveyor system, for the dumper drivers as well as for the higher number of mechanics. With conveyor transport, 60% of the driving in the tunnel is unnecessary. It is also possible to ventilate with less air volume. The cost of conveyor belt operation is about 25% less than with dumpers, without considering the reduction of construction ventilation.

The investment for a road tunnel of 4 km length is for

- a conveyor belt system approx. 2.3 million Euro
- vehicle transport with dumpers approx. 1.5 million Euro

Further savings are possible if the transport for the pneumatic stowing and for the pea gravel and fill in the tunnel invert can be dealt with by appropriate adaptation of the return belt.

88   5 Material Transport

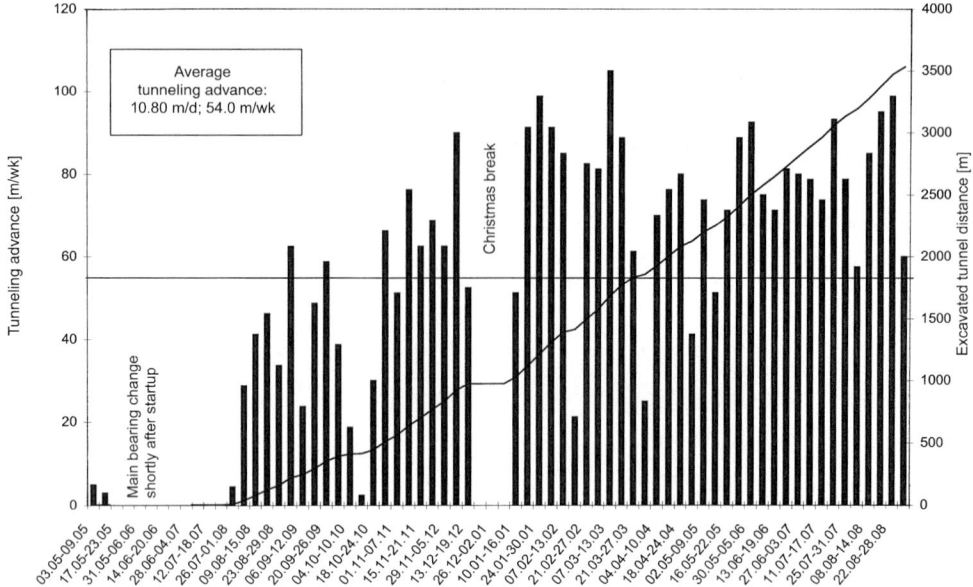

Figure 5-16
Weekly advance diagram TBM drive, Bözberg tunnel with muck transport by dumper

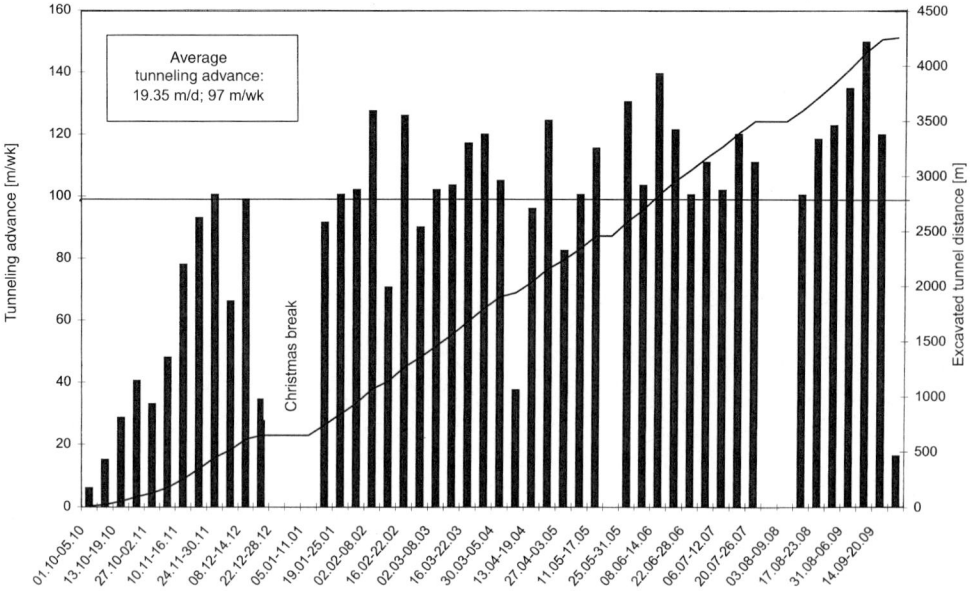

Figure 5-17
Weekly advance diagram TBM drive, Murgenthal tunnel with muck transport by conveyor belt

6 Backup Equipment

Mechanical excavation with a tunnel boring machine and the necessary rock support, at least in the forward direction, necessitate systems to ensure supply, control, waste disposal and also working platforms for the various activities.

A steel construction serves as the framework for this system. This is described as the backup, with all the installations required for the tunnel drive. Figure 6-1 shows the equipment, which has to be fitted into the backup.

Energy supply TBM	TBM steering
High-voltage cable drum Transformers Electrical cabinets Hydraulic systems Compressed air	Steering cabin

Material transport
Conveyor belt Material transfer station Belt assembly and extension Temporary muck storage (stacker)

Rock support
Segment erector Segment storage (invert segment or complete rings) Grouting of annular gap Drill equipment Ring erector Wire mesh erector Installation material (transport, storage, installation)

Components
Invert segments Waterproofing materials De-watering system Cable canal Invert fill Transport roads

Ventilation
Main ventilation, air ducts Ventilation TBM area Dust suppression system

Figure 6-1
List of equipment in the backup

6.1 Backup Concept

Backups with steel construction run on rails or slide on skids (Fig. 6-2), usually without a drive unit but pulled by the TBM. Figure 6-3 shows a typical backup for a gripper TBM with all essential plant. In the mostly very restricted space available, the transport of invert segments, with their mostly relatively large dimensions, becomes a real problem.

The train with the muck cars is loaded in the backup by the extendable loading conveyor (Fig. 6-4). In this tunnel, with its excavation diameter of 5.2 m, the rock support is dealt with locally over long stretches with large quantities of shotcrete and steelring installation, this being the simplest method of construction.

90   6 Backup Equipment

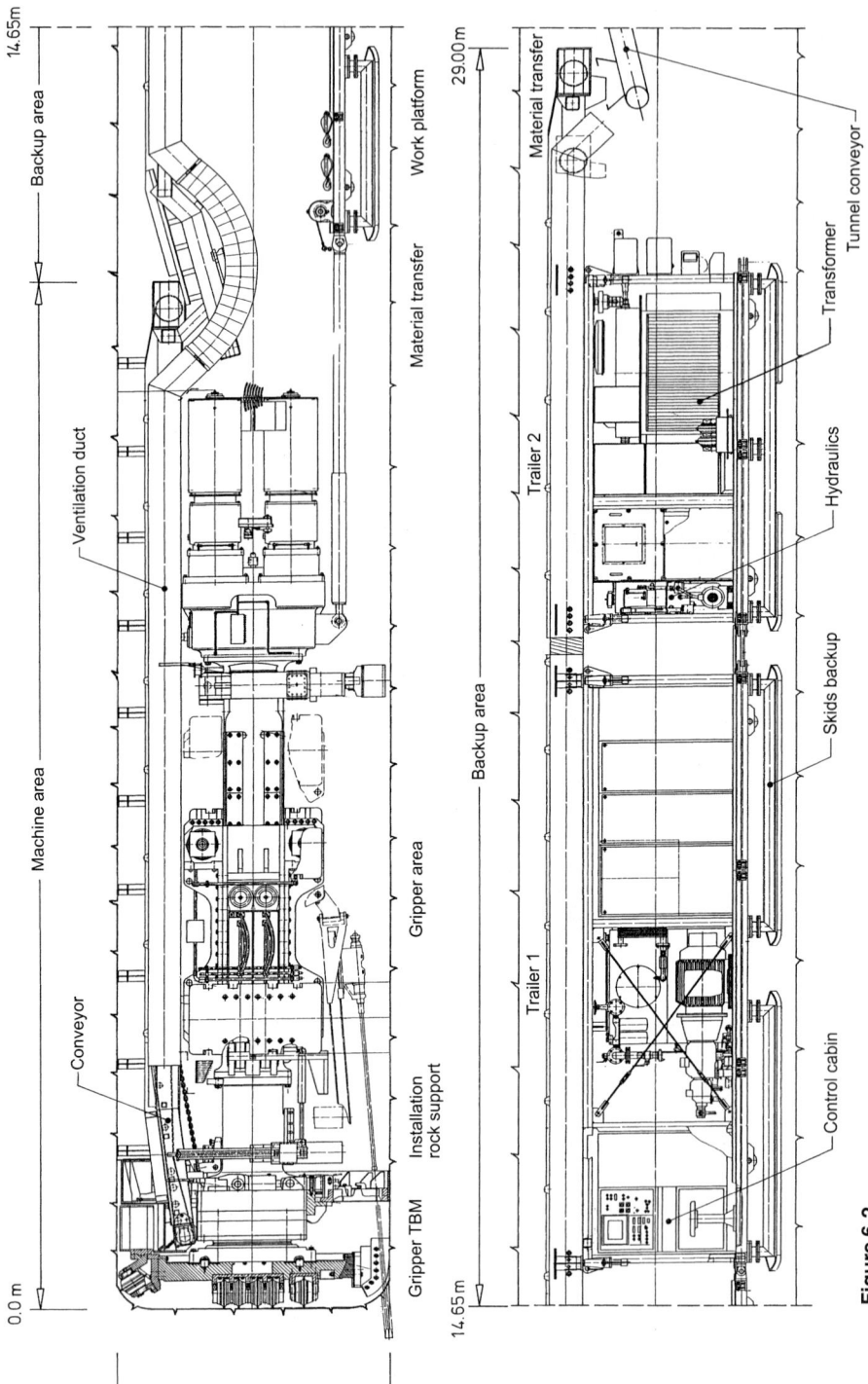

Figure 6-2
Gripper TBM S-96 with backup (Herrenknecht), Kompakt TBM 3000, ⌀ 3,03 m

6.1 Backup Concept

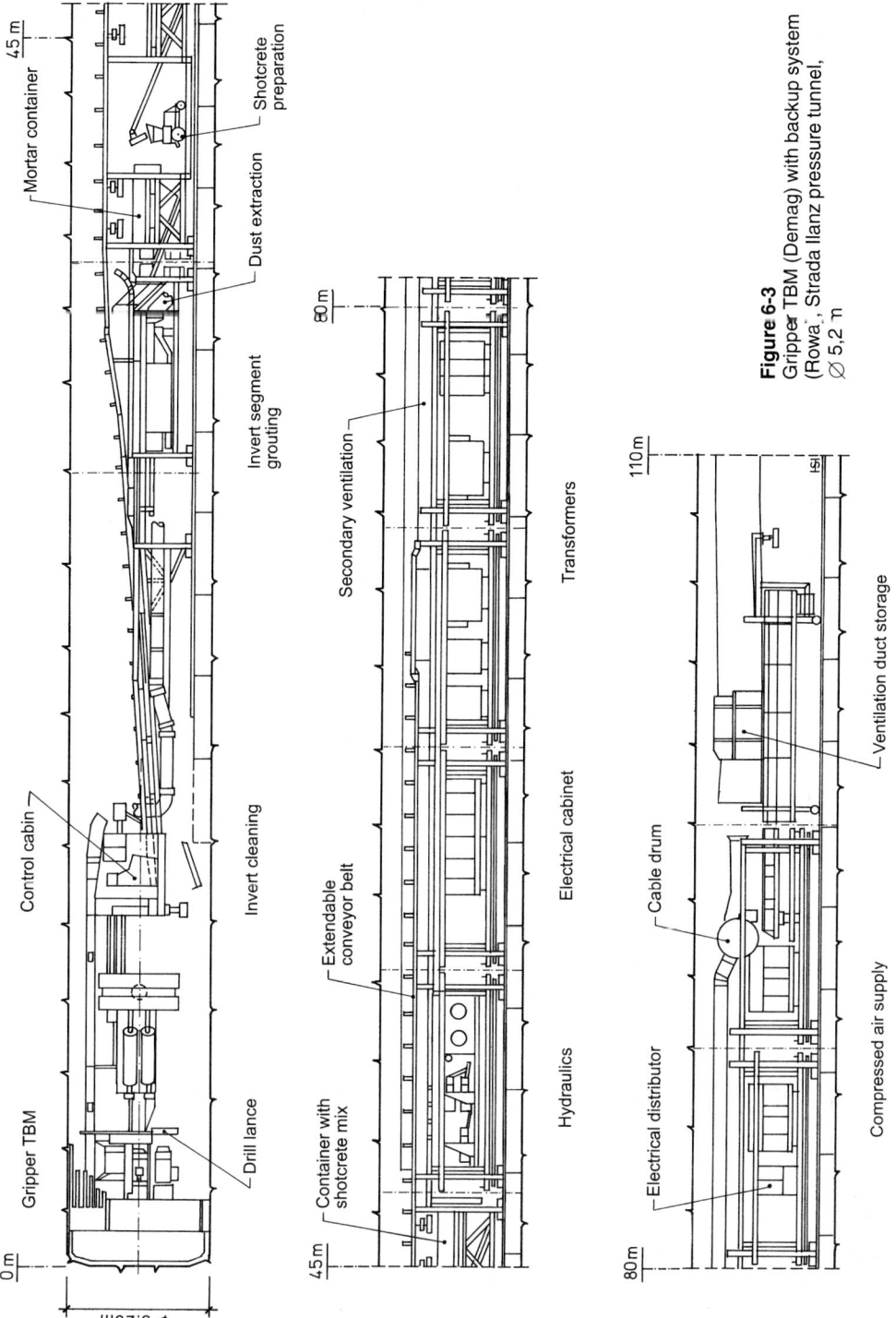

Figure 6-3
Gripper TBM (Demag) with backup system (Rowa), Strada Ilanz pressure tunnel, ⌀ 5.2 m

Figure 6-4
Loading conveyor in the drive equipment,
Strada Ilanz pressure tunnel

In large tunnels for two-track railways or for roads, the structure of the invert is often installed in the backup. This produces a much wider area for transport in the tunnel. It is possible for vehicles to pass anywhere. The illustration of the backup (Fig. 6-5, see p. 94) shows as an example the entire setup for the Zürich-Thalwil shield TBM with the gantry over of the working area road.

The part labelled as trailer 1 carries all electrical and hydraulic supplies as well as the temporary store for segments for a complete ring. The pea gravel pumps in the lower deck of the backup, located directly under the gravel silo (Fig. 6-6), transport the gravel pneumatically to the injection location behind the shield tail.

The main dust removal plant, mostly a dry dust remover, blows the air extracted and cleaned from the excavation area of the TBM out through the outlet at the end of the supply backup.

Figure 6-7 shows the supply trailer 1 at the transition to the first backup trailer with the outlet of the dust removal plant (left side of illustration).

Works on the road include the laying of the main drainage and the spreading of the crushed stone fill for the carriageway and its compaction (Fig. 6-8). These activities require a certain working length. In order to achieve the greatest possible independence of these operations from the actual driving of the tunnel, a spatial separation from the extensive transport to and from the TBM is necessary. Backup gantries ensure this.

The arrangement of the supply backup, labelled as backup 8 in Fig. 6-5, is determined by the choice of transport system in the gallery. Rail or road operation with dumpers necessitates large temporary storage in order to be able to compensate the changeover from the continuous operation of the conveyor to the intermittent operation of the spoil wagons or even dumpers without negative effects on the TBM drive. For conveyor belt transport, suitable belt extension equipment will be required to extend the length of the conveyor belt (see Fig. 5-14).

Whichever method is chosen of transporting the spoil away, measures will have to be installed for the handling of deliveries. Examples of deliveries to be handled here are segments as complete rings or as invert segments, steel inserts, wearing and replacement parts for the entire driving equipment.

The high-voltage cable drum sits at the end of the backup. The storage capacity will often be many hundreds of metres of high-voltage cable, which is rolled out as the boring advances and rolled in again after the gallery cable has been installed.

The air duct for the main ventilation during the drive is continuously drawn out from the air duct storage stationed at the end of the backup frame and suspended under the crown with cables.

6.2　Design Specifications

Backups are too often built taking mechanical and technical considerations into account and neglecting the operational requirements. The space required for all the necessary mechanical equipment is determined, the sight line for the laser steering is held open. After these decisions, free space has to be found for the ventilation ducts, often consequently entailing many changes of direction and cross-section. This increases the pressure loss to be compensated by the fan to such an extent that the resulting quantity of air lost leads to a unacceptably low rate of air changes.

Backup constructions have to ensure first of all economic and safe tunnel driving. The priorities must be set accordingly in the construction process.

The following should, therefore, be determined first:

- Safe walking routes, including emergency routes
- Tidy pipework and cables without bends and kinks
- Alley for the conveyor belt
- Window laser
- Working areas including safety considerations

After these points have been decided, the devices and equipment can be placed in the remaining spaces. Compromises are unavoidable in small diameter tunnels. It can, however, simplify the situation if the steel construction can be used to carry cables. Instead of using steel sections, tubes of sufficient diameter can serve as structural frame and also carry cables.

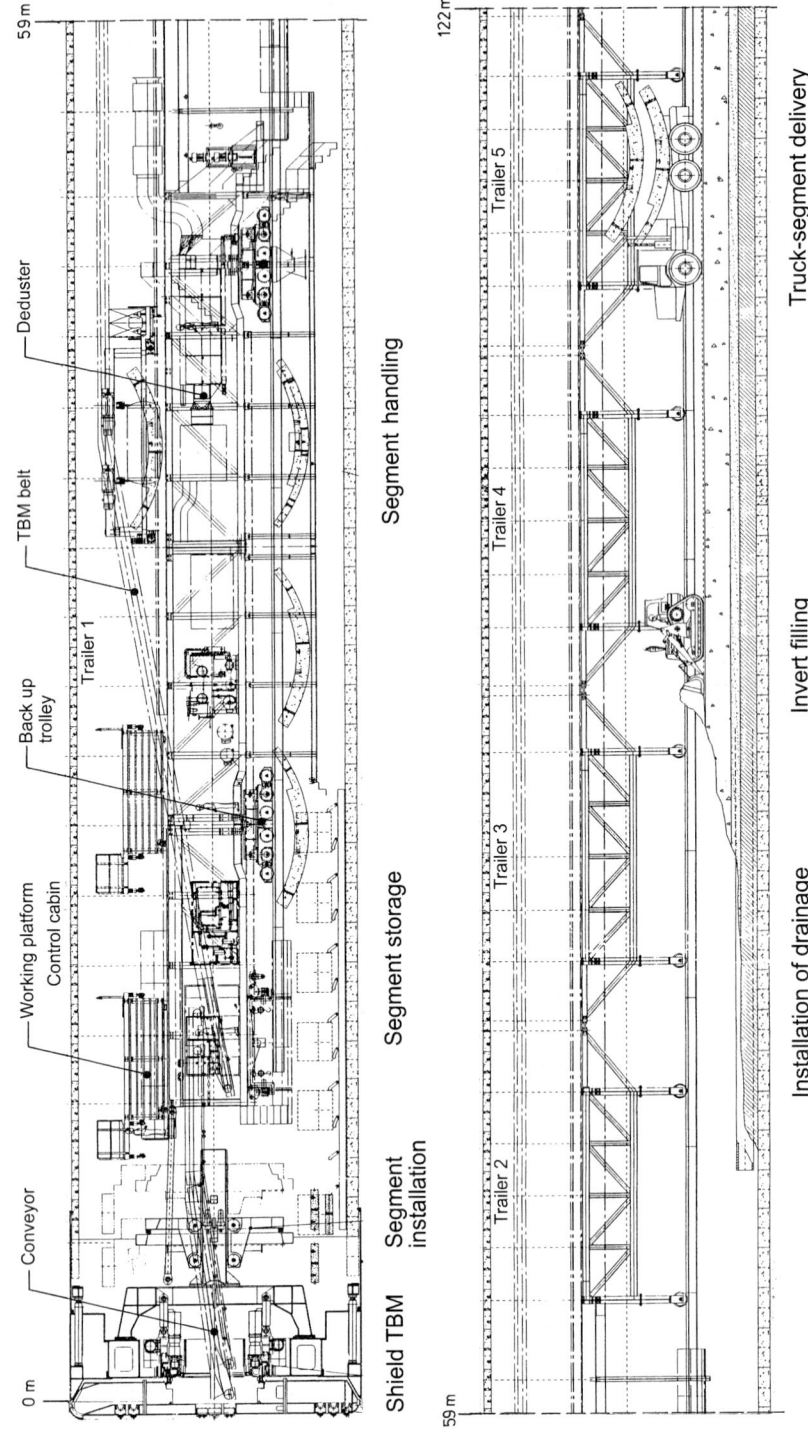

6.2 Design Specifications

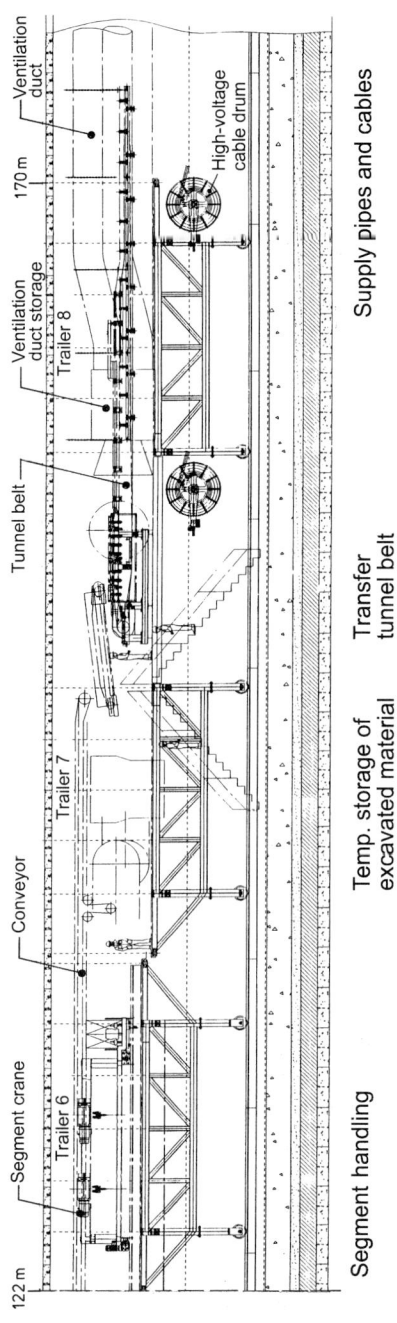

Figure 6-5
Shield TBM S-139 (Herrenknecht) with backup and the associated equipment, Zürich–Thalwil tunnel, ⌀ 12.35 m

Figure 6-6
Pea gravel pumps

Figure 6-7
Rearward part of supply trailer 1, seen from the working road, Murgenthal tunnel

Figure 6-8
Delivery and compaction of the gravel sand for the carriageway under the bridege backup, Murgenthal tunnel

6.2 Design Specifications

Figure 6-9
Backup trolley with bearing rollers and support rollers rolling diagonally on the segment (Herrenknecht)

Figure 6-10
Backup on rails with special bearing construction, starting situation, Zürich–Thalwil tunnel

Backups normally slide on skids or travel on rails. The simplest backups usually slide on skids or are pulled on rails. A very simple track results from using the edge of the invert segments at the arch abutment (Fig. 6-9).

Rails with special bearing details (Fig. 6-10) have to resist a remarkable dynamic loading. The compaction of the invert fill (Fig. 6-8) leads to considerable vibration in the segments, which is then transmitted to the rail bearings. Low vertical loading can lead under the resulting vibration to the bearing coming loose from the segment if the bearing is insufficiently fixed.

7 Ventilation, Dust Removal, Working Safety, Vibration

7.1 Ventilation

7.1.1 Danger

The mechanical excavation of rock with the partial pulverising of the rock under the disc cutters produces a lot of dust. At the same time, the mechanical energy is converted into the destruction of the rock and also to a considerable degree turned into heat. In order to ensure good air at the working site – this is a precondition for respectable human productivity – the dust has to be intercepted and removed and also the considerable heat produced has to be cooled by suitable measures.

The gases, which occur much more frequently in rock than is commonly supposed, are a significant source of danger. Mostly this is methane and other higher hydrocarbons with a rather low lower explosion limit (UEG), or sulphur dioxide in combination or alone; more unusual are carbon dioxide and radon.

7.1.2 Ventilation Schemes, Ventilation Systems

Basically, there are two different zones during TBM excavation:

- The area of the entire TBM equipment
- The area at the rear

The relevant national regulations apply for the ventilation of the area at the rear, the SUVA for Switzerland, the TBG for Germany and the AUVA for Austria. The necessary ventilation system and scheme is not different from drilling and blasting (see Tunnelbau im Sprengvortrieb *(Tunnel Building by Drilling and Blasting)*, Springer 1997) [96]. The standard SIA 196 "Ventilation in Underground Construction" (1998) [153] explains concepts and calculations for ventilation.

A new requirement of the SUVA from summer 2000 or 2001 for diesel vehicles is the fitting of a particle filter to eliminate the fine soot particles, which can cause cancer.

Exceptions to this rule are:

- Vehicles with auxiliary diesel motor, e.g. drilling jumbos, which are only driven with the diesel motor.
- Vehicles, whose motors are in use for less than 2 operating hours per day.
- Vehicles, whose motors are under the power limit of 50 kW.

For TBM operations, particular ventilation measures have to be taken at the working face. If the dust production caused by the disc cutters in the excavation area alone is considered, then the partitioning between the excavation area and the machinery area, which is integrated in all machines, would be sufficient as long as the dust extraction was adequately proportioned. But if there is a danger of methane emission, whether by opening the pores in the rock during boring or from a joint in the rock, which in an extreme case could lead to the excavation area being flooded, then the extraction of the excavation area by mechanical fans is no longer sufficient for larger machines. The excavation area must be ventilated through. Fans built into the partition can make sure, together with the dust extraction, that such gases are thinned sufficiently.

In Fig. 7-1, the main fan 1 drives the fresh air into the machine area with positive pressure. The flushing ventilation of the excavation area extracts the dust-laden air around the stator of the machine and feeds it to deduster 2. The outlet of the dust remover feeds it back through to main duct. A flush ventilator 3 ensures sufficient fresh air in the backup area up to the TBM. The flush ventilation extracts from the excavation area to the dust remover. Flush ventilations 2 and 3 are permanently installed in the backup. The main ventilation consists normally of an air duct, or for larger tunnels and long drives two parallel air ducts.

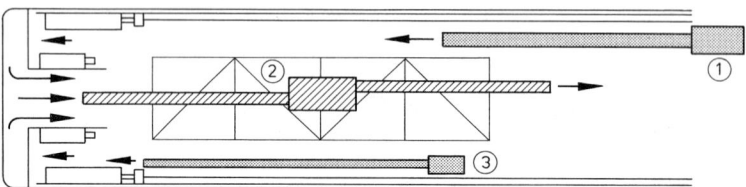

Figure 7-1
Ventilation scheme for a TBM drive [138, 141]

7.2   Dust Removal

The high concentrations in the excavation area exceed many times the permissible values for workplaces. To ensure that the concentration of contaminants in the air is not sufficient to endanger health, the excavation area must be partitioned off and the air containing dust must be so efficiently filtered by suitable dedusters so that the concentration in the tunnel itself lies under the permissible value.

The TBM diameter determines the dust removal capacity directly. With a good partitioning of the excavation area, a minimal opening for the removal conveyor – this opening can also be made smaller with tarpaulins or plastic curtains – the dust removal capacity detailed in Fig. 7-2 will be adequate. These values are based on experience for various types of rock.

There are dry and wet deduster devices on the market today, even for small sections. Dry dedusters are easier to operate, so the SUVA prefers such devices.

High air suction speeds in the air duct lead to not only dust being extracted, they also carry grains of sand and fine gravel. The result is very high wear on the filter hoses and the dust removal mechanism of the deduster (Figs. 7-3 and 7-4). It is a good idea in such cases to install a coarse deduster before the actual deduster. This increases the en-

7.2 Dust Removal

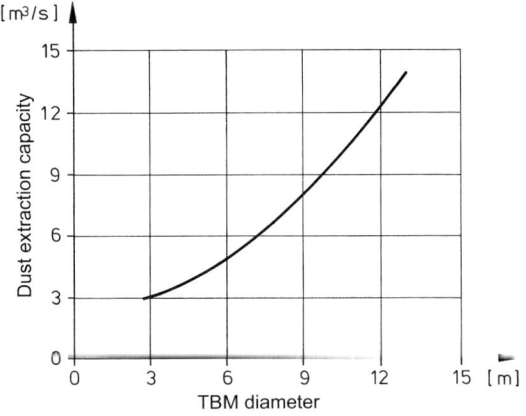

Figure 7-2
Dust removal performance against TBM diameter

Figure 7-3
Dry deduster of the gripper TBM S-155 (Herrenknecht), Tscharner [61]

ergy requirement, but considerably reduces the wear. To minimise sedimentation in the air duct, the minimum air speed in the air duct should be about 18–20 m/s.

Although dry dedusters are preferred because of their simple and safe operation, they are less suitable for TBM tunnelling where shotcrete is used for rock support. The dust containing cement would quickly clog the filter and lead to incrustation [50]. In this case, wet dedusters are used, which bind the dust with added water. These always need a water supply.

102   7 Ventilation, Dust Removal, Working Safety, Vibration

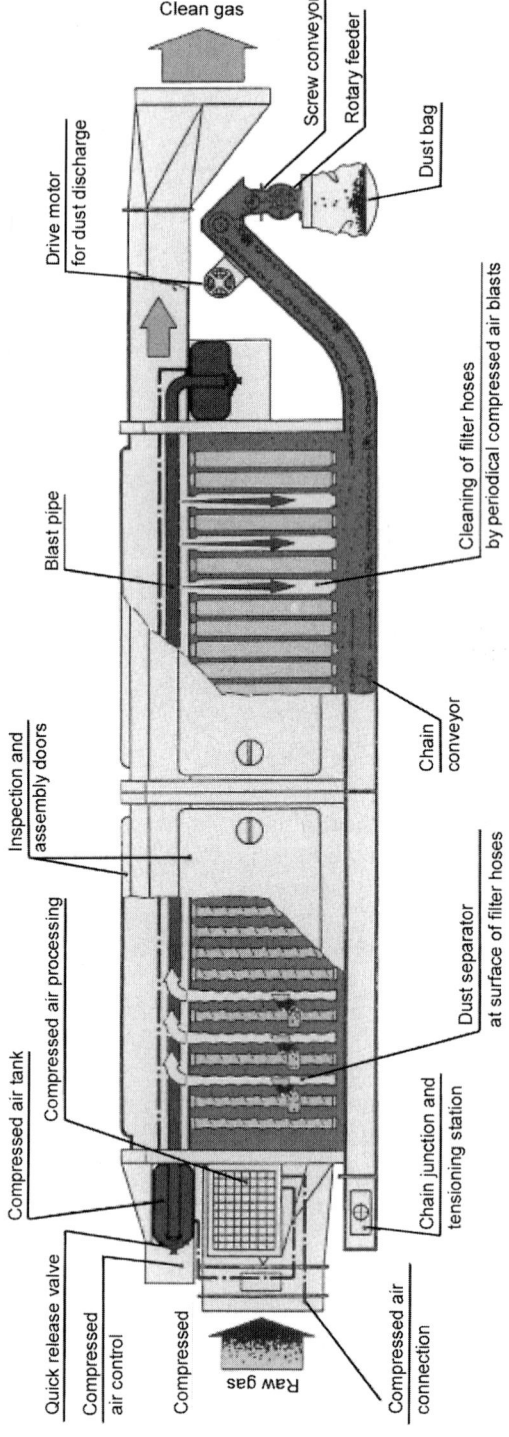

Figure 7-4
Functional principle of a dry deduster with blowing-off of the filter hoses by blasts of compressed air [171]

7.3 Occupational Safety and Safety Planning for TBM Operation

7.3.1 General

The term safety is defined as a condition of being without threat, which results objectively from the presence of protection measures or the absence of danger and is felt subjectively by individuals or social groups as a certainty about the reliability of safety and protection measures. Safety considerations thus presume the recognition of dangers and the risks regulating from them.

Safety cannot refer to the structure alone. Safety in an integrated sense (Fig. 7-5) includes three essential sections, which all have to be included in the considerations.

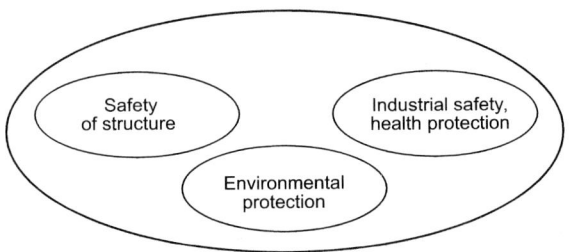

Figure 7-5
Integrated safety

Especially in underground work, there is an increased risk of workplace accidents. The reasons for this are:

- Danger from the mostly heterogeneous ground
- Lighting problems, although artificial lighting is available
- Concentration of moving machines in a restricted space

Basically, every person involved in construction in a constitutional state is responsible for safety as part of his duties. The safety of the structure itself lies within the classic responsibility of the engineer. It is properly required that, even during the design phase, the consideration of safety is included, whether this means the choice of a section or the determination of a construction process or schedule. This is what first makes it possible for the contractor to correctly organise the necessary protection measures for accident prevention and health care.

Experience shows that mistakes are often made during the preparation of work protection measures and the safety-relevant equipment. Possible danger points have not been considered in the design phase of a project or only insufficient protection measures are planned. The necessary safety measures are not expertly tendered, so that protection measures are not appropriately provided for in the construction contract. Not last, the safety equipment on site is not available at the correct time or in the correct quantity. International regulations at European and national level are trying to correct these failures.

7.3.2 International Guidelines and National Regulations

7.3.2.1 International Guidelines

The European Union deals with the areas of "Occupational Safety" and "Health Protection" in various directives, although for tunnelling with a TBM, there are basic regulations.

The member countries of the EU are obliged to adopt all directives into national law. Countries, which have joined CEN are also obliged to adopt European EN standards.

There is a difference here between guidelines based on:

- Article 100a and
- Article 118a of the EEC contract

The former are intended to remove technical hindrances to trade between the individual countries. The member states of the EU are not allowed, as with all guidelines according to article 100a, to introduce regulations stricter than the guideline because this would have the same effect as a trade restriction.

The guidelines according to Article 118a of the EEC law refer to the area of the protection of employees. These are minimum regulations, which may not be exceeded by the individual countries. Stricter regulations in this area may, however, be introduced; but existing regulations, which are more favourable to the employee, are not affected.

The following are essentially of importance in the area of tunnel building:

- Directive 89/391 on the introduction of measures to encourage improvements in the safety and health care of workers at work [117]; all the general ground rules of employee protection are fixed in this directive.

- The 8th individual directive 92/57 on the implementation of minimum health and safety requirements at temporary or mobile construction sites (Building site directive) [121].

- The standard EN 815, which covers TBMs without shield [41].

7.3.2.2 National Regulations

Germany: There are various current national laws, standards and guidelines regarding occupational safety. Of special interest, however, are the relevant regulations affecting tunnelling, especially the accident avoidance regulation (UVV), regulations (BGV or VBG), safety rules (BGR) and information (BGI) from the *Employer's liability insurance associations (BG Bau)*.

In Germany, the *building site regulation (Baustellenverordnung)*, which came into effect on 1 July 1998, implements European law into German law. This is intended through special measures to considerably improve the safety and health protection of the employees on the building site. The measures introduced include the appointment of a coordinator and the creation of a safety and health plan *(SiGe Plan)*.

7.3 Occupational Safety and Safety Planning for TBM Operation

Coordination on the building site means that the work of various employees or contractors, which is usual in tunnelling, are arranged according to one progress schedule in order to ensure safe and economic construction progress. To implement this requirement, the client must name one or more coordinators for building sites where employees work for various employers. It is, however, also possible that he undertakes this task himself or appoints a third party to do it.

The safety and health plan *(SiGe Plan)* is required so that safety technical and occupational health measures have already been sufficiently considered in the design phase of a project, functionally planned, expertly tendered and regulated in building contracts in an appropriate manner. This can ensure that measures are available at the right time and in adequate quantity for the construction phase. The SiGe Plan must contain at least occupational health and safety measures and provisions for the relevant building site: it is developed from the danger catalogue of the employer's liability insurance association for the building industry or from a SiGe Plan from another building project. In the danger catalogue, all relevant regulations (like DIN, UVV etc) are related to the risks and measures. The responsible employee is appointed and the planning schedule fixed with a bar chart. Samples for the creation of SiGe Plans are available from the employer's liability insurance association for the building industry.

The safety of tunnel sites and the occupational protection are regulated in:

DIN EN 815	Safety of tunnel boring machines without shield and shaft boring machines without drill stem for use in rock, 11/1196
BaustellV	Regulation concerning health and safety protection on building sites (Building site regulation), 06/1998
DruckluftV	Regulation concerning working in compressed air, 06/1997
BGV A1	General regulations, 1998
BGV C22	Building work, 04/1993
VGB 119	Mineral dust dangerous to health, 10/1988
BGR 160	Safety rules for underground construction work, 10/1994
BGI 504–1-1	Selection criteria for special occupational medical care according to the employer's liability insurance association guideline Gl. 1, Part 1: "Dust containing quartz", 1998

Austria: The construction work coordination law came into effect on 01 07 1999. This law obliges clients and project managers to appoint coordinators for safety and health protection, where employees from more than one company are employed on a constructions site at the same time or one after the other. For larger building sites, like in Germany, a safety and health protection plan (*SiGE Plan*) is to be created. These client obligations belong to the duties of the employer according to building worker protection regulations. The law serves to implement those regulations of the European building site directive, which have not yet been introduced internationally, with the aim of clearly reducing the accident rate on building sites.

The occupational safety of employees on Austrian building sites is regulated according to the following rules, regulations and laws:

ÖNORM EN 815 Safety of tunnel boring machines without shield and shaft boring machines without drill stem for use in rock, 12/1996

BauV Building worker protection regulation, 04/1998

BauKG Building worker coordination law, 07/1999

AStV Workplace regulations, 01/1999

AschG Employee internal protection law, 06/1994

The guidelines and pamphlets of the AUVA are also to be observed.

Switzerland: Many laws and standards regulate occupational safety in Switzerland:

UVG Accident insurance law

STEG Federal law regarding the safety of technical equipment and devices

Standard SIA 118 General regulations
Art. 104 This standard refers quite briefly to the duties of those involved according to the building directive EEC 92/57, especially however to the duties of the client and his representatives

Standard EN 815 Safety of tunnel boring machines without shield; has the status of a Swiss standard

Regulation concerning the safety and health protection of employees engaged in building work, 03/2000

7.3.3 Integrated Safety Plan

7.3.3.1 The Safety Plan in the Environment of Management Plans

Safety as a complete theme is included in the earliest implementations of the client's aims in design. The consultant is therefore obliged to consider the interests of safety in the general project, at the latest in the preliminary project. This makes safety a part of quality management, alongside the usability, economics and environmental compatibility of the works.

The integrated safety plan is therefore embedded in the operation plan and the quality management plan of the project. It should give information about the areas:

- Safety aims
- Description of dangers and risk analyses
- Action Plan containing measures to prevent or limit damage
- Emergency rescue concept with the emergency services

7.3 Occupational Safety and Safety Planning for TBM Operation

7.3.3.2 Safety Aims

Safety aims are generally considered to have been reached when a hazard of any type is reduced to the relevant acceptable limit. The degree of acceptance is a public interest measure.

The safety aims regarding occupational safety, health protection and environmental protection are to be dealt with in the safety plan. Rock support takes a special place, on the one hand described and evaluated in the operation plan as part of the construction work, and on the other hand the effectiveness of the support work has high significance for occupational safety, for example the early strength of shotcrete in the first hours.

7.3.3.3 Description of Dangers and Risk Analyses

The recognition of danger is the most important factor in the entire risk analysis, because a danger that has not been recognised is the greatest danger.

In order to estimate the danger rationally, the corresponding danger patterns are selected, which mostly comes from practical experience, ability to put things together, intuition and an ability to recognise discrepancies. A rather analytic, systematic route leads to error trees, event trees and cause-effect diagrams [146].

A risk is a danger evaluated at least according to the consequences and probability (Fig. 7-6). According to this definition, all dangers can be categorised as a risk according to the product of the consequences with the probability of it occurring.

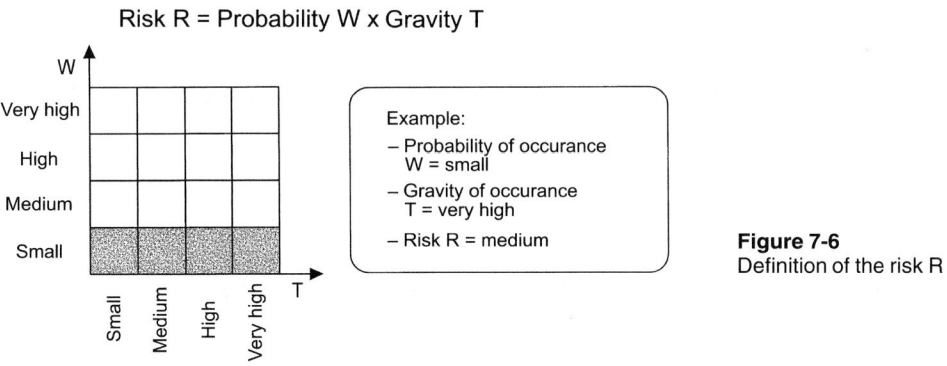

Figure 7-6 Definition of the risk R

Risk-reducing measures, taken purposefully, have the ability to reduce the evaluated risk to an acceptable, mostly small remaining risk. In the listing of the dangers and their evaluation it may be assumed that the client employs a capable contractor, whose employees are familiar with the recognised rules of technology for underground work.

Danger descriptions for the following areas are to be included in the safety plan:

- **Building site setup**
  - Traffic and transport underground
  - Natural dangers like rock fall, flood and avalanches
  - Noise, related to the construction site itself and to the environment
  - Safety of electrical equipment

- **Ventilation**
  - Ventilation breakdowns due to faulty fans, power failures
  - Gas problems, especially the flooding of the TBM with natural gas, special regulations for switching off and on the entire electrical system
  - Insufficient amount of fresh air to adequately thin toxic concentrations
  - Rise of natural gas concentration through increased quantity of gas
  - Damage to the air ducts in the rearward area
  - Concentration of toxic substances in ventilation and after breakthrough

- **Fire in the tunnel**
  - Single vehicle on fire or two vehicles after an accident with smoke through the tunnel or burning of the air duct
  - Fire caused by construction work (building chemical products) also in the rearward area or from noise insulation in the backup
  - Fire after breakthrough

- **Transport equipment for material and personnel transport**
  - Hitting and running over of persons at the face and in the rearward area (reversing of wheeled vehicles, insufficient sight for locomotive driver in railway operation)
    - Driving into working scaffolding
    - Rolling stock running away
    - Falls in the shaft due to insufficiently secured access routes and working platforms

- **Lighting**
  - Injuries due to insufficient lighting of the workplace, the transport routes or persons outside the workplace

- **Temporary electricity supply**
  - Injuries due to inexpertly installed high-voltage equipment (e. g. cable runs in transport area)
  - Fires or explosion
  - Results of power failure (pumping, ventilation, measurement and monitoring equipment)

- **Excavation and support**
  - Injuries due to caving or rock burst (early strength of shotcrete in the first hours)
  - Injuries from rock fall (protection roof on the TBM)
  - Water and mud ingress
  - High concentration of dust during excavation

- **Gas emissions**
  - Explosions due to excessive concentration of natural gas
  - Radiation injury due to excessive concentration of radon

These areas are to be investigated on each site for possible danger patterns.

7.3.3.4 Action Plan

Safety measures of an organisational or material type serve the complete or at least partial defence against the danger, and reduce the risks to the acceptable level. For damage, which nonetheless occurs, a rescue concept is to be integrated with the relevant local emergency services is to be included in the action plan.

The national organisations SUVA, BG Bau, AUVA have published pamphlets, guidelines and regulations, which help each consultant to create the action plan containing measures against risks (e.g. Tunnel Construction, "Working Safely" from the BG Bau) [151, 168].

The rescue concept would ideally be in three- parts with structural, material and personal measures. These measures are explained with examples below.

- **Structural measures**
  These serve to ensure communication and measures in case of fire:
  - Communications
  - Alarm organisation
  - Fire outbreak with alarm and extinguishing measures
  - Rescue plan with escape routes, transport system on the construction site, helicopter pad etc.

- **Material measures**
  These include measurement equipment, rescue material and fire extinguishing material:
  - Measurement devices, mobile and fixed for gas occurrences, which could be expected, including oxygen
  - Rescue material, self-rescuer, circuit devices for rescue, first aid material
  - Fire extinguishing equipment like hose, hand fire extinguisher etc.

- **Personnel measures**
  These serve to train the emergency helpers, the rescue squads:
  - General first aid and rescue training
  - Training in use of a self-rescuer
  - Training of rescue squads in use of breathing apparatus and gas detectors

7.3.4 TBM Details and Specifics Regarding Natural Gas Danger and Rock Support

Two safety goals have a special importance in TBM excavation:

- Ventilation at the face because of natural gas and the particular equipment to deal with it.
- Rock support with protection measures for personnel and securing cavities.

7.3.4.1 Natural Gas Danger

All countries have introduced regulations for the prevention of accidents during the construction of works in rock strata containing natural gas (e. g. for Switzerland, SUVA guideline 1497, for Germany BG Bau guideline BGR 160).

The following points are of particular importance for TBM excavation:

- In contrast to mining, it is never possible in construction to keep plant fully explosion-safe.
- The electrical equipment of all elements (motors, fuses, switches etc.) should therefore be designed so that switching off without danger is always possible after a partial or total switch-off.
- Measurement devices at various locations of the boring equipment, including all trailers, give information about the gas concentration curve at the measurement points, with an alarm at the maximum permissible workplace concentration <1,5% vol. CH_4, which should set off a gas alarm. The advance is stopped. All except the monitoring team leave the tunnel.

If the concentration reaches the limit value of 1.5% vol. CH_4 through a continuous increase or by sudden flooding, then the electrical equipment is switched off, except that equipment, which is absolutely safe and part of the fresh air shower (gas-free) of the main ventilation.

Because the many switching and fuse cabinets can never all be permanently constructed to explosion-safe standards, a way of flooding them with gas-free air is required before the systems are switched on again.

Decision procedures are to be included in the action plan, which should act as a guideline for the personnel in case of:

- Switching on after an automatic cutout of the drive resulting from a gas concentration of 1.5 vol.% and functioning main ventilation after its shutdown.
- Switching on after a power failure in the excavation area without main ventilation being switched off.
- Switching on after a power failure with switch-off of the main ventilation, measures from the portal are necessary here.

7.3.4.2 Rock Support

This is intended on the one hand to make the workplace safe and ensure protection of persons, and on the other to secure cavities. Under poor conditions, like for example in geology with pronounced faulting, TBMs without a roof as protection, under which support can be installed, are extremely dangerous.

Collapses with large blocks falling over the body of the machine and resulting difficulties to support the rock lead to interruptions of the advance and major danger to the crew and the machine (Fig. 7-7).

Figure 7-7
Blocks hanging in the support from a collapse starting immediately behind the TBM stator

Under these conditions, regarding the requirements of EN 815 and the national standards, support can only be undertaken under the protection of a roof projecting at the back, whether a full or a trailing finger roof. Similarly to segments, only systematically precast elements of steel rings and nets or similar support measures seem sufficient. The profiles of the steel rings and the ring width of the elements should be adapted to the geotechnical conditions.

Tunnel boring machines with support equipment designed like these comply with EN 815.

7.4 Vibration

The process of a TBM boring produces considerable vibration, which could have unpleasant consequences near to buildings. If the client is obliged to keep the vibration, which lead to noise in buildings, within close limits, this can lead to a working time limitation for boring. The vibration, measured at the foundation of a house above the tunnel during a TBM drive are shown in Fig. 7-8.

112   7 Ventilation, Dust Removal, Working Safety, Vibration

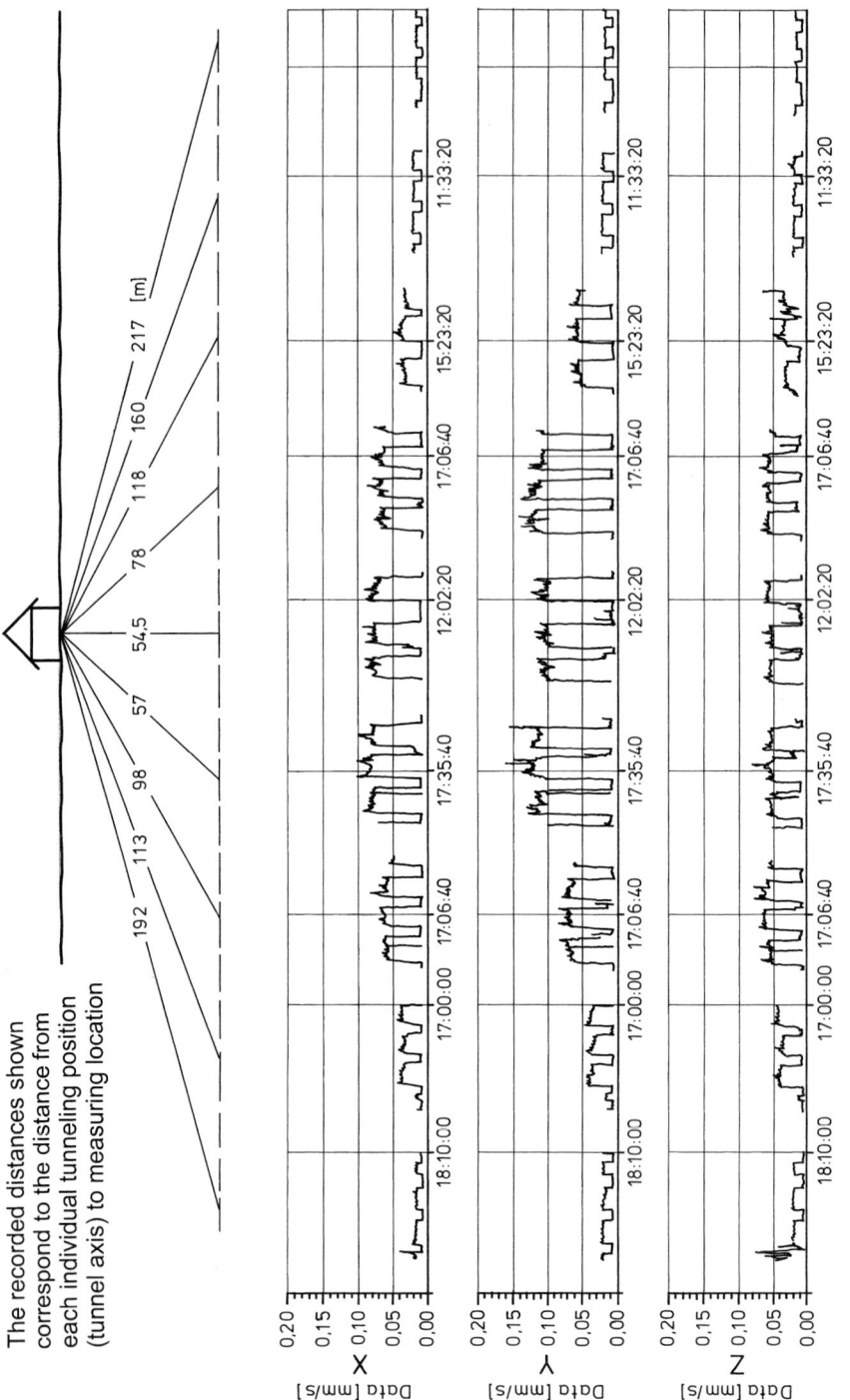

Figure 7-8
Vibration at the foundation of a house during TBM advance, Murgenthal tunnel

7.4 Vibration

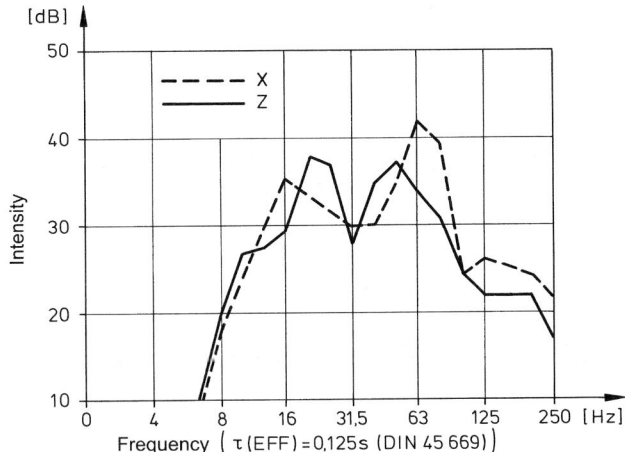

Figure 7-9
Third band spectrum as transfer spectrum of the TBM from Fig. 7-8

The boring times and still periods for rock support as segmental lining in a single shield are recognisable over the 400 m long stretch of tunnel monitored.

Figure 7-8 shows the vibrations in the three main axes x, y, z at the foundation of a house at the relevant distance from the cutter head of the TBM. The vibrations at the house with a cutter head distance of 192m are shown at the left hand side of the illustration. These vibrations rise to a maximum shortly before the TBM travels under the house They reduce rapidly and strongly until they have come down to very low values at a distance of about 217 m.

The measurements show that there are not only negative consequences. The local vibrations caused by the TBM, at least for larger tunnel boring machines, lie in the spectrum area of a goods train travelling fast through the tunnel and certainly under those of a high-speed train travelling through. The transfer spectra at the foundation of a property situated over the tunnel (Fig. 7-9) also correspond to those of modern trains, and are often more than these.

8 Additional Equipment

8.1 Investigation and Improvement of the Geological Conditions

However good the geological investigation for the tunnel may be, there is always more or less uncertainty, making the TBM excavation a journey into the unknown. Known fault zones, suspected and above all known geological events lead to uncertainty, which can often cause substantial interruptions in case of a sudden discovery.

Tunnel boring machines should therefore by equipped as standard for:

- Investigation drilling over the periphery or also around a cavity (also to check for gas).
- Equipment for ground improvement. Ground improvement should be possible through the investigation hole, in order to be able to inject a few decametres ahead of the machine, and on the other hand it is necessary to be able to consolidate broken ground, ground, which might collapse, or ground, which has already collapsed immediately in front of the cutter head.
- Drainage drillings.

Investigation by drilling enables a reliable prognosis, as described above. Seismic investigation from the cutter head produces, even today, no conclusive results. In addition to investigation drilling, which is best carried out with hammer drills and with full-scale recording of drilling data, drill hole photography enables a direct view into the geology, which is to be driven into (see Fig. 12-2).

Drillings within the tunnel cross-section are indeed possible, but there is always the danger of the breaking of the drilling rod, which results in an almost completely impassable hindrance for the TBM.

It is therefore of advantage to use an in-the-hole hammer, which has a closed casing, which does not project forward. The positioning of the drilling rig is very difficult with a gripper TBM. With a shield TBM, it is easily possible to erect the drilling rig on the segment erector. This makes every drilling position accessible through the shield.

The drilling of the injection holes for the consolidation of the ground immediately in front of the cutter head can no longer be done through the stator of the TBM with machines of large diameter. The only possibility is to drill the holes for the injection lances in a close pattern out of the cutter head itself. This requires small drills (Fig. 8-2a) and drilling equipment, which also functions as injection lances. These represent no hindrance for boring with the TBM after the injection measures are complete, or catch up in the cutter head, which would tear out the rock again with the rotation. Very suitable tools for this purpose are extension tubes made of glass fibre reinforced plastic with a throwaway drilling head (Fig. 8-2b). These extension tubes, only 50 cm long, can be excavated by the cutter discs.

116   8 Additional Equipment

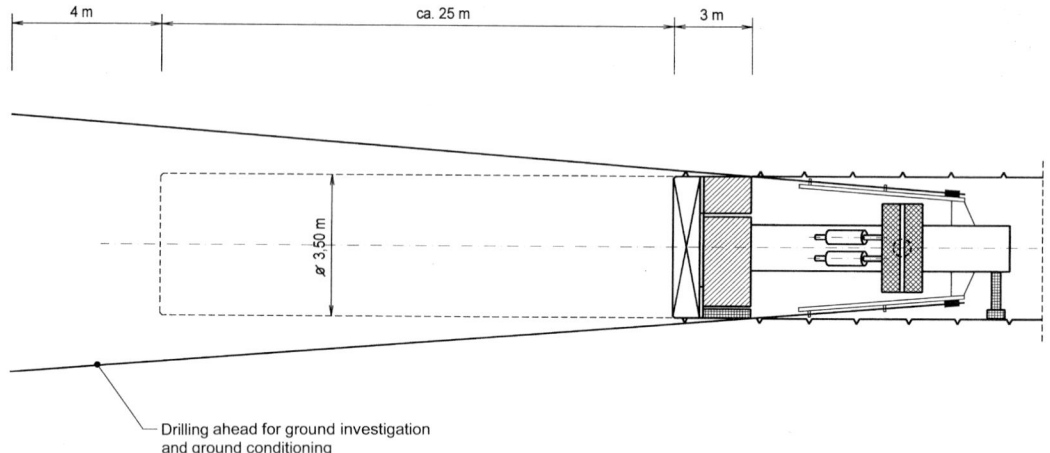

Figure 8-1
Positioning the drilling equipment for ground investigation or for drilling injection holes with a gripper TBM

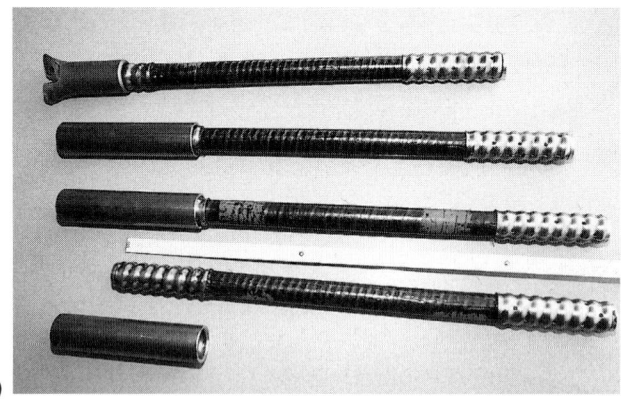

Figure 8-2
Drilling equipment
a) Lumesa drill for extension tubes of 0.5 m length
b) Weidmann GRP drilling extension rods

8.2 Equipment for Rock Support

Injections through the cutter head are best carried out with relatively rapidly hardening artificial resins (PU and Acrylic resins). The cutter head does not stick in this and the injected substances do not run down to the invert of the excavation area of the TBM.

8.2 Equipment for Rock Support

Gripper TBMs are often fitted with equipment for drilling anchors of medium to large diameter, also with equipment for erecting perimeter lining with arches. Mechanised mesh installation with a mesh erection device is being tried out at the moment (Fig. 8-3).

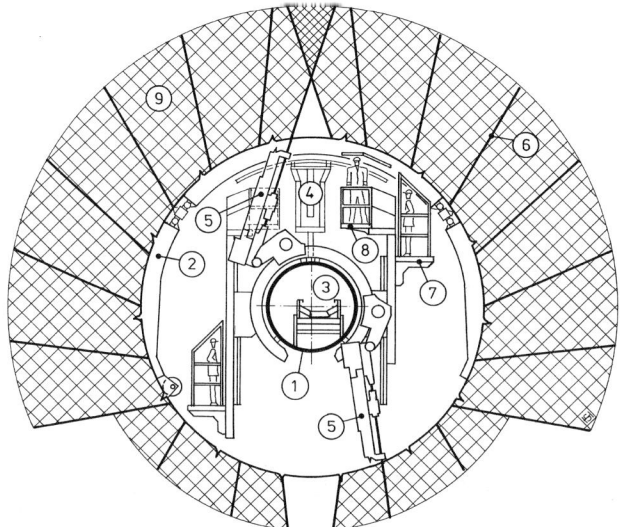

① Machine carrier
② Mesh erection equipment
③ Conveyor belt
④ Erection equipment for expanding segments
⑤ Drilling guide
⑥ Anchor
⑦ Working platform with protection roof
⑧ Working basket with protection roof
⑨ Area in which anchors can be installed

Figure 8-3
Equipment for rock support, Gripper TBM S-167 (Herrenknecht), Lötschberg base tunnel, Steg section, ⌀ 9,43 m [61]

8.2.1 Anchor Drills

Anchor drills, arranged between cutter head and gripper unit at one clamping level or between the gripper units where there are two, can drill holes for anchors in a limited range determined by the geometry (Fig. 8-3). Each anchor drill consists of a telescopic drilling rig, a travel device and an extending working basket, and they can be moved independently of each other.

In order that anchor holes can be drilled at the same time as the TBM is boring, the anchor drill rig can be moved to the extent of the stroke of the TBM. Anchors in the upper area from about 10 o'clock to 2 o'clock can be placed systematically but not strictly radially.

8.2.2 Steel Ring Equipment

Immediately behind the dust wall of a gripper TBM, there is generally a ring erector, with which closed steel rings can be installed early under the protection of the trailing fingers (Fig. 8-4a). The complete ring is moved to the installation location and braced against the tunnel walls. The smallest possible ring separation is determined by the construction and arrangement of the grippers.

The steel ring devices in use are today almost always limited to erecting single steel rings. The time required for the erection of the rock support increases correspondingly, which inevitably leads to a reduction of the overall advance rate.

8.2.3 Mesh Installation Equipment

The installation of mesh was done until now by hand and took place in the unsecured area. Mesh erection equipment is now being tried out on site (Fig. 8-4b), making it possible to draw rock fall protection netting or reinforcing mesh onto the telescopic bearing frame of the mesh erection device within the already-secured area of the tunnel. The mesh prepared thus is driven forward to behind the slat roof with the anchor drills

a) b)

Figure 8-4
Mechanical equipment for rock support, Gripper TBM S-167 (Herrenknecht),
Lötschberg base tunnel, Steg section [61]
a) Anchor drilling and installation equipment
b) Mesh erection equipment

8.3 TBM Steering

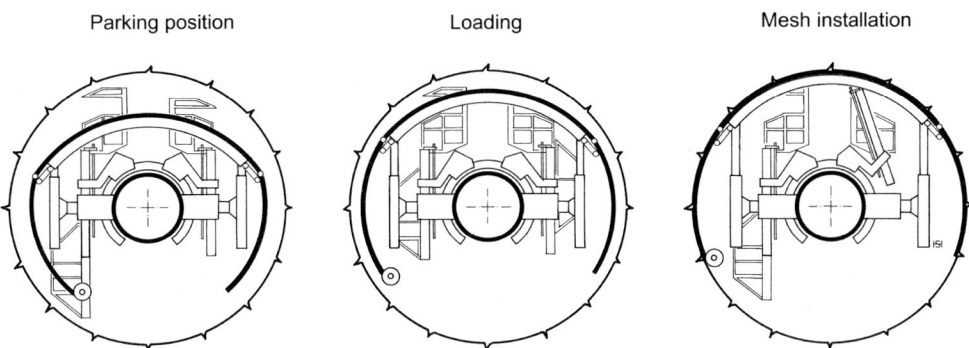

Figure 8-5
Mesh installation with mesh erection equipment (Herrenknecht) [61]

pulled in, braced with the cage-shaped bearing frame and anchored with the anchor drills (Fig. 8-5).

8.2.4 Innovation Aims

In order to achieve high advance rates even in fractured rock, innovations are required to permit the installation of area support elements, as is done with segments in a shield TBM drive, for example steel ring mesh units. This would avoid improvisation with material and the resulting damage (Fig. 8-6). Corresponding high advance rates would, however, entail higher costs for the manufacture of the support elements.

In comparison, with shield TBM excavation the erection of the segments with gripper or vacuum erectors has long been standard.

8.3 TBM Steering

The steering is intended to lead the TBM as exactly as possible along the intended route. The various types of tunnel boring machine behave completely differently when it come to steering. The different types are (see also Chapters 2 and 4):

- Gripper TBM with single clamping
- Gripper TBM with X-type clamping
- Single shield TBM
- Double shield TBM with gripper clamping

8.3.1 Steering the Gripper TBM with Single Bracing

This type of TBM slides forwards on the invert shoe during advance. The clamping unit, also thrust forward by the main beam, remains stationary during a stroke (Fig. 8-7). The steering is done with this clamping unit, by lowering the main beam of the TBM or

120   8 Additional Equipment

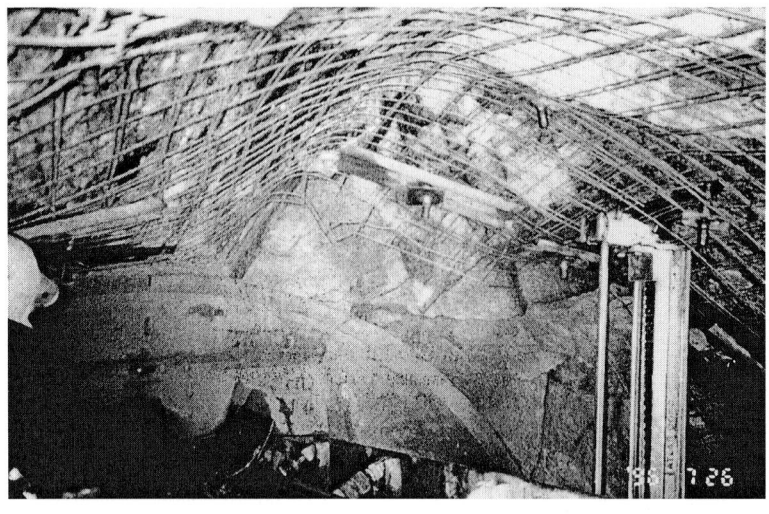
a)

b)

Figure 8-6
Comparison of support work directly behind the machine
a) Improvised rock support
b) Mechanised segment installation

8.3 TBM Steering

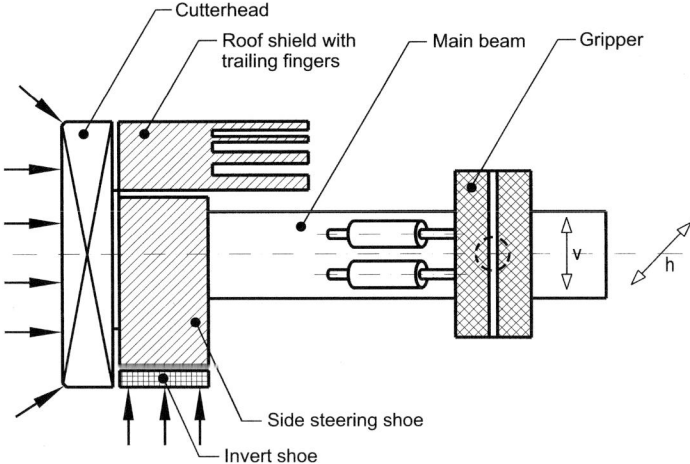

Figure 8-7
Steering of a Gripper-TBM with single clamping

moving it sideways before beginning to bore a new stroke. The invert shoe, which slides along the tunnel floor, can be used to lift or lower the TBM. The up and down steering of the machine is done like this if the rock is sufficiently hard. The side steering shoe, which is often fixed with a link to the invert shoe and can be thrust upwards, serves to stabilise the TBM.

Slow or constant extension or withdrawal of the opposite side of this side steering shoe can also be used, in addition to the adjustment of the clamping unit, to influence the sideways movement of the TBM (Fig. 8-8). It is naturally clear that no absolute accu-

Figure 8-8
Cutter head of a Gripper-TBM with slewed side steering shoe

racy of steering is possible with such simple operation, although efficient in practice. The TBM therefore has to be newly directed with the clamping unit according to the tunnel laser after every stroke.

8.3.2 Steering the Gripper TBM with X-Type Clamping

This type of TBM does not change direction during a stroke. The machine has to be newly directed before each stroke. This is done by the horizontal and vertical movement of the main beam within the rear bracing unit (Fig. 8-9). The TBM thus drives a spatial polygon course.

When it comes to steering, the gripper TBM with two clamping units is very good. The cutter head drifting off when the rock on one side is glancing into harder rock, or subsiding in very weak rock strata in the invert, do not happen with this type of machine.

The new direction of the TBM can, however, only take place while the cutter head is rotating, otherwise impermissible loading is put on the cutter head, in the cutter head bearing and in the Kelly bar.

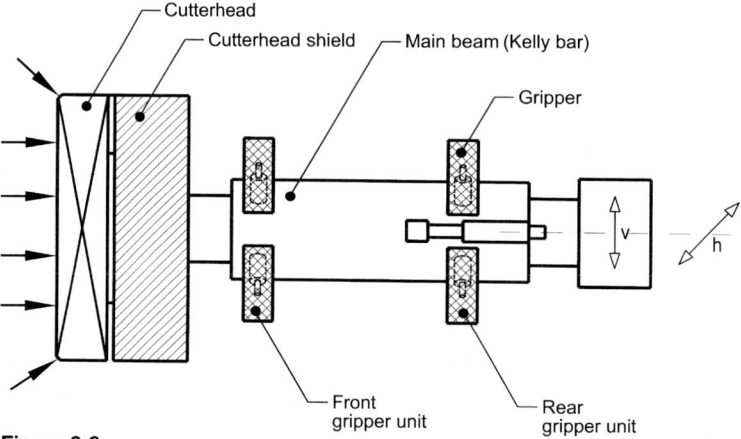

Figure 8-9
Steering of a Gripper-TBM with X-type clamping

8.3.3 Steering a Single Shield TBM

The single shield TBM is thrust against the continually installed segment linings [95]. The resulting system of forces can be seen in Fig. 8-10. The cutter head is moveable in all directions and turns in the shield. The force exerted on the cutters amount together to the resultant cutter head thrust force F_B. Presuming consistent geological conditions, this lies in the axis of the cutter head. The cladding friction RM results from a function of the stability of the ground (rock). Under stable geological conditions, there is in the lower part of the shield, in the bearing surface, a friction force. While advancing in fractured rock with ruptures, friction arises over the entire body of the shield; this is,

8.3 TBM Steering

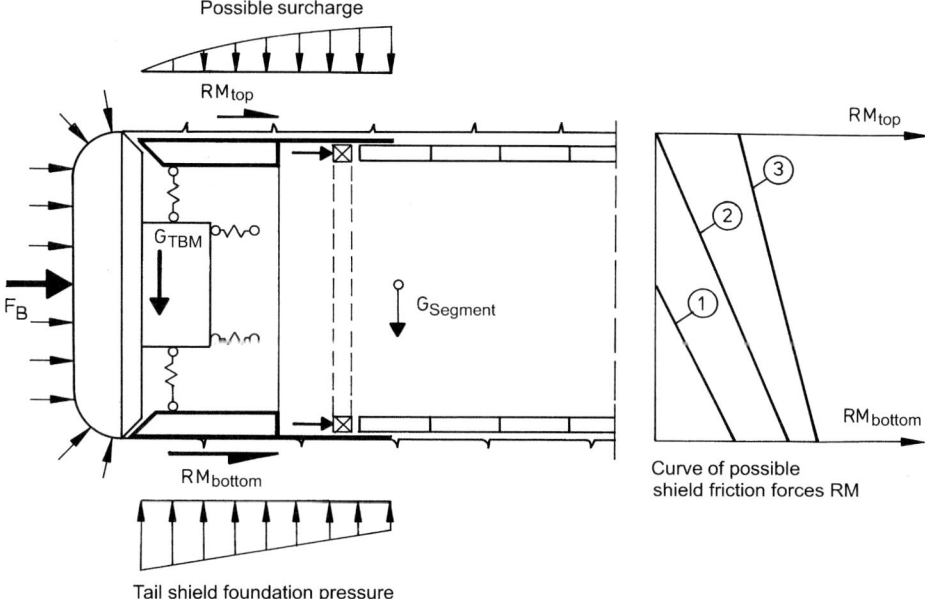

Figure 8-10
System of forces for a single shield TBM with friction forces and shield foundation pressure as well as a possible surcharge

however, less at the crown than at the invert. According to the stability of the rock, the surcharge may also load onto the shield. This is determined by the amount of loose material resting on the shield. The resulting shield cladding friction can in an extreme case for a large TBM vary between 0 and 100 kN/m^2.

The shield cladding friction RM is shown qualitatively in Fig. 8-10. Case 1 represents a shield in a practically stable rock. In case 2, a slightly fractured rock, a little material falls out of the rock structure between the bored cross-section and the shield. In case 3, a very easily fractured or fractured rock, a considerable cladding friction arises because of the surcharge.

The shield bearing force is a reaction to the total force system. A relative centre of the shield lies, through the steering control of the thrust jacks, approximately in the tunnel invert at the bearing of the pressure ring onto the segment ring. The absolute centre of the relevant momentary shield movement lies many 10 diameters outside the tunnel. In order to achieve the different thrust forces with the propelling jacks, which have to be controlled by travel, these are grouped into press groups. The total forces, which have to be provided to propel the TBM, require force concentration in the lower area, so normally two press groups are located in the lower quarter of the shield. In the sides and the crown, one group each is sufficient.

The vertical steering of a shield TBM can be easy or difficult according to the geological conditions. If the rock is stable, lifting the cutter head can produce a gradual steady lifting. If, however, the rock is less stable, for example a molasse marl or broken sandstone, the lifting of the cutter head over the line of the lower shield skin has the opposite effect, because the cutting edge presses into the ground and shears off weaker layers. The slow lifting of the shield TBM can only be achieved by producing resistance at the top of the shield by extending steering fins.

A shield TBM reacts sensitively to glancing meetings with hard stone strata (Fig. 8-11). The high thrust forces on the cutter discs in the hard strata let the shield TBM drift. This drifting continues until the cutter discs at the front of the cutter head create a righting moment, which overcomes the drifting. To deal with this, in some cases only a step-by step approach works: boring with a standing shield and extending the cutter head for a short distance and immediately pushing the shield after it with the cutter floating.

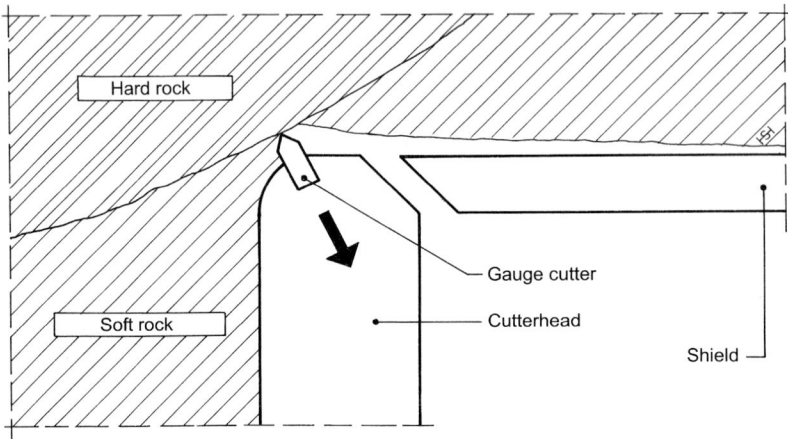

Figure 8-11
Drifting of a shield TBM after a glancing drive into a hard rock stratum during excavation in soft rock

The steering of a shield TBM generally becomes harder, the greater the risk of instability in the ground is. Driving in problematic geological conditions can be optimised by [143]:

- Controlling the penetration and revolution speed
- Material quantity checks to recognise a collapse
- Checking the thrust force and the revolution speed continuously while driving
- Checking the settlement of the ground onto the shield with effects on the control of the cutter head cut-out and the lubrication of the upper surface of the shield with bentonite suspension
- Carry out steering of the cutter head as the cutter head rotates
- Adaptation of the steering to ensure the axial precision when the rock on one side is considerably harder

8.3 TBM Steering

The shield TBM pushes itself from the segment lining. For corrections up and down or left and right, and also for driving curves, the cylinder of the segment linings formed by the last ring, should be as nearly as possible at right angles to the advance direction. This is only possible using correction segment rings, which are shorter at one side of the lining than at the opposite side. This has the result that the supporting plane for the shield is newly directed. For segment systems with differing wall thickness of the individual segment rings, correcting rings are needed for all four sides. Rings symmetrical about the centre can generally be installed as correcting rings. One cylinder-cutting plane is inclined to the axis; when they are turned against each other, the drive direction can be straightened, and any required direction is possible by putting them together or by a partial rotation.

The rotation of the cutter head leads to a contra-rotation of the shield, the shield rolls. This forced rolling is resisted by a light twisting of all the thrust jacks on the bearing face of the segments. The light twisting leads to a force direction away from the axis, with which the rolling can be resisted or, if too large, can be turned back.

8.3.4 Steering a Double Shield TBM

The actual double shield TBM with gripper clamping normally props itself off the clamping unit in the rear shield, the gripper shield. In very poor geological conditions, where the clamping is not very effective, a combination of the gripper clamping and propping off the segments can be used. The front shield with the cutter head is normally supported from a firm abutment, similar to the open TBM with simple clamping.

The V-shaped arrangement of the thrust jacks, as connection between the shields and as thrust unit (Fig. 8-12), allows precise steering of the front shield.

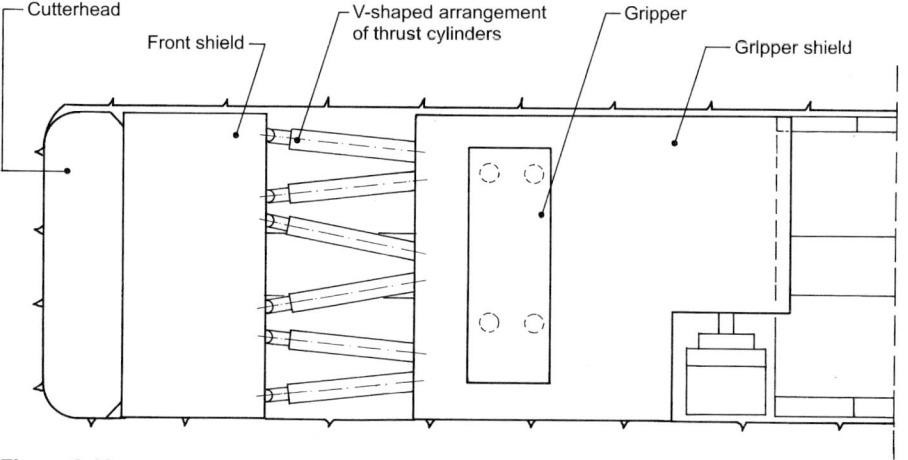

Figure 8-12
Diagram of a double shield TBM with front and gripper shields and V-shaped arrangement of jacks

The continuous control of such inclined thrust jacks during the thrust phase can only be computer-controlled. As with the single shield TBM, the segment rings are to be arranged so that they are connected to form a closed ring behind the drive.

The use of a double shield TBM in very fractured rock can lead to serious difficulties, because fallen rock between the shields hinders or even prevents the withdrawal of the gripper shield after a completed stroke. In some cases this is described as the shield TBM becoming trapped. A better description is that it is seized. Being trapped would require considerably greater overburden, in order that a correspondingly large deformation of the ground could occur. It does, however, make clear the real danger of using a double shield machine at great depths or in less stable conditions.

8.4 Surveying

A tunnel boring machine can never be driven exactly on the intended spatial line. Gripper TBMs allow quite rapid corrections; shield TBMs in contrast have lethargic reaction behaviour.

Surveying should ideally be capable of showing the position of the TBM relative to the intended line at any time, i.e. every half metre of advance. This aim has first been made practical with computer-controlled position fixing. Such devices are the current state of technology [95].

Surveying as an overall concept also contains the potential for errors. This conjectural error potential should be taken into account by the consultant in the layout of the standard profile. With modern surveying technology, the directional accuracy of a TBM is about 8–10 cm in all directions.

Overall, the main sources of errors in tunnel surveying are as follows:

- Basis surveying errors
- Surveying errors
- Setting out errors
- Errors in the TBM advance

There is the tendency to calculate the probable error, according to the theory of propagation of errors, as the square root of the sum of the squares of the individual errors. This assumption does not, however, correspond to the real conditions. A surveying error caused by an incorrect laser direction beam leads to an incorrect theoretical boring axis, which no longer lies within the bounds of such theoretical error considerations.

The stated boring accuracy of 8–10 cm in all directions can be taken as an achievable dimension for tunnels up to 5 km long. In longer tunnels, however, the basis surveying error increases; this has to be corrected at the target portal by adapting the route or taken into account by increasing the section of the tunnel, a substantially more expensive measure.

8.4.1 Surveying the Position of the Tunnel Boring Machine

A modern surveying system enables permanent monitoring of the position of the TBM. This is achieved by receiving a laser light beam at an active receiver system. The installation of this receiver system is included in the construction of the TBM. It is fixed to the shield. It determines the longitudinal position and the rolling of the TBM at the same time.

The exact angle of yaw is determined using intelligent sensors as the deflection of the target table longitudinal axis to the laser beam. The differences in horizontal and vertical directions of the point where the laser beam passes through the target table to the intended position are communicated to an industrial computer, together with the roll angle, pitch angle, and yaw angle, and used to determine the position and level. The result is displayed graphically and numerically related to the planned tunnel route or the correction curve on a monitor at the control position. This enables the TBM driver to continuously take into account the reaction of the machine while steering. This guarantees that the steering operations are always related to the position and that the course is continuously led back to the intended route.

The distance between laser and target table is normally measured electrically-optically and updated with a new tunnel survey for each relocation of the laser. Any commercially available type of tunnel laser can be used. This must, however, be adjusted from time to time, especially when driving curves. This altered position of the laser must be entered into the system manually.

8.4.2 Forward Calculation of the TBM Route

After the position of the TBM has been determined, for a gripper TBM after the last driven advance and for a shield TBM after the last installed ring, the ideal route ahead for the TBM can be calculated. If there are only minor deviations, then the intended route remains the future drive axis. If there are significant deviations, a number of centimetres, a modern computer system can calculate the correction curve needing to be driven. This correction curve starts from the most recently actually driven curve and leads within the capabilities of the TBM slowly back to the intended curve. Too-sudden corrections mostly lead to deviating from the ideal axis on the other side.

9 Tunnel Support

9.1 General

Support against collapse should use suitable means to maintain the stability of the rock, installed at the right time, and protect the workers and equipment from injury and damage by falling rocks or collapse.

It is only possible in very rare cases to excavate do without support. Fresh rock is mostly disturbed or faulted by the stresses, which form mountains, and thus have insufficient stand-up time without support; also the employer's liability insurance associations (SUVA, BG Bau, AUVA) demand rock-fall support for cavities above a limited height.

The materials and the type of rock support in TBM tunnelling have hardly or insignificantly altered in comparison to tunnelling by drilling and blasting or driving with roadheaders. Even the language used with TBMs remains the same, with the expressions head protection and a light to heavy support.

Connected with this, though is the traditional, often very temporary way of installing the support with adaptation as precisely as possible to the seemingly recognisable requirements of the rock. This way of working came into use when the material costs for the support were more significant than the wage costs, which were lower.

Today, the wage costs are far ahead of the material costs. The main aim is to advance the tunnel at a fast rate. The tunnel builder is required to develop innovative new support systems, which, in contrast to the high proportion of the advance time taken up with support work as shown in Fig. 3-33, make more time available for boring.

The support mostly has a final character or at least partially final character. In comparison to drilling and blasting, as is often stated, the rock is treated gently but the very high clamping forces often produce a quite different picture.

There is, however, an essential difference that the rapid advance, which is aimed at requires bearing capacity much sooner than with drilling and blasting. Neither normal mortar anchors nor shotcrete reach the required bearing capacity for the daily advance rates, even with tunnels of up to 20–40 m diameter.

The experience so far with shotcrete support applied near the machine shows, in addition to the general problem of incompatibility between machine and shotcrete, a serious fall-of of advance rate even with systematic use (Ilanz I and Heitersberg East). In both cases the advance rate fell strongly, to less than 4 m in the large section of the Heitersberg tunnel ($\varnothing$ 10.65 m) and to less than 5 m in the smaller power station heading Ilanz I with 5.2 m diameter.

With the primary purpose being higher advance rates, support systems are required, which can also be used with a gripper TBM similar to segments with the shield TBM.

Support elements can no longer be categorised as temporary materials, they should be used as parts of the construction.

Shotcrete, with its superb properties like bonding to the rock and remarkable ductility, should not be abandoned. The necessary completion of the support as a whole can be done with shotcrete, though first 30 m behind the cutter head.

9.2 Support Systems and Advance Rates

The strong fall-of of advance rate with increasing requirement for support has already been shown in 1980 in the SIA 198 (Fig. 9-1).

The excavation classes of the national standards (see chapter 13) basically represent nothing else than the function of advance rate against the support measures. The *Swiss master builders' association (Schweizerische Baumeisterverband)* [152] also shows the fall-off in advance rate depending on the support measures (Fig. 9-2).

The illustration shows clearly how the advance rate with traditional support methods sinks strongly with increasing support, and also the objective of improving the advance rate using innovative support systems. The advance rate for a shield TBM remains at a very high level whatever class of rock is encountered.

The common types of TBM today were too often built with the aim of achieving high performance in rock, which is hard to bore through, and this in rock requiring practi-

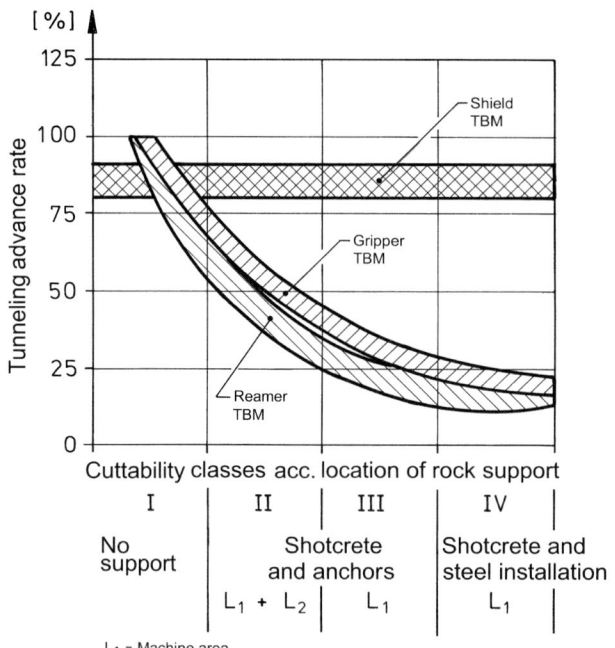

Figure 9-1
Advance as a function of the support work for various TBM types [135]

9.2 Support Systems and Advance Rates

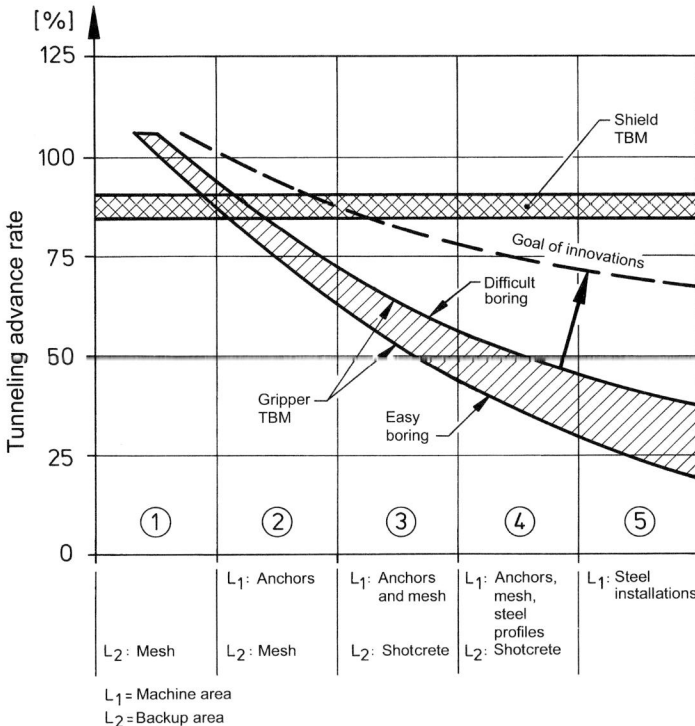

Figure 9-2
Graph of the advance rates for a gripper TBM with traditional support system and a shield TBM [152]

cally no support. The radial speed of the disc cutters is reaching limiting values; any faster and the wear on disc cutters and roller housings would rise over proportionally. The increase of cutters pressure has almost reached the material-technological limit. Cutters bursting in hard rock demonstrate this, because the actual cutter loading increases significantly above the calculated long-term values. With long-term loading of 250 kN, the actual loading can increase in the short term to 450 kN. An increase in the overall advance rate can therefore only be achieved by shortening the time required for the installation of rock support. Such an increase is only possible within limits, e.g. by using anchor drills with improved performance. What is required is innovations in the type and the installation of rock support in various excavation classes.

The aim of such innovations must be to improve the performance curve for gripper TBMs in rock, which necessitates more support work. If support systems similar to segments could also be installed with gripper TBMs, then the advance rate would increase considerably, even in hard rock and hard excavation classes, with a resulting lowering of costs.

Figure 9-3 shows the cost of tunnelling with a gripper TBM against the support work, calculated from standard prices from the *Swiss master builders' association (Schweize-*

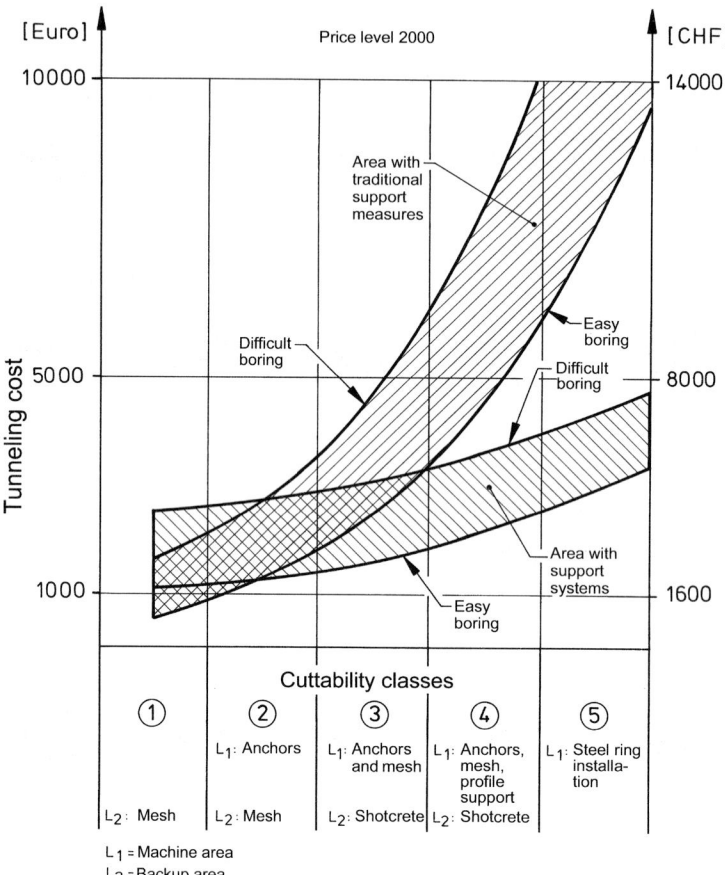

Figure 9-3
Advance costs (excavation and support, year 2000 prices in CHF and Euro) with a gripper TBM ⌀ 5 m against the rock support and the level of boring. The support system corresponds to Fig. 9-2

rische Baumeisterverband). It is insignificant here whether the prices are set higher or lower, because the relationships are maintained. With the performance improvement through innovative support systems according to Fig. 9-2, the overall costs for the advance sink sharply. Higher cost of the support materials leads to lower overall costs, even with a relatively moderate total support requirement.

The precondition for the efficient installation of rock support is the corresponding design of the TBM. As with a shield, it should be shield to install such systems under the protection of the trailing fingers or a protective roof shield, even a shield with side steering shoes. These support elements are not in contact with the rock until the TBM drives forward, and according to this system the expanding would take place after the forward stroke in each case.

9.3 Systematic Support at the Machine

Systematic support at the machine should be able to do duty in supporting the rock immediately (see also Chapter 8). Support elements, which have a delayed development of bearing capacity, are not suitable; for example shotcrete, which only develops sufficient strength hours after spraying, or mortar anchors.

The good effect of slack full-contact anchors is already generally known from experience with drilling and blasting. The use of these is, however, very questionable with the high advance rates of tunnel boring machines, unless the mortar in the deepest depths of the drill hole has accelerator added.

The following are suitable as systematic support at the machine:

- Steel ring installation, both in the upper part and closed ring, also in combination with yieldable connections
- Liner plates
- Segments: segments as all-round lining; segments as invert segments, in smaller cross-sections also as carriers for the main drainage and as roadway; segments in the invert, often combined with temporary drainage and an invert filling for large cross-sections

9.3.1 Steel Arch Support

Steel installations out of rolled sections are capable of being installed behind the stator of the machine and take loading from the rock. With a TBM without a roof shield, however, the rock must have sufficient stand-up time until the room to install the steel arch has been created by the forward movement of the machine. With a roof, the support can be installed as single elements, or better as support elements, with better protection against rockfall.

Figure 9-4 shows a steel installation with reinforcement nets inserted in the crown. The steel installation element, in this case made of HEB sections (I beams), is installed as a rigid support.

In addition to the common wide-flange rolled profiles (HEA, HEB, HEM), bell sections are also suitable. These are designated TH-sections. They enable a yieldable support at first, which allows remarkable convergence. Figure 9-5 shows a connection of steel arch elements with yieldable support.

If the profile is bent with its opening upwards, then the later application of shotcrete can be completed without cavities. Table 9-1 shows the geometrical and structural values for common bell profiles.

Bell profiles also have the considerable advantage that it is possible, after the installation under the protection of the trailing fingers and the advance of the TBM, to hydraulically expand them to make contact with the rock.

Bell profiles were successfully installed as yieldable support in the Viktoria and Haus Aden coalmines in 1977–1982 in fault zones with an overburden of about 1000 m. The

Figure 9-4
Rigid steel installation in the Walgau heading [69]

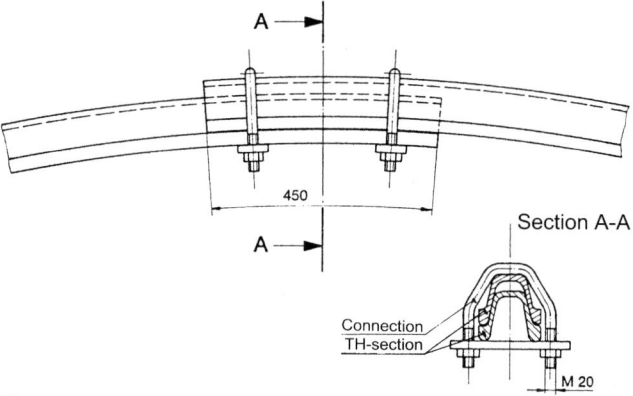

Figure 9-5
Bell section coupling in a yielding support with the bell opening to the cavity

gripper TBMs (Demag) in the Viktoria mine with 6.1 m and in the Haus Aden mine with 6.5 m diameter achieved remarkable average advance rates of 13 to 14 m per day with steel profile support throughout in sand slates, slates and sandstone, even with squeezing in places.

For light steel installation elements, above all for small to medium diameters, U profiles are used, either as closed ring or described as crown arches: Crown arches first become bearing elements in combination with rock anchors (Fig. 9-6). Friction anchors are frequently used, although the corrosion resistance is very limited. It should be noted with these that the final support can be loaded on one side if single anchors fail.

9.3 Systematic Support at the Machine

Table 9-1
Bell profiles

Steel strength 32 Mn 3 Tension strength 60/70 kg/mm^2
or E30-2 Yield limit 40/45 kg/mm^2
naturally hard Strain 20/24%

Profile	Minimum bending radius		Area	Weight	Resistance moment		Moment of inertia		Dimensions of the profile		
TH	r	max			Wx	Wy	Ixx	Iyy	Height	Width	Thickness
	m	L = m	cm^2	kg/m	cm^3	cm^3	cm^4	cm^4	mm	mm	mm
13/48	1.5	9	16.3	13	32	30.6	137.1	150.1	85	98	9.5
16.5/48	1.5	9	21	16.5	42.8	47	190.5	266	88.3	114	10
21/48	2.0	12	26.8	21	57.8	63	293.5	385	98.5	122	12
25/48	2.0	12	31.8	25	77.2	76	428.6	466.3	111	123	14.5
29/48	2.0	12	37	29	99.6	107	600	799	119.4	149	13
36/48	2.5	15	45.8	36	137.3	148	926.5	1265.5	132.3	171	14.5
44/48	2.5	15	55.6	44	179	179	1323.7	1558.8	144.3	174	17
21/58	2.5	15	26.8	21	61.3	64	341	398	108	124	14
25/58	2.5	15	31.8	25	80	83	484	560	118	135	15
29/58	3	15	37	29	96.8	106	627	783	124	148.5	16
36/58	3	15	45.8	36	138	151	977	1270	138	169	17

Figure 9-6
Installation of crown arch support with mesh reinforcement and anchor drill [182]

Steel ring elements of whatever type need to be matched to the size of the gripper shoes of the TBM. It must be possible to apply the grippers between the steel arches. If they are not matched, then it would not be possible to apply the clamping forces and on the other hand the arches would be fully deformed, so losing their bearing capacity as support.

Lattice girders, which are very suitable in drilling and blasting and also in tunnel driving with roadheaders, mostly sprayed in both applications with shotcrete right in the working area, are not very suitable for TBMs. They are not capable of taking rock pressure until they have been sprayed at a distance of 30–60 m behind the cutter head.

9.3.2 Liner Plates

Liner plates are curved sheeting plates on the tunnel wall with a raised edge: they can be screwed together through this raised edge. Figure 9-7 shows support with liner plates in a small TBM tunnel.

Liner plates are a very good solution in small cross-sections.

Figure 9-7
Liner plate sheeting, San Leopoldo tunnel [108]

9.3 Systematic Support at the Machine

9.3.3 Segments

9.3.3.1 Invert Segments

Modern TBM excavations without the use of invert segments would no longer be imaginable.

The invert segment forms the temporary and the final support of the invert and contributes to cost-effectiveness through an increase of performance. At the same time, it also serves to provide a reliable rail track bed for material transport from and to the backup and for the drainage. Invert segments are installed using a crane in the back part of the machine at the segment erection station (Fig. 9-8).

Figure 9-8
Installation of an invert segment with central drain, Vereina tunnel

There are appropriate recesses in the invert segments for the ring arches at a particular spacing. It is important to install the steel arches to the correct dimensions so that the recesses in the segments to be installed later fit.

9.3.3.2 Segmental Lining

The segment ring is a typical support system, used widely today, including in TBM tunnelling (see also Chapter 15). The segments consist almost entirely of reinforced concrete, seldom of cast iron or steel. Concrete segments are built in very varied forms, as curve segments, as cassette segments or a simple block segments.

The discussion as to whether it is better to build with a single shell or double is mostly in terms of cost effectiveness. It is a good idea to design the support system segment as the final tunnel lining. In pressurized face tunnelling with mixshield or earth pressure TBMs, this is today standard technology.

For TBM tunnelling, however, there are some disadvantages. Single-shell support is only cost effective under particular conditions. A comparison of tenders for many larger traffic tunnels in Switzerland has shown that two-shell construction is still 5–10% cheaper. The reasons for this are:

- Simple concrete block segments as temporary support are cheaper to produce, with less reinforcement, less precision, simpler formwork and operation (Fig. 9-9).

- Greater advances are possible with the simple Swiss type of concrete block segment because the ring installation time, max. 15 minutes for 12 m diameter, is only 30–40% of that of a single-shell ring.

- The simple unsealed concrete block segment can be bedded with a fine gravel using pneumatic supply. The single-shell, watertight ring with sealing profiles in all joints needs external pressure to preserve the sealing because of the relaxation on leaving the shield. Grouting of the annular gaps, as is usual in soft ground, is therefore essential.

- The inner shell often requires no reinforcement; the two-shell construction is therefore very economic to construct (Fig. 9-10).

The design of segments can sensibly only be carried out as a detail-sensitive calculation. It is usual to use a bar construction model for the bedded ring and increasingly FE calculations are being used. These last have the advantage that the rock can be considered not only as loading but also as bearing.

A comparison of the required reinforcement shows very large scattering (Table 9-2). The wide range results less from the various assumptions about the rock than from the use of reinforced concrete design codes. The higher reinforcement areas result from the strict application of rules for minimum reinforcement. Precisely with a two-shell sup-

Figure 9-9
Segmental lining with simple, unreinforced concrete block segments, Murgenthal tunnel

9.3 Systematic Support at the Machine

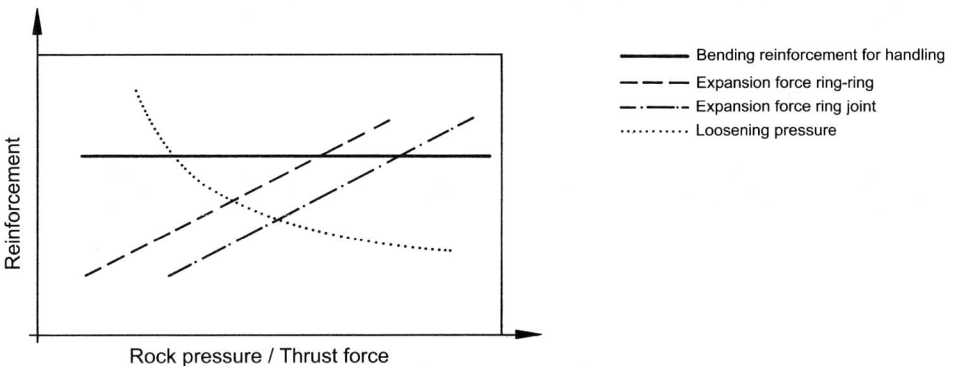

Figure 9-10
Qualitative diagram of the shares of the segment reinforcement

Table 9-2
Scattering of the segment reinforcement in various tunnels in Switzerland, all in two-shell construction in the road area, or in the case of symmetrical construction, two-shell also in the invert

Tunnel	Geology	Ring length	Thickness	$\varnothing$ m	Steel kg/m^3	Steel kg/m^3
Gubrist	Upper sweetwater molasse	1.2 m	28 cm side/crown 40–58 cm invert	11.65	98.6	1010
Zürichberg	Upper sweetwater molasse	1.2 m	25 cm side/crown 40–53 cm invert	11.65	74	745
Bözberg	Jurassic limestone, marl	1.25 m	40–70 cm 29 cm symmetrical	11.87 11.87	90 100	1110 1055
Murgenthal	Lower sweetwater molasse	1.5 m	28 cm side/crown 45–56 cm invert	12.03	51.2	656
Zürich-Thalwil	Upper sweetwater molasse	1.7 m	30 cm side/crown 50–60 cm invert	12.28	88.5	978
Sachseln	Marl, lime flint	1.25 m	30 cm symmetrical	11.76	66	712
Arrissoules	Lower sweetwater molasse	1.25 m	30 cm symmetrical	11.76	54	583

port, which has only meagre demands on its water tightness, the system should be able to help itself rather with an additional joint, formed by a crack under extreme loading, than with a corresponding relaxation.

The necessary segment reinforcement consists of many parts. The diagram in Fig. 9-10 shows the qualitative distribution of the main sections against the ground pressure and the thrust force of the TBM.

The attempts to use steel fibres as reinforcement could be helpful in meeting the requirement for the best possible ductility of the segments. Steel fibres alone are not economical for large diameter segments, but their use in combination would be sensible, particularly where complicated segments are required.

For segmental lining of small diameter, however, steel fibre reinforcement is very suitable. (Fig. 9-11).

Resistance Against Aggressive Water

Concrete segments as invert segments only or as a complete lining can be damaged by ground water or by alkali sensitiveness of the aggregate.

a)

b)

Figure 9-11
Segmental lining with steel fibre reinforcement, Sörenberg gas heading

9.3 Systematic Support at the Machine

Damage occurs through corrosion in the concrete, mostly in the cement binder, resulting from:

- Soft water, especially when it also contains carbonic acid
- Water with a high content of aggressive carbonic acid
- Water containing sulphates as sodium, magnesium or calcium sulphate
- Water containing chlorides (in rare cases), which would be a problem for the reinforcement
- Chloride from salt spread on roads and water contaminated by agriculture or other civilisation influences
- Alkali reaction of the aggregates

Sulphate water damage occurs often in descriptions of damage. The solution has generally been to use cement low in C_{3A} for sulphate resistance. The fact that concrete with such cement does not have sufficient resistance is shown by the damage to the San Bernardino tunnel (Fig. 9-12).

Figure 9-12
Sulphate water damage in the eastern vault abutment and in the tunnel drainage channel, San Bernardino tunnel

In contrast to the assumption that sulphate attack can be avoided by the formation of calcium sulphoaluminate (Ettringite) by low-C_{3A} cement, large-scale trials by the SBB [137, 140] have shown that sulphate resistance of concrete can only be achieved by:

- A very dense concrete structure with high-quality Portland cement
- A w/z factor of < 0.45

- A microsilicate admixture of 8–10% by weight of cement
- An appropriate, high-quality super-liquefying agent

Aggregate liable to alkali attack can lead to the breakdown of the concrete structure. Non-crystalline, so-called amorphous silica can convert to sodium silicate at normal temperatures with alkali [79].

9.4 Shotcrete Support

9.4.1 Shotcrete Support at the Machine

The application of shotcrete in the machine area leads, through the inevitable rebound, to fouling and damage of the TBM. The provision of a travelling shotcrete robot competes for the available space with the other devices for installing rock support (drilling guides etc).

The young shotcrete support is partially destroyed by the clamping forces applied by the grippers.

Regular use of shotcrete is therefore normally carried out in the area of the backup. Shotcrete at the machine is rather used for local support in front of and over the cutter head under difficult geological conditions. Figure 9-13 shows the steps, which were necessary for 2,400 m of drive (46.8% of the entire drive length) of the Gossensaß pilot heading ($\varnothing$ 3.5 m). The shotcrete was applied using the dry spraying process in layers 3 m thick [119].

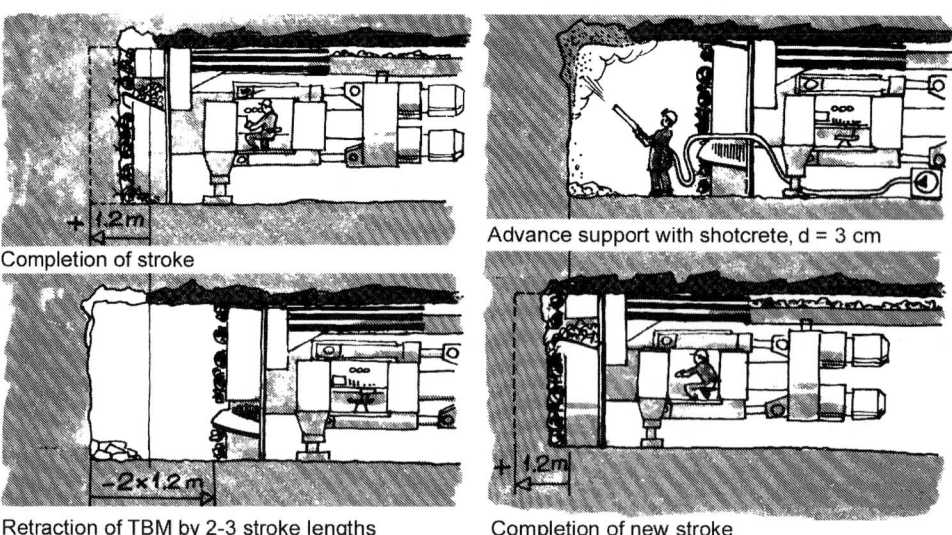

Completion of stroke

Advance support with shotcrete, d = 3 cm

Retraction of TBM by 2-3 stroke lengths

Completion of new stroke

Figure 9-13
Application of shotcrete at the machine, Gossensass tunnel [72]

9.4 Shotcrete Support

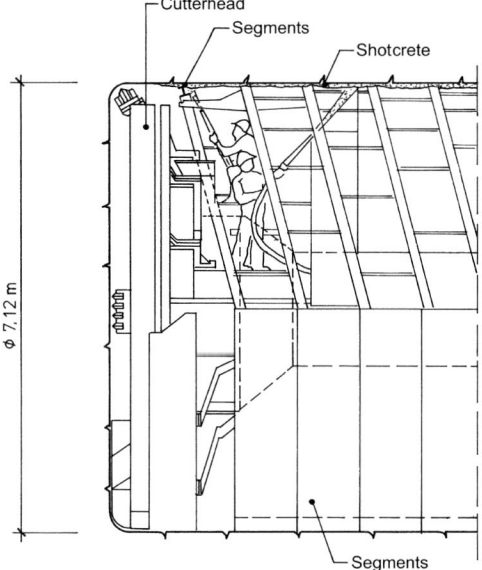

Figure 9-14
Application of shotcrete at the machine, Murla project, Australia [126]

An example of the systematic use of shotcrete in the machine area is given by Robbins, the Murla project in Australia (Fig. 9-14) [126]. To what extent the problems described with the use of shotcrete occurred is not further reported.

Robbins sees, however, further development potential in the use of shotcrete with anchors. He points out the possibility in this connection of producing flexible support, which follows the basics of shotcrete in order to reduce the high rock pressures by allowing deformation.

9.4.2 Shotcrete Support in the Backup Area

Because of the generally known incompatibility of shotcrete and tunnel boring machines, the only option normally is to apply the shotcrete systematically in the backup area. The hoped-for bond between the shotcrete and the rock cannot always be achieved when the shotcrete is applied so late. The many stress cycles caused by the TBM clamping with the gripper shoes often lead to spalling of rock chunks. This results in the rock being loose behind the reinforcement. The shotcrete can only be applied to a "loose" rock layer.

Not only the incompatibility of the shotcrete and the TBM compels the application of shotcrete at a separation of 30–60 m from the cutter head. With an advance rate of 20–40 m per day, the sprayed shotcrete only reaches the required early strength with high accelerator content, in order to be able to function as load bearing support.

There are also problems spraying shotcrete in a TBM advance with workplace health and safety (accelerator, dust), making the machine dirty and by cleaning the hoses when blocked. In this case, the dry spraying process has additional disadvantages to overcome.

Nonetheless, the development should be continued in this area in order to use the advantages of shotcrete support known from conventional tunnelling for TBM excavation as well. Further details about shotcrete support can be found in [89, 96].

9.5 Localised Support

Localised support is carried out according to the geotechnical conditions that are encountered. In the Alps, conditions can alter very rapidly. This poses the question of decision competence, decisions being mostly required immediately for support in the machine area of gripper TBMs.

The following are suitable for localised, mostly sporadic support:

- Anchors, also in combination with reinforcing mesh (Fig. 9-15)
- Single arches

Figure 9-15
Temporary support out of anchors, nets and shotcrete

9.5.1 Anchors and Mesh

Rock anchors or rockbolts are generally regarded as an important means of rock support in underground construction. Methods proven as successful in drilling and blasting, with moderate advance rates compared to TBM tunnelling, are not necessarily also suitable for high-performance tunnelling.

The mode of operation of these short anchors depends on their task, the type of anchor and the length:

9.5 Localised Support

- Fixing of individual blocks of rock (coffin lid) or packets of rock
- Prevention of rock burst
- Production of a rock-bearing ring in collaboration with shotcrete

In order that the anchors work best, they should be installed soon after the cavity has opened. The worse the condition of the rock and the quicker and larger the deformation is, the quicker the anchors have to be installed.

It can be easily recognised here that the demand for an effective rock-bearing ring shortly after excavation by the TBM cannot be achieved in practice [136, 158].

The type of anchor used should be matched to the required lifetime:

- Permanent anchors with long-term compatibility of the materials with each other and against outside influences. The lifetime of these anchors can be considered the same as that of the entire structure.
- Provisional or temporary anchors. These do not fulfil their function over the entire life of the structure; their task is superseded by the installation of another support system.

It is observed all too often that anchors fail to produce the required or hoped-for fixing force. This problem occurs mostly because the anchor type or version has not been selected according to the type of rock.

The problems develop due to:

- Unsuitable choice of the drill hole diameter. For glued anchors and artificial resin cartridges, the drill hole diameter should never exceed the dimension of the anchor bolt plus 10 mm. This results in, according to anchor bar, 30–35 mm holes. Such small drill hole diameters can scarcely be drilled with high-performance hammer drills.
- Drill holes not stable
- Drill holes poorly filled with mortar
- Incorrectly stored artificial resin cartridges

Table 9-3 shows the advantages and disadvantages of the common anchor systems for support work in TBM tunnelling.

When combining the anchors with steel arch-support, great care should be taken with the durability of the anchors. A single broken anchor, deprived of its bearing capacity through corrosion, can lead to extreme loading on the support or the tunnel waterproofing.

9.5.2 Arch Support

Propping with single arches or also partially with support elements running all the way round (liner plates) (Fig. 9-16) are certainly possible individual cases. These are, however, less cost-effective solutions outside a systematic collapse classification.

Table 9-3
Anchor systems for use in TBM tunnelling

Anchor system/anchor type	Bearing behaviour General	Bearing behaviour Temporary	Cost effectiveness and suitability for construction work	Durability
Full contact anchor, slack • Mortar anchor: steel or GRP	Good	Often too late	Good	Usable – bad (chemistry of the water)
• GRP resin anchor	Good	Mostly good	Good – optimal	Good – optimal
Anchor tube, slack • Friction anchor (e.g. Swellex)	Good	Good	Good	Bad
Anchor tube, tensioned • Steel expanding anchor • GRP expanding anchor • Mortar bond length with Plastic mortar Steel GRP	Bad Moderate	Immediately load-bearing Quickly load-bearing	Bad Bad	Very bad Bad
Anchor tube, tensioned, with later full bonding • Mortar anchor with accelerator to the end of the drill hole • Expanding anchor with later resin injection Steel GRP	Good Good Good	Mostly too late Good Good	Bad Usable Usable	Usable (chemistry of the water) Good Good

9.6 Stabilisation Ahead of the Cutter Head

As explained in Chapter 8.1, a TBM being used in unstable rock should be equipped with the means of stabilising the rock in front of the cutter head. Stabilisation should not limit the continuing work of the TBM. Interruptions of the excavation with a realistic scheduling of the investigations for the advance should be included in the construction schedule, based on an appropriate risk analysis. Prepared stops, or at least planned in the schedule, are clearly shorter and the use of the equipment is significantly more efficient.

Synthetic resin injections have shown themselves as particularly effective in stabilisation. Resin components cannot be sufficiently mixed in the drill hole. The range of in-

9.6 Stabilisation Ahead of the Cutter Head

Figure 9-16
Partial rock support with liner plates, partly all-round, San Leopoldo tunnel [108]

jections with epoxy or PU resins is therefore strictly limited. Longer-range injections can be performed with acrylic resins with special systems. Very thin pressure hoses (internal diameter 8–10 mm), formed at the end of 2 m as a sleeve pipe, fit together to the total injection length. With a 24 m measured overall injection length, about 10 pressure hoses with sleeves formed at the ends will be needed. Such bundles of pressure hoses can be simply inserted into the drill hole and filled with mortar. The solubility of PMA resin in water permits, in the first minutes after the first injection, immediate flushing of the pressure hoses with a quickly inserted flush pipe in running water. This method enables a second injection process to be carried out.

In addition to ground improvement in front of the cutter head with injections, ground freezing with liquid nitrogen can lead to the required stability in faulted ground saturated with water, not however when the water is flowing (see Chapter 17, example San Pellegrino tunnel). This is an expensive process, which is suitable for short fault zones, and the use of liquid nitrogen requires corresponding ventilation of the working area.

10 Gripper TBM and Shield Machine Combinations

Two particular problem areas have to be considered in the detailed design of a TBM. Firstly, to secure the excavated cavity early in accordance with the geological conditions encountered. Secondly, the thrust required for the boring process must be produced and the support of the cutter head must be provided reliably.

In addition to these design details and the solution concepts resulting from them, this chapter discusses further tendencies in development. The classic shield for soft ground is to be equipped for work in hard rock.

If loosening of the rock structure, as is to be expected in lightly fractured rock, is expected during TBM excavation, then the support has to be installed as near as possible to the working face. That is, immediately behind the cutter head. This requirement is, particularly with small and medium diameter tunnels, a contradiction to the available space around the tunnelling machine. The consequence is a serious restriction of the advance rate.

These collapses can be dealt with by the provision of movable support in the form of a protective shield. Various special types of tunnelling machines have been developed and built for this purpose.

According to the layout of the machine, these special types can also cope with supporting the reaction forces from boring. Although the weight of the machine can be carried directly to a radial support by the rock, the thrust force and the torque can only be transferred to the rock either indirectly through a radial clamping (gripper) or directly against a segmental lining.

The question of the adequate transfer of thrust forces is just as significant as the question of support around the cutter head when deciding which solution to follow.

Various basic types can be categorised from the multitude of individual solutions, which approach the classic shield tunnelling machine through the choice of temporary support or represent a combination with it. This extends to the installation of roofs and side shields, through the use of a cutter head shield, and on to the use of fully shielded hard rock tunnelling machines as single or double shield.

In addition to the machine layouts and special constructions resulting from the question of rock support or clamping, further innovative combinations of TBMs with shield machines are discussed. These represent extensions of the classic shield for soft ground, which are equipped for use in hard rock.

10.1 Roof Shields

Isolated falls of rock and frequently, in addition, deformations of the tunnel section are to be expected in every tunnel excavation. TBMs are therefore equipped with roof shields, which are intended to provide protection against falling rock (Fig. 10-1 a).

a)

b)

Figure 10-1
Roof shields of gripper TBMs
a) Roof shield (Herrenknecht)
b) Roof shield extended with trailing fingers (Robbins)

These will not, however, exert a stabilising effect on the rock with a higher supporting function. These roof shields are often extended backwards with trailing fingers, which more or less bridges the distance between the stator of the TBM and the installed rock support in fractured rock as a load-bearing element (Figs. 10-1 b and 3-3 a). A favourable property of this construction is the elasticity of the individual fingers, which are capable of holding large blocks of rock (Fig. 10-2). The roof shield is also continued as far as possible forwards in order to protect the cutter head from falling rock.

Articulated load bearing protection roofs over the entire area of the machine, which were produced in the 1970s by various TBM manufacturers like, for example, the gripper TBM for the Schwelme heading (Fig. 10-3 a), were often damaged by falling rock [120]. Hydraulically adjustable multi-part protection roofs (Fig. 10-3 b), which were pressed against the crown like with the gripper TBM in the Kielder tunnel, are also no longer used in this form. Multi-part protection roofs can today only be found in special designs like the Mobile Miner (see Fig. 3-38) and the Continous Miner (see Fig. 3-40 b).

10.1 Roof Shields   151

Figure 10-2
Trailing fingers of a gripper TBM (∅ 5.2 m) in a fractured phylitic Verrucano, Ilanz II

a)   b)

Figure 10-3
Multi-part crown shield construction over the machine area
a) Gripper TBM 134-153 with static roof (Robbins), Schwelme heading, ∅ 4.0 m [120]
b) Gripper TBM TVM 34-38 with hydraulically adjustable roof (Demag), Kielder tunnel, ∅ 3.5 m [67]

10.2 Roof Shield and Side Steering Shoes and Cutter Head Shields

While boring continues, the cutter head exerts a propping force on the working face through the attack points of the cutting tools. This propping force is, however, is only effective as long as the cutter tools attack the rock, and not at all while the machine is re-gripped, during servicing, maintenance and repair work in front of the cutter head.

A propping force can be exerted on the rock around the cutter head of a gripper machine by the roof shield and the side steering shoes or cutter head shield, whose segments can be extended radially. This propping force stabilises the tunnel wall locally and also assists the stabilisation of the working face. The cutter head is protected from falling rock around the gauge cutters and the blocking of the cutter head is prevented. The shield surfaces additionally serve to clean the invert and as a dust shield for the effective functioning of the dust removal equipment.

Side steering shoes are protection units, which are arranged behind the cutter head of a main beam TBM (see Chapter 4) and are similar to a shield. When they are arranged to cover the whole surface, the support consists of an invert shield, also called slide shoe, the side shoes arranged on rockers and the roof shield (Figs. 10-4a and 3-4b). The side steering shoes can be radially slewed and thus serve, like the height-adjustable invert shoe, to maintain the position of the cutter head during the boring operation.

Cutter head shields were originally only found on Kelly TBMs. Mounted on a linkage, they served only as a protection for the cutter head, with the maintenance of the posi-

a)   b)

Figure 10-4
Full-surface protection equipment in cutter head area
a) Roof shield and side steering shoes main beam TBM S-155 (Herrenknecht), Tscharner,
  $\varnothing$ 9.53 m, 1999 [61]
b) Cutter head shield Kelly TBM 880 E (Wirth), Qinling tunnel, $\varnothing$ 8.80 m, 1997 [182]

10.3 Walking Blade Gripper TBM

tion of the TBM and the cutter head being primarily performed by the Kelly construction with the double clamping unit. Because of the front-heaviness of today's larger Kelly machines caused by the cutter head, the height-adjustable invert element of the cutter head shield also serves to bear the weight, so that the originally different constructions regarding support around the cutter head have converged again (Fig. 10-5).

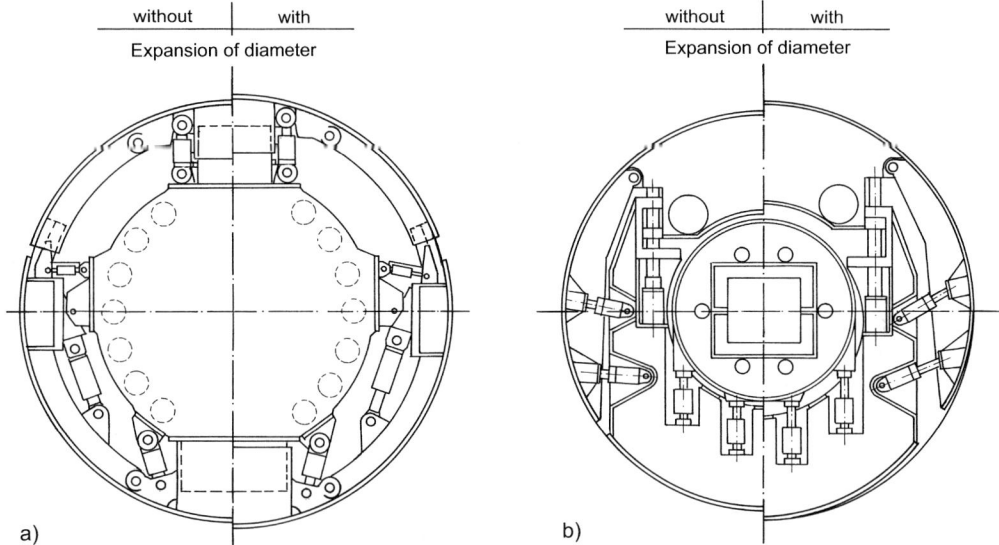

Figure 10-5
Alteration of diameter
a) Roofl shield and side steering shoes main beam TBM S-155 (Herrenknecht) Tscharner, ⌀ 9.53 m, 1999 [61]
b) Cutter head shield Kelly TBM 770–850 E (Wirth), Vereina, ⌀ 7.70 m, 1994 [181]

Depending on the diameter of the tunnelling machine and the resulting amount of space available, support measures like anchors or steel arches can be installed directly behind the side steering shoe surface or the cutter head shield.

10.3 Walking Blade Gripper TBM

In addition to the various types of gripper TBM discussed, special constructions have sometimes been built to meet the demands of individual projects. Out of the multitude of special constructions, the walking blade gripper TBM will be discussed in more detail here, because the use of this type of machine is often proposed for use in tunnelling through squeezing rock.

This machine has repeatedly been suggested by Robbins [123, 124, 126] for tunnelling in squeezing rock. This type of machine has so far been used only for the Stillwater tunnel in Utah, USA, and for the investigation heading for the Freudenstein tunnel.

At the Stillwater tunnel, the excavation was started with a double shield TBM ($\varnothing$ 2.91 m). After the machine encountered an area of slate clay cut up by faults, re-gripping was hindered 14 times by the blocking of the gripper shield, and, at 1000 m into the drive, the machine could proceed no further; the project was tendered again. For the second attempt, the machine was rebuilt underground into a walking blade gripper TBM. The expanding blades of the shield were controlled in two groups. While one group clamped into the rock to transfer the reaction forces from the boring process, the blades of the other group were pushed forward with the TBM, and after the maximum stroke of the expanding blades of the first group had been reached, the TBM was clamped with the second group of blades and the expanding blades of the first group were pushed forward

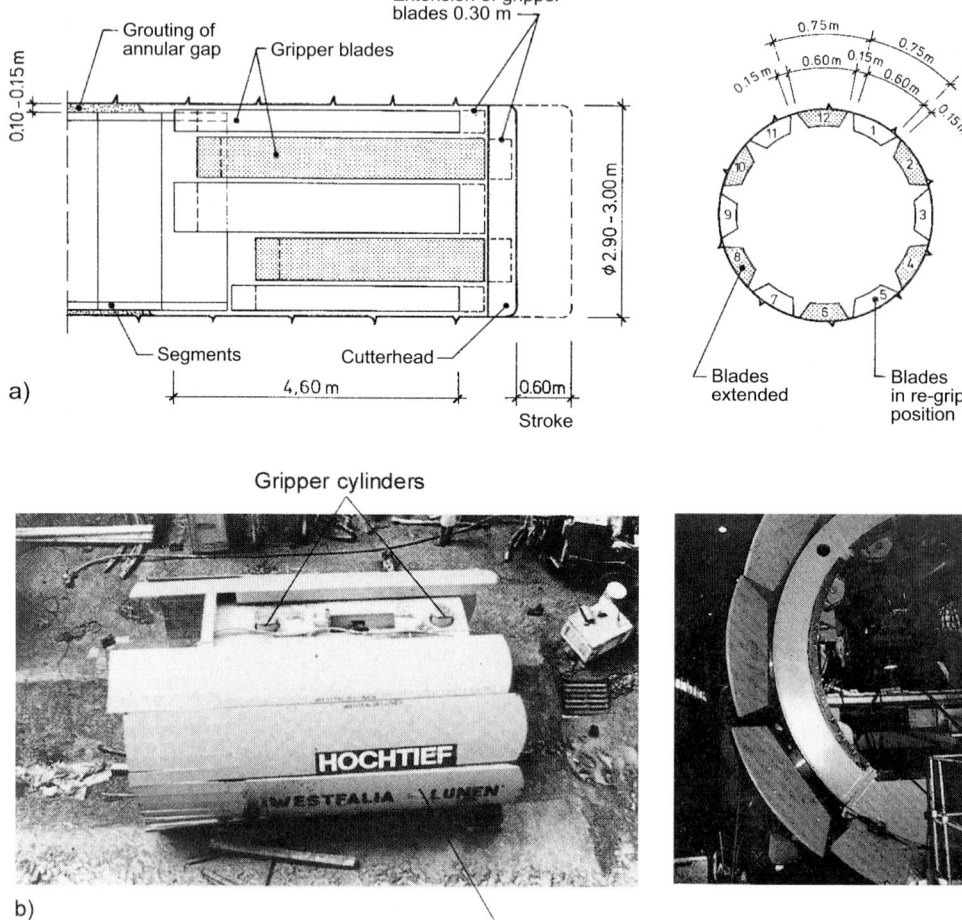

Figure 10-6
Walking blade gripper TBM
a) Blade-TBM 92-192 (Robbins), Stillwater tunnel, $\varnothing$ 2.9/3.0 m, 1983 [161]
b) Blade-TBM (Westfalia Lünen), Freudenstein tunnel, $\varnothing$ 5.4 m, 1994 [62]

with the machine. The average advance rate of this TBM was 9.1 m/d, while the gripper TBM with roof shield (∅ 3.2 m, 1.8 m long roof schield with 1.5 m long trailing fingers extension at the crown) used in the advance in the other direction achieved under the same unfavourable geological conditions an average advance rate of 41 m/d [129, 148].

The blade TBM (∅ 5.4 m) for the Freudenstein tunnel reached an average daily advance of 2.5 m in leached gypsum Keuper. The machine turned out to be extremely susceptible to water. As soon as a certain water content of the excavated rock was exceeded, the closed arrangement of the cutter head resulted in the cutter head and the bucket chutes agglutination of the muck. Tedious cleaning work with retracted cutter head and the many resulting collapses hindered the advance. Only sinking of the groundwater and additional compressed air operation made it possible for the machine to reach its end station in non-leached anhydrite rock [74]. Because of the poor advance rate and other construction delays, the use of the machine in non-leached gypsum Keuper was abandoned and the tunnel section was excavated using an additionally purchased gripper TBM with cutter head shield (∅ 5.64 m), which achieved an average advance rate of 20 m/d [82].

The experience of tunnelling in both applications of this TBM has not been able to fulfil the expectations for the machine layout. Especially the steering problems with the TBM excavation at the Stillwater tunnel, which were caused by loose rock in the unsupported area (0.15 m) between the blades (l = 0.6 m), show the difficulties of applying this type of machine. The tunnelling concept of a short-shield machine with immediate support seems more favourable under such geological conditions, according to existing experience.

For the development of mechanised tunnelling systems with yieldable support for squeezing rock (see Chapter 15.2), the machine layout of a TBM with a shield adjustable in diameter, which to a certain extent can withdraw from a squeeze situation, seems to be the only practical method, even if the advance rates of the prototypes have not been convincing.

10.4 Full-Face Shield Machines

10.4.1 Developments

The idea of tunnelling in hard rock with low stand-up time or in fractured rock with a shielded tunnel boring machine was essentially influenced by experience in Switzerland. Gripper TBMs were used there at the start of the 1970s to excavate tunnels in the diameter range > 10 m and only achieved low advance rates in the sections of the sweet water molasse, which tends to collapse. A support for the rock with steel arches was necessary and hindered the advance. The basic idea was now to develop a machine for such geological conditions, with which the dependence between the utilisation rate of the machine and the stability of the rock is minimal. This idea was implemented technically with the single shield TBM with segmental lining (Fig. 10-7). The fact that the advance process is mostly separated from the rock support makes a consistently high grade of mechanisation possible and this makes for high advance rates, even in changeable

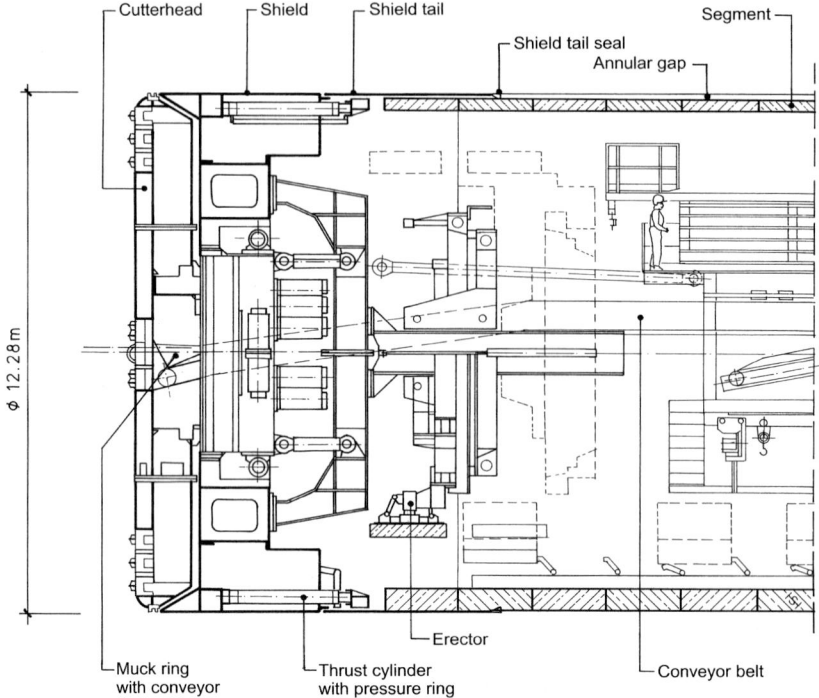

Figure 10-7
Single shield TBM S-139 (Herrenknecht), Zürich–Thalwil, ⌀ 12.35 m, 1998 [61]

geological conditions. This tunnelling method has meanwhile reached a state of perfection in Switzerland, which shows in the increased advance rates. For example, the shield TBM in the southern section of the Zürich–Thalwil tunnel (1999) achieved an average advance of 26 m/d in comparison with 11.90 m for the driving of the second bore of the Gubrist tunnel (1980). The use is in no way restricted to easy-to bore rock like molasse or jurassic. Applications in alpine rocks or in gneiss have also been successful.

This experience in the range of diameters of over 10 m is certainly valid for other countries.

10.4.2 Special Characteristics

Shielded tunnel boring machines completely avoid the use of shotcrete in rock support, in contrast to the gripper TBM. Segmental lining, well known in soft ground, is used with adaptations for the situation in hard rock. The rock is never open. Geology with a tendency to collapse can usually be managed and does not cause any delay with the excavation.

This machine type also exhibits some characteristics in comparison to shielded machines for use in soft ground, apart from the method of working face support, and these will be discussed here in greater detail.

10.4 Full-Face Shield Machines

10.4.2.1 Cutter Head and Shield

The diameter of the cutter head (see Fig. 3-3 c and d) of a shield TBM is usually slightly larger than the shield. This is intended to produce an overbreak, which prevents the blocking of the cylindrical shield. The cutter head axis is placed slightly higher than centreline of the shield. This creates a small gap between the exterior surface of the shield and the crown (Fig. 10-7). In order to stabilize the cutter head without side support of the shield during tunnelling, and to ensure the disc cutters can run in their tracks without offset, two hydraulic stabilising shoes (Fig. 10-8) are located in the upper half of the shield, which can be pressed against the side of the tunnel through an opening in the shield. The use of stabilisers in tunnelling originated in double shield technology (see Section 10.5).

Figure 10-8
Stabilising shoe shield TBM S-160 (Herrenknecht), Metro Porto, ⌀ 8.7 m, 1999 [61]

10.4.2.2 Thrust Ring

When shielded machines are used in loose rock, the face support requires continuous contact to the segments, which is applied by the remaining thrust cylinders during the ring installation process. In contrast, shielded hard rock TBMs with a five-part segment lining and one keystone in the invert (see Chapter 15.2) have a thrust ring between the thrust cylinders and the last segment ring installed (Fig. 10-9 a), which all cylinders press against. After completion of the stroke, the thrust cylinders and thrust ring are retracted, making space in the shield tail to install the next segment ring. The shield TBMs used in Switzerland have further devices for ring installation, like swivelling carrier rollers for placing the side segments and expanding devices for the installation of the key stone in the invert (Fig. 10-9 b and c).

Figure 10-9
Auxiliary device for ring installation, shield TBM, Zürich–Thalwil
a) Thrust ring
b) Swivelling carrier rollers
c) Expanding mechanism for installation of the keystone in the invert joint

10.5 Double Shields

10.5.1 Developments

Alongside the successful development of the single shield TBM in Switzerland, a further new machine concept was developed, combining a shield and a gripper; the double shield TBM. This type of machine, developed by Carlo Grandori in 1972 [52], was designed with the objective of obtaining high advance rates even in bad and very variable rock conditions. As with the single shield TBM (see Section 10.4), the installation of support is separated from the excavation. Further advantages over the gripper TBM are better steerability in soft rock formations and the possibility of variable installation of segments. In contrast to the single shield TBM, the double shield machine has alternative possibilities for thrust and clamping; the double shield TBM has both the gripper system of the open gripper TBM and telescopic and thrust cylinders aligned along the tunnel.

Shield machines of this type are currently used very successfully for the excavation of water tunnels in the diameter range between 3.8 and 7 m. In particular, the development of special segment systems, like the hexagonal or honeycomb segment (see Chapter 15), for the excavation of water tunnels with double shield machines has been successful. The short construction periods demanded for fully lined long pressure tunnels can only be achieved by double shield machines, even with rock characteristics ideal for gripper TBMs, since excavation and lining can be implemented simultaneously.

Under ideal conditions, double shields in the diameter range 5 to 7 m can reach average advances of 35 to 70 m/d. The cycle time for boring and installation of a segment ring (hexagonal segment, 1.3 m long) for 5 m excavation diameter is set at 15 min. The TBM is re-gripped twice (every 1.5 min.) and the assembly of a segment stone is performed in approximately 3 min. [174].

10.5.2 Functional Principle

A double shield TBM consists of the front shield with cutter head, main bearing and drive as well as a gripper shield with clamping unit (gripper plates), tail shield and auxiliary thrust cylinders. Both parts of the shield are connected by a section called the telescopic shield with the telescopic thrust cylinders, which operate as the main thrust cylinders (Fig. 10-10).

The basic principle is that the machine clamps itself radially at the tunnel wall with the grippers of the gripper shield, with excavation and installation of the segmental lining being performed at the same time. The cutter head and front shield are pushed forward by the telescopic cylinders. The auxiliary thrust cylinders in the tail shield are only to hold the installed segments. On reaching the end of the stroke of the telescopic cylinders, the clamping of the gripper shield is released and the gripper shield is pulled forward towards the front shield. At the same time, the auxiliary thrust cylinders are extended in order to hold the last segment ring in position. The support during the re-gripping of the gripper shield is provided by the stabilising shoes, the shield of the front

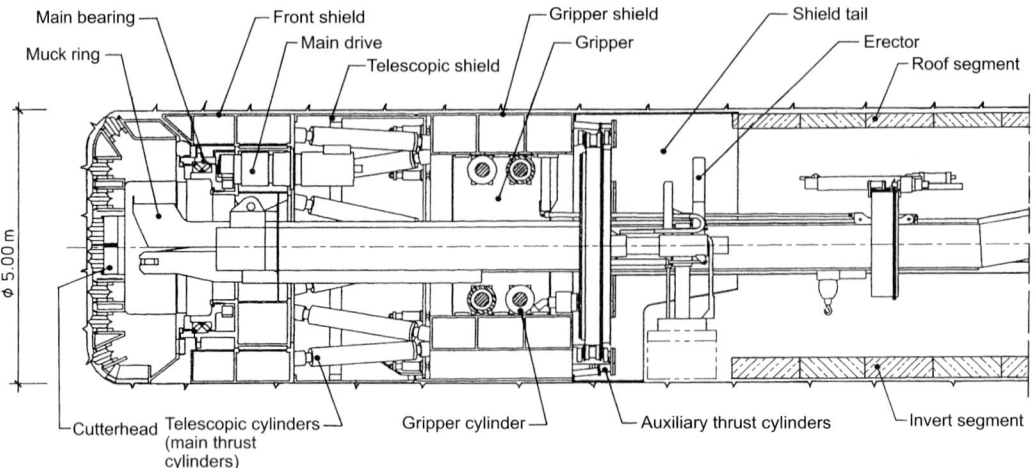

Figure 10-10
Double shield TBM ⌀ 5.0 m (Robbins) [128]

shield and the auxiliary thrust cylinders. The grippers can be clamped sideways; this is, however, generally done at an upward angle of 45° in order to push the gripper shield downwards. This type of clamping is enabled by the arrangement of the grippers on a swinging arm. It achieves more stable clamping and enables the vertical forces from the front shield to be resisted.

If it is not possible to clamp radially through the grippers, then the necessary thrust forces can either be provided by the telescopic cylinders (stationary gripper shield) or by the auxiliary thrust cylinders. In the first mode described with the telescopic cylinders, the auxiliary cylinders only transfer the thrust forces on to the segmental lining.

In the second mode, also called single shield mode, front and gripper shield form a rigid unit, the telescopic joint is completely closed and the cylinders in this area are retracted. The auxiliary thrust cylinders produce the necessary forward thrust. Simultaneous tunnelling and building of rings are no longer possible and the advance speed is reduced accordingly.

10.5.3 Special Cases

10.5.3.1 Shield and Bentonite Lubrication

The double shield TBM is, due to the machine concept with its long shield, in danger of jamming if it encounters squeezing rock. The problem is not generally the jamming of the TBM along the entire shield length, but rockfalls near the telescopic joint hindering the retraction of the gripper shield. This possible danger to the double shield TBM is answered from a design point of view by graduated diameters and vertical offsets of the longitudinal axes of cutter head, front and gripper shield. Bentonite lubrication systems on the shield reduce the shield friction during the stroke as well as when pulling up the gripper shield.

The high advance speeds attainable with a double shield TBM can be considered favourable if squeezing rock is expected. This advantage naturally only exists if the machine is constantly moving forward and not standing.

10.5.3.2 Telescopic Shield

The design of the telescopic shield requires special attention, since lateral shield movements for steering and longitudinal movements for the stroke are overlaid. The spoil collects here, which is formed into a "hard cake" by the front edge of the gripper shield and clogs the telescopic joint and the annular gap between shield and tunnel wall. The angle between the shield components produces a one-sided gaping of the telescopic joint. Based on the first experiences excavating with double shields, a divided telescopic shield (Fig. 10-11), moveable with its own cylinders, was designed to seal the telescopic joint and fitted with muck scrapers [51, 53]. The sealing of the telescopic joint still poses a problem to be solved in the design of double shield TBMs, because of the intention of preventing mud and water getting into the shield when water ingress occurs.

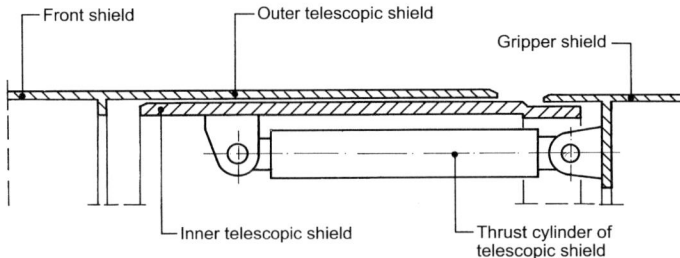

Figure 10-11
Telescopic joint of a double shield TBM (Herrenknecht) [61]

The internal telescopic shield of a modern double shield TBM is adjustable in length from 600–800 mm and thus permits an opening of the TBM in exceptional cases and provides a way out for the crew to perform any support work necessary above the cutter head. Such work was necessary at the "Ginevra" TBM drive for the Evinos Mornos tunnel [54]. After a collapse at the tunnel face in a flysch section, the area above the TBM and in front of the face had to be stabilised. Figure 10-12 shows the sequence of the working stages.

Probe drilling equipment and drill channels for the creation of a pipe umbrella above the TBM are taken into account in this design and are usually located in the tail shield area of the double shield, so that the entire machine area, but in particular the necessary open telescopic area, can be secured by by the pipe umbrella.

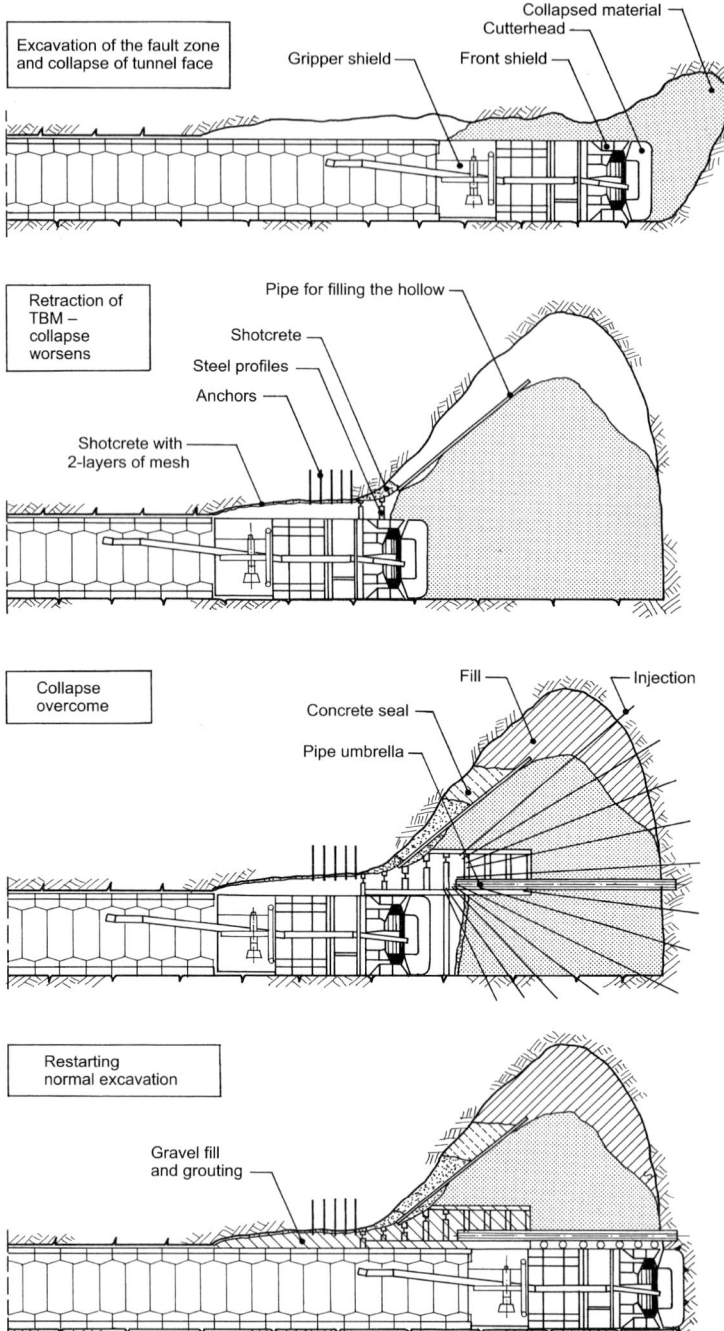

Figure 10-12
Overcoming the collapse of the DS-TBM „Ginevra" of the Evinos-Mornos tunnel, ⌀ 4.04 m [54]

10.6 Slurry Shield Machines

10.6.1 Developments

The common equipment for muck transport from TBMs is the conveyor belt. While tunnel boring machines, except for short interruptions for re-gripping, bore continuously and simultaneously transport the excavated material away from the machine area, most follow-up transport mechanisms only work continuously to a limited extent.

One possible continuous transport system for spoil removal in tunnels and shaft is hydraulic transport. The criteria for the selection of the follow-up spoil transportation are the traffic volume in the tunnel, traffic safety, possible hold-ups at tunnelworks and the cost of ventilation. Hydraulic transport, as an option for the implementation of high transport volumes with small diameters, enables high advance rates in long tunnels with small diameters in which spoil trains cannot pass [58, 130]. In comparison with today's high-performance TBM advances, with the common use of conveyor belts in the tunnel and pocket conveyors in shafts, the disadvantage of the process that the spoil has to be separated and tipped should not be forgotten.

A slurry circuit for a TBM excavation becomes worth considering again where tunnelling has to be be accomplished in problematic geological and hydrological conditions like faulted rock or tunnelling under waterways, where high water penetration is to be expected, requiring a complete sealing of the cutter head area while excavating.

These requirements can be certainly be dealt with by a mixshield with hard rock cutter head. This machine type represents another combination of TBM and shield machine. Successful examples for the use of these are the drives in Muelheim [95] and Sydney. Such a tunnelling machine is preferred for applications in soft ground where hard rock boulders can be anticipated.

Combination shields of the types mixshield with pressurised face and single shield TBM respectively were also used for the excavation of the Grauholz tunnel [95] and section 2.01 of the Zurich–Thalwil tunnel [17] (see also Chapter 16) [95].

10.6.2 Working Principle

With a gripper TBM with downstream hydraulic transport system (Fig. 10-13), the excavated material is transferred from the buckets into the cutter head and on to the TBM conveyor. From there it is transferred to the intake box for the hydraulic conveying system. The material then has to be mixed with a liquid, generally water, to create a suspension, which can be pumped. The material is transported by rhe slurry circuit to the separation plant outside the tunnel. After solids have been separated from the liquid, the carrier liquid is pumped back to the TBM.

The combination of a mixshield with pressurised face and hard rock cutter head of a TBM, in contrast, has separate pressure chambers, filled with suspension and divided by a submerged wall into working and excavation chambers (Fig. 10-14). The hard rock cutter head, equipped with cutters, rotates in the suspension with resulting higher wear.

164   10 Gripper TBM and Shield Machine Combinations

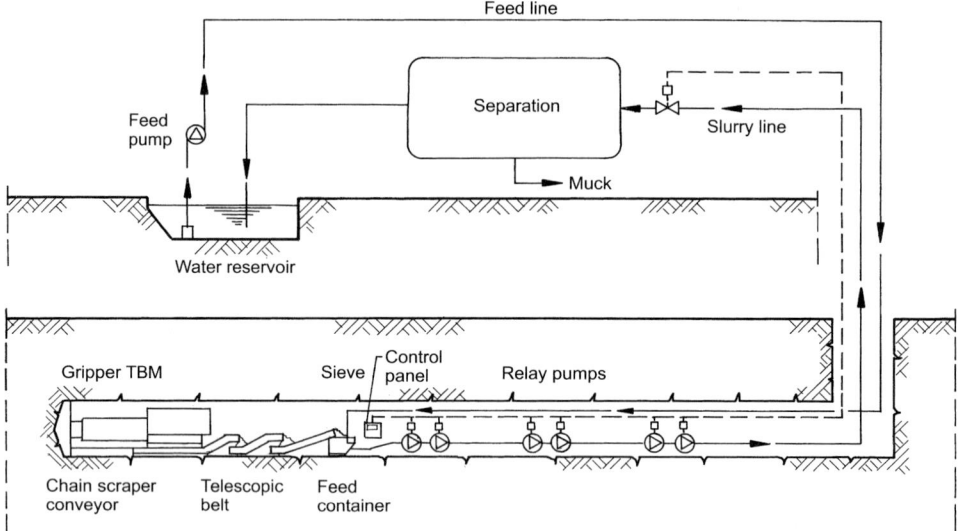

Figure 10-13
Hydraulic spoil removal during the excavation of the Radau header, 1977 [58]

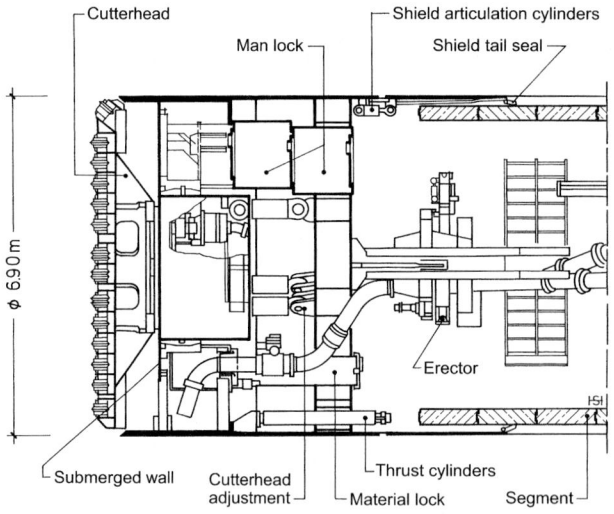

Figure 10-14
Mixshield S-52 (Herrenknecht) with hard rock cutter head, Muelheim metro, Ruhr crossing ⌀ 6.9 m, 1989/91 [95]

This results in special requirements for the cutter head design to encourage better hydrological properties of the buckets and spoil chutes, sealing of cutter bearings and wear characteristics of the cutter discs even when boring in medium strength rock

After passing through the bucket openings in the cutter head, the spoil sinks, more or less quickly depending on chip size, into the suspension rotating with the cutter head and slowly down towards the submerged wall opening, where it is mechanically pushed backwards before it is caught by the bentonite nozzles and stirred up. Then the mixture of spoil and slurry is sucked out of the working chamber and fed through the slurry line in the tunnel to the separation plant. Here, the spoil is separated from the slurry. It often suffices to use water as the carrier liquid. To reduce wear on the downstream conveying systems, pumps and pipes, the use of bentonite as an additive can be considered and also implemented, even if it is not required as face support.

10.7 Shields with Screw Conveyors

10.7.1 Developments

The screw conveyor presents a very interesting alternative to belt and slurry circuit systems for transport of excavated material from the excavation chamber.

This type of shield has the advantage above all that the process can be changed from open to closed mode without any substantial mechanical or time-consuming modifications being required. Shields with screw conveyors for use in rock are essentially very similar to EPB shields in their basic construction. Instead of the usual soft ground excavation tools fitted to the cutter head of an earth pressure balance shield machine, the cutter head is equipped with disc cutters for operation in hard rock. On account of the mostly heterogeneous operating conditions, the cutter heads are often equipped with a combination of cutting knives or ripper teeth. Shields with screw conveyors were already in use for the construction of the Channel Tunnel at the end of the 1980s, which combined dry and hydraulic transport combinations (Fig. 10-15). In some cases, however, the limits of the method were reached; additional measures like injections became necessary and caused substantial extra costs [95].

The possible scope of application for shields with screw conveyor and rock cutter head are projects, which exhibit a section of loose rock suitable for a EPB and a pure hard rock section, both to be accomplished with the same tunnel boring machine. An application could also be considered in jointed rock formations with water flow and limited stand-up time. In loose rock, the application of a hard rock cutter head may become necessary if boulders or thin to medium size rock layers obstruct the cross section.

10.7.2 Working Principle

Rock is excavated at the tunnel face by tools on the rotating cutter head as with the already described types of TBMs and shields. The thrust force is applied directly to the excavation tools in open operation.

Using a shield with screw conveyor in unstable, water-bearing rock or in loose rock, loss of stability at the tunnel face must be avoided by applying a supporting pressure. In this case part of the thrust force can be transferred by the pressure wall to the medium in the excavation chamber when operating in closed mode, that is earth pressure balance operation. With the EPB shield, in contrast to other shield tunnel methods, it is possible to do without a secondary support medium (compressed air, suspension, mechanical face support), since the excavated soil serves as supporting medium. In order to use the soil as a supporting medium, the excavated soil must, however, fulfil the following requirements [98]:

- Good plastic deformation characteristics
- Pulpy to soft consistency
- Low internal friction
- Low permeability

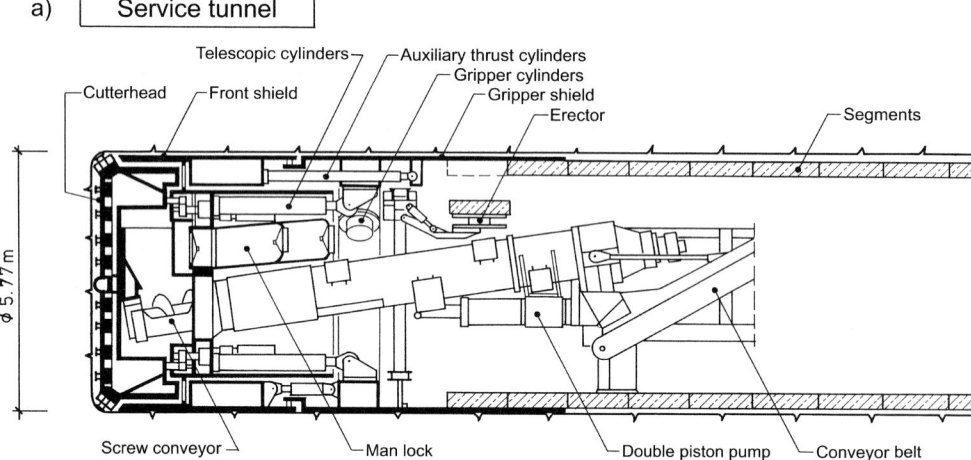

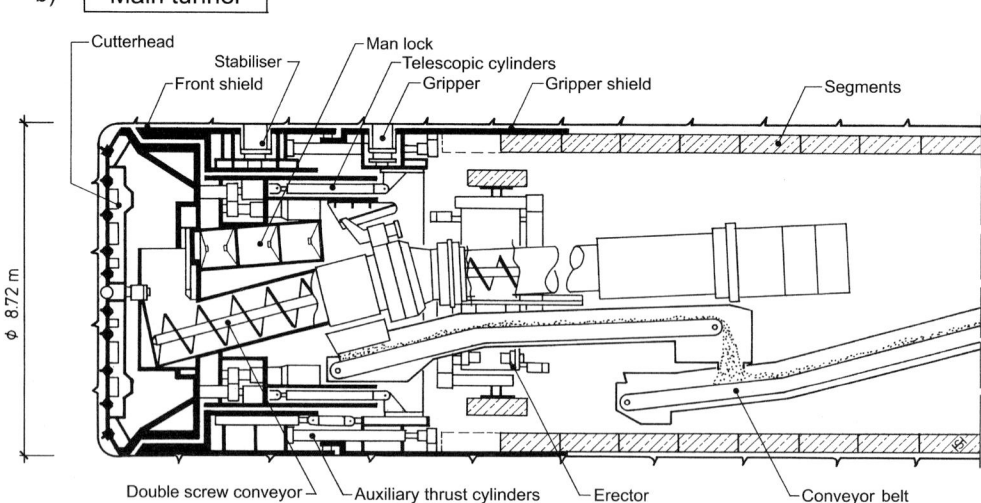

Figure 10-15
Shield with screw conveyor, Channel tunnel, French side, 1988 [95]
a) Double shield TBM, service tunnel, ⌀ 5.77 m (Robbins)
b) Double shield TBM, main tunnel ⌀ 8.72 m (Robbins/Kawasaki)

10.7 Shields with Screw Conveyors

If the required characteristics are not available in the excavated material, then the material must be conditioned. The conditioning process selected depends on the in-situ soil type and so depends on the soil parameters grain curve, water content w (%), plastic limit w_L (%), degree of plasticity (I_n) and consistency number (I_s) [104]. These can be influenced as follows:

- Addition of water
- Addition of bentonite, clay or polymer suspensions
- Addition of foam

When operating in rock, none of the characteristics specified above are present. Ground conditioning only offers, if at all, chances of success in extremely weathered rock formations, the operation of a shield with screw conveyor in EPB mode is therefore limited in rock and requires additional equipment.

The further transport of the excavated material through the tunnel can be done by bulk material transport, (conveyor, rail/truck) or, adding a liquid suspension, through a hydraulic conveyor with piston pumps, The diameter must be limited as well as the proportion of stones in order to ensure the functioning of the conveyor screw. This must already be taken into account in the design of the cutter head. In contrast to mixshields with pressurised face no stone crusher can be installed within the excavation chamber.

10.7.3 Machine Types

The principle machine types can be categorised according to the method of spoil removal (Fig. 10-16).

10.7.3.1 Open Mode (Screw Conveyor – Conveyor Belt)

The majority of shield machines with screw conveyor are operating in rock in open mode, i.e. without support against ground or water pressure. The cutter head can be designed and equipped as cutter wheel with cutter discs or in combination with cutting knives. In water-bearing zones it is possible, during the shut-down after closing off the screw conveyor gate valve, to apply air or water pressure in the excavation chamber.

10.7.3.2 Closed Mode (Screw Conveyor – Conveyor Belt)

The type of machine described above can be operated without modification in closed mode with half lowering. This presumes, however, low permeability of the soil in the lower part of the excavation chamber. The upper area of the excavation chamber is filled with compressed air, displacing joint and pore water and reducing the water flow. The application in connection with drum cutter heads or cutter wheels with carrier buckets is not recommended, since the ground plug in the invert area of the excavation chamber as well as the necessary high level of filling of the conveyor screw are not attainable. Difficulties with the process result from uncontrolled blow-outs from the conveyor screw. Sufficient sealing to produce a closed system is only possible in clay

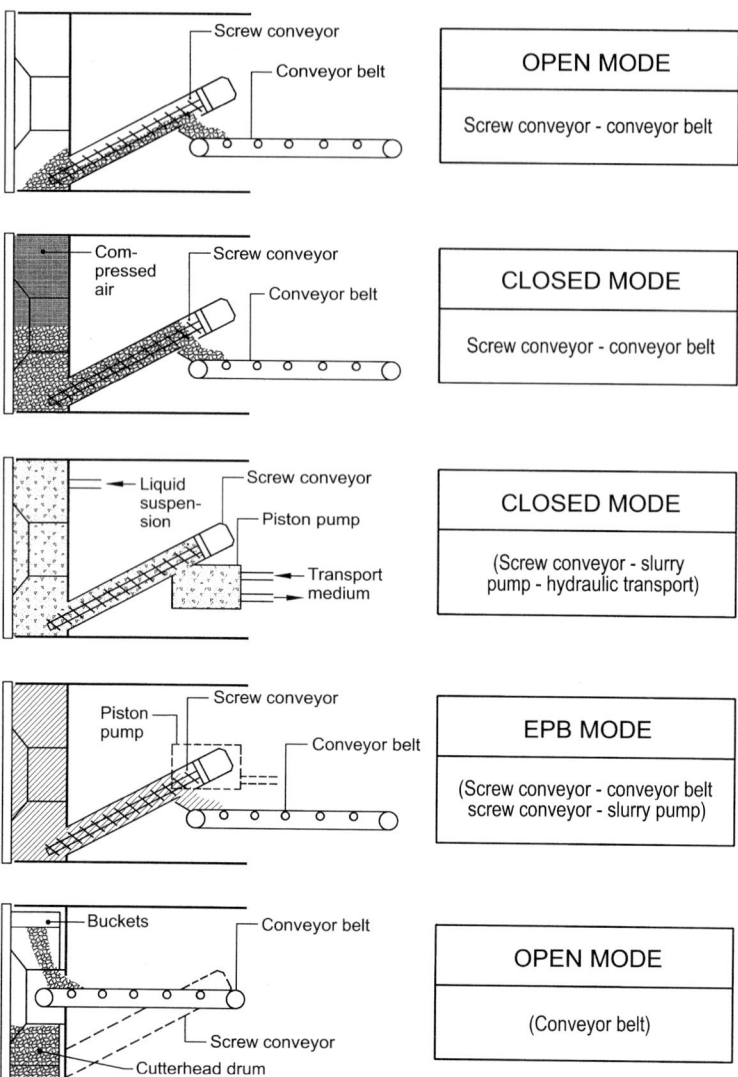

Figure 10-16
Machine types according to type of material transport

or strongly weathered rock. The future for these type of machines could be the use of foam.

10.7.3.3 Closed Mode (Slurry Circuit)

The danger of blow-outs at the screw outlet can be avoided by connection to an enclosed conveyor system. Sealing and lock systems for dry operation, such as transfer boxes, rotary feeder locks or even double screws have proved extremely problematic in practice. Hydraulic conveyor systems with conveyors piston pumps and also with slurry pumps have already been used successfully, however. The slurry can either be fed directly into the pressurised excavation chamber or into the slurry box directly behind the screw conveyor. The cutter head and the screw must be matched to each other depending on rock strength, abrasiveness and ease of boring of the rock. The additional use of a rock crusher at the connection to the screw to the conveyor pump should be considered at the design stage.

10.7.3.4 EPB Mode (Screw Conveyor – Conveyor Belt or Screw Conveyor/ Slurry Pump)

The majority of shields with screw conveyor, which have been used in rock (e. g. Channel Tunnel, Athens Metro etc.), were described as Earth Pressure Balance shield, but never actually operated in EPB mode.

The operation of a shield with screw conveyor in EPB mode in rock requires extremely complex conditioning measures (see Section 10.7.3.2) [102] due to the inadequate deformation characteristics.

EPB shields seem predestined for use in fine-grained soft ground with pulpy to soft consistency (Fig. 10-17) [104].

For this reason, the spoil transport and spoil disposal have to be included in the decision-making process and/or in the choice of the best excavation process, as well as the support of the tunnel face.

In EPB mode, the soil is loosened at the tunnel face by tools mounted on the rotating cutter head. The excavated spoil does not fall into the excavation chamber as with a slurry shield or an open shield, but is pressed into the excavation chamber through openings in the cutter head and mixes into the earth slurry there. The thrust load is transferred through the pressure wall onto the earth slurry and prevents uncontrolled ingress of the soil from the tunnel face into the excavation chamber. A steady state is reached when the earth slurry in the excavation chamber is no longer consolidated by ground and water pressure. If the supporting pressure of the earth slurry is increased beyond the steady state, the earth slurry in the excavation chamber and the ground are compressed further, which could cause heaving of the ground in front of the shield. If the ground pressure is reduced, the in-situ soil could force its way uncontrolled into the earth slurry in the excavation chamber and so cause settlement on the ground surface.

The material is cleared out of the excavation chamber by a screw conveyor. This process must be controlled to avoid even short term reduction of the earth pressure in the excavation chamber and resulting settlement. Further transport through the tunnel can be done by bulk material transport (conveyor belt – rail/dumper) or also, after the addition of a liquid, by hydraulic conveyor with piston pumps. There are special references to soil conditioning in [95, 98, 104].

AREA	CONDITIONS	CONDITIONING MATERIAL
1	Ic Support medium = 0.4–0.75	Water Clay and polymer suspension Tenside foams
2	k < 10 E-05 m/s Water pressure < 2 bar	Clay and polymer suspensions Polymer foams
3	k < 10 E-05 m/s No ground water pressure	High-density-slurrys High molecular polymer suspensions Polymer foams

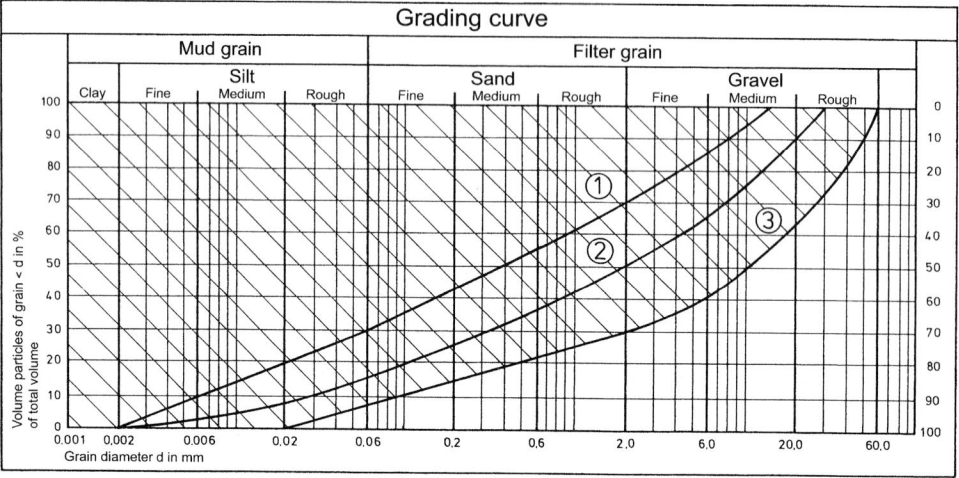

Figure 10-17
Scope of applications of earth pressure shields [104]

10.7.3.5 Open Mode (Conveyor Belt)

Systems under development from various manufacturers offer the possibility of operating the shield either in closed mode with screw conveyor or in open mode with direct belt conveyor out of the central area. The open operating mode corresponds essentially to the full face shields in function and machine construction. Open mode can be used where the rock possesses sufficient stand-up time and if controllable water ingress is expected while tunnelling. When the machine stops, the conveyor can be retracted in a short time and the centre can be sealed pressure-tight. Relief valves and pump systems make it possible to resume tunnelling in open mode.

The cutter head is often designed as a rotary drum and is equipped with plates or buckets, which carry the spoil into the center. The excavated rock then falls by gravity through the muck ring located in the centre onto the conveyor belt. The drum design of the cutter head has disadvantages for EPB mode, however, since the conditioning of the soil is made considerably more difficult.

10.7 Shields with Screw Conveyors

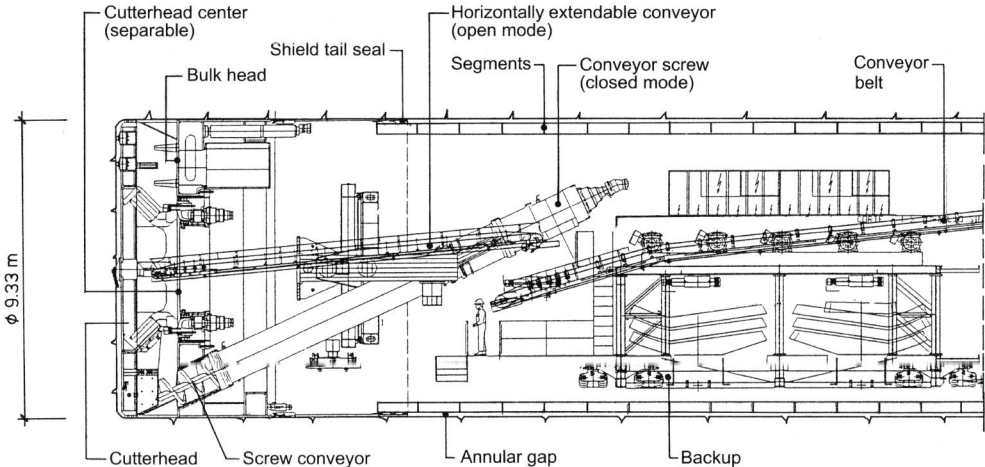

Figure 10-18
EPB Shield S-165 (Herrenknecht) in open mode, Metro Madrid, ⌀ 9.33 m, 1997 [61]

Many manufacturers today offer combinations of the excavation and conveyor systems described above. Figure 10-18 shows the EPB shield for a Metro project in Madrid from Herrenknecht AG. The machine can work both in open and in closed mode. The changeover time from belt to screw conveyor only takes a few minutes.

The Lovat system, with the Muckring conveyor constellation typical for the manufacturer (Fig. 10-19), is considered extremely durable and has often proved itself in practice. The patented "pressure relief gates" offer additional possibilities of closing and material dosing, permitting, in effect, closed operation. The changeover to screw conveyor from the excavation chamber is also possible but with long rigging time. This system requires soil with good deformation characteristics for regulated precise earth pressure operation.

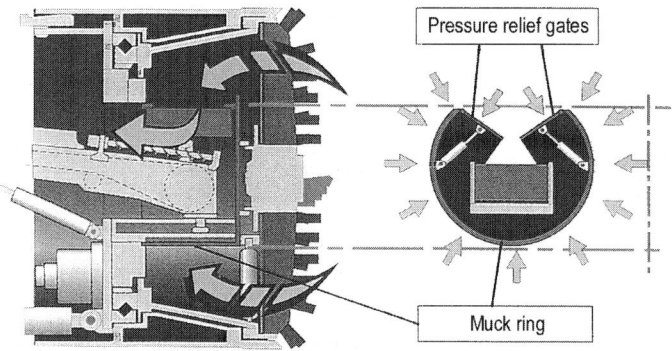

Figure 10-19
EPB – Muckring – conveyor belt constellation [86]

The planning of the construction process for the use of shields with screw conveyors requires a detailed analysis of the rock conditions. Particularly the operation in closed mode causes problems in rock because the possibilities of successful conditioning are extremely limited.

Practical experience has been positive with screw conveyors in open mode. The problem of wear in abrasive rock, however, should be combated with appropriate preventive measures.

10.8 Micro Machines for Hard Rock

The development of tunnel boring machines with smaller diameters has been driven forward by the installation of water supply pipes, sewers, gas and electricity supply as well as telecommunication cables in hard rock areas, where open cut or blasting cannot be carried out.

At present there are two general lines of development that can be recognised of micro machines in hard rock:

- The downsizing of tunnelling machines for hard rock based on standard TBM technology (Mini TBM).
- Shield tunnelling processes in pipe jacking with cutter heads specially equipped for rock tunnelling.

10.8.1 Mini TBM

Mini TBMs are made by various manufacturers in a range of diameters around 2 m. They are essentially a gripper TBM and machines have been made with both single and double gripper systems. The application of such machines was intended for stable rock, because the already restricted space is further reduced by the body of the machine. Because of the reduced diameter compared to normal TBMs, an extension of the stand-up time has been observed when using Mini TBMs in fractured rock, but in general static protection roofs are still used as head protection over the machine and backup areas.

An advantage of the small dimensions of Mini TBMs is the earlier availability, because larger machines, on account of the restricted space in the approach route, often have to be commissioned in stages.

A relatively new development is the application of Mini TBMs as double shield TBMs with Gripper clamping in the diameter range of 2 m. In the comparison to pipe jacking, double shield operation has the advantage of easier tunnelling of curves, the avoidance of intermediate jacking stations and that it is no longer longer necessary to dimension the pipes to be built in to accept thrust pressure.

The double shield TBM developed by Boretec has a smallest shield diameter of only 1600 mm (Fig. 10-20). With this decrease of diameter, apart from the necessary power supply, provided by a hydraulic drive, problems of ventilation and working temperature

10.9 Micro machines for Hard Rock

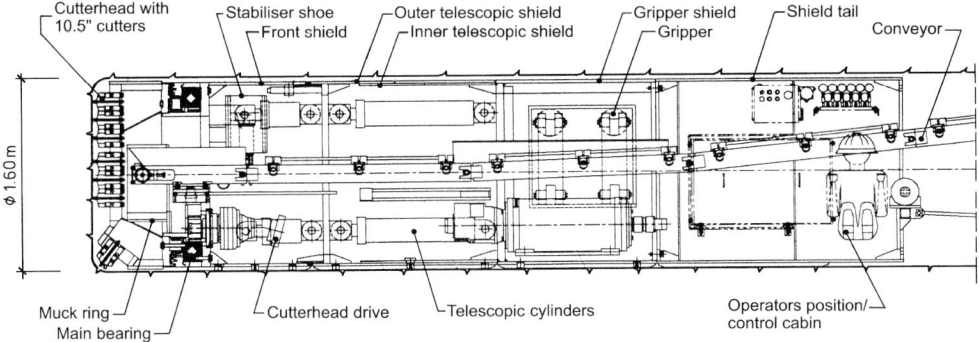

Figure 10-20
Mini double shield TBM Knoxville, ⌀ 1.60 m, 1996 (Boretec) [178]

Figure 10-21
Shield tail with control cabin of the mini double shield TBM Knoxville, ⌀ 1.60 m, 1996 (Boretec) [178]

also have to be solved, since a shield driver located in the tail shield is required to steer this TBM (Fig. 10-21). There have also been developments in the excavation tools, with a decrease in cutter disc diameter to 10.5″ because the standard discs of large TBMs could not be used and drag picks did not provide the anticipated tunnelling performance in hard rock.

Depending on the geological conditions encountered, average advance performance of these tunnel boring machines is between 9 and 18 m/d.

10.8.2 Pipe Jacking

Pipe jacking operations in hard rock are enabled by special hard rock cutter heads. The possible pipe-jacking processes are:

- Press boring pipe jacking
- Shield pipe jacking

Both tunnelling procedures start from a shaft or a start excavation and push either final pipes (single-phase tunnelling) or temporary pipes (two-phase tunnelling), with the help of a hydraulic ram device, through to the destination.

10.8.2.1 Press-Boring Pipe Jacking

In the press-boring pipe jacking process, a pipe line is pushed forward by thrust jacks with the simultaneous excavation of the face by a cutter head. The continuous spoil removal is done by a screw conveyor. The procedure is particularly suitable for the excavation of shorter distances because of the low amount of machinery required.

For applications in large-grained soil, the cutter head can be equipped with disc cutters as well as crushers, which grind up the excavated material, meeting the grain size restriction on the transport through the screw conveyor. The scope of applications of pipe jacking systems can even be extended into rock. The development of special mini discs permits a reduction of diameter of the pipes to a minimum of 600 mm.

10.8.2.2 Shield Pipe Jacking

Pipe jacking with shield is performed by pushing forward a temporary or final pipes while simultaneously excavating the full face using a cutter head, with continuous hydraulic muck transport.

In rock, cutter heads with special excavation tools (rock cutter heads) are used, which can be applied in ground classes 6 and 7 according to DIN 18300. Rock cutter heads (Fig. 10-22) are available, starting with a diameter of 400 mm.

Micro tunnel machines equipped like this can master hard rocks up to 100 MN/m (from diameter 1600 mm). Operation in rock compression strengths over 100 MN/m^2 is possible from a diameter of 800 mm. Special attention needs to be paid to the problem of wear.

From a machine diameter of 1200 mm, a door can be included in the micro tunnelling machines for access to the tunnel face. Obstacles along the tunnel route can be removed and cutters can be changed. The length of stretches in rock can be extended to 500 m, reducing the number of intermediate shafts required.

While boring through non-homogeneous subsoil with rock sections, the necessary rock cutter heads of micro machines with hydraulic spoil removal can be equipped with an additional high-pressure flushing system, to flush the entries into subsoil openings during excavation in cohesive soil using the rock cutter head.

10.9 Micro machines for Hard Rock

a)

b)

Figure 10-22
Rock cutter heads of micro machines (Herrenknecht) [61]
a) Cutter head ⌀ 600 mm fitted with 7" disc cutters
b) Cutter head ⌀ 1500 mm fitted with 12" disc cutters and toothed cutters

Table 10-1
Rock classes according to DIN 18300

Class 6: Rock, which is easy to excavate and comparable soil types
Rock types having an internal, minerally bound consistency, but very jointed, fractured, brittle, friable, slatey, soft or weathered, as well as comparable consolidated non-cohesive and cohesive rock types.
Non-cohesive and cohesive soil types with more than 30% by weight of stones of over 0.01 to 0.1 m^3 volume.
Class 7: Rock, which is hard to excavate
Rock types having an internal, minerally bound consistency and high joint strength and only slightly jointed or weathered.
Solidly bedded, un-weathered slate, molasse conglomerate strata, slag heaps of smelteries and similar.
Boulders over 0.1 m^3 volume.

[1] 0.01 m^3 volume corresponds to a sphere with diameter of about 0.30 m. 0.1 m^3 volume corresponds to a sphere with diameter of about 0.60 m.

Articulated Steel Pipe Shield

The articulated steel tube shield, first developed by Dyckerhoff and Widmann in 1984, has the distinctive feature of two-phase boring. In the first phase, a length of steel pipe (external diameter 860 mm) is thrust forward with the shield, which is later replaced in a second processing step after completion of the actual tunnelling works by the actual product pipe with outer diameter of the steel pipe, but with any internal cross section.

a) b)

Figure 10-23
Articulated steel pipe shield (Dyckerhoff & Widmann)
a) Cutter head ⌀ 860 mm fitted with scraper head and toothed cutters [40]
b) Steel tube ⌀ 860 mm with integrated supply pipes and cables [160]

The length of steel pipe consists of sections 2 m long, in which all necessary supply lines for hydraulic spoil removal, oil circulation, air, water and electrical supply are already installed (Fig. 10-23 b).

The excavation of the soil is performed by a combination of a rotating scraper head with toothed cutters (Fig. 10-23 a). With this cutter head configuration, the independent advances of the scraper head and toothed cutter enables different methods of excavation to be selected. The result is quick adaptability to changing ground conditions during excavation and the particular suitability of this process for non-homogeneous, changing geological conditions. The articulated steel pipe shield can be used in all types of water-bearing soil with boulders of any size or shape. When tunnelling in cohesive soils, the advance performance reduces due to blocking of the cutters.

In solid rock, the concept of the articulated steel pipe shield offers operating advantages through the use of steel protecting pipe anchored to each other. Higher thrust forces can be achieved compared to product pipes and the correction of rolling is also simple.

Stretches of up to 250 m can be accomplished with this technically very complex machine concept. Tunnelling advances of more than 20 m/d are possible, although the performance in rock can fall off to 4–6 m/d [80]. Another disadvantage of this system is the restricted diameter of 860 mm.

11 Special Processes: Combinations of TBM Drives with Shotcrete

Under particular conditions particular to individual projects, a combination of mechanical and conventional tunnelling methods can be preferable, considering the properties of the rock and economically.

Construction processes combining TBM tunnelling with shotcrete work are outlined below.

11.1 Scope of Application

Combined construction processes can be used for deep tunnels as well as those near the surface. The method also makes it possible to produce non-circular cross sections. Generally, a partial section within the entire profile is excavated with a TBM and then expanded to the final cross section with shotcrete support. Figure 11-1 shows, as an example, the Amberg motorway tunnel, where this process was used. The Pfaender tunnel, which is part of the same motorway, was driven the same way [96].

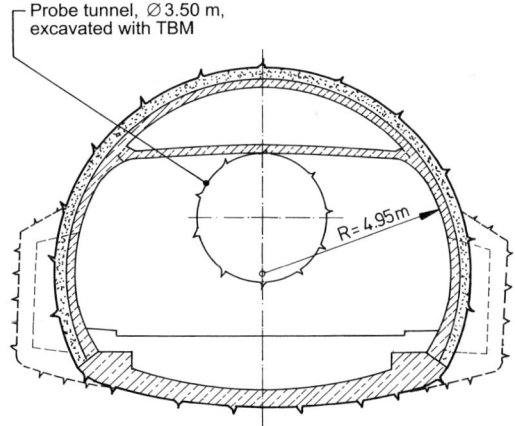

Figure 11-1
Overall section with location of the probe heading, Amberg motorway tunnel (Austria) [57]

The combination of economic TBM tunnelling with the flexible conventional method provides the following substantial advantages:

- Early discovery of the geological and hydrological ground conditions with a resulting reduction of the uncertainties and risks due to the ground, which affect the subsequent enlargement.
- The possibility of using available tunnel boring machines.
- Avoidance of environmental intrusion by reducing the quantity of blasting – less noise nuisance for the inhabitants and less vibration during the enlargement by drilling and blasting.

- Improved facilities for ventilation and drainage during the drilling and blasting work.
- Reduction of the overall cost risks.

11.2 Construction Options

Depending on the particular conditions for the overall project, possible implementations of combined construction processes can be categorised as follows:

- Probe or investigation headings, to be completed in advance for the almost complete investigation of the existing subsoil, which are used mainly for tunnels with deep overburden. These are generally tendered as a separate project in advance of the construction of the main tunnel and can be used for the entire route or only sections of it. The information gained can then be used to plan the method of tunnelling the main tunnel and for the tender documents.
- Pilot headings, which are bored first and generally lie inside the later profile of the tunnel, belong to the construction contract of the main tunnel. Pilot headings are arranged from the viewpoint of construction technology and integrated into the construction process of the main tunnel. At the same time, they offer the advantages of a probe heading.
- Cross section enlargements, which are used along the mechanised bored stretch for various purposes, like construction of stations, sidings, points or machine halls.

11.2.1 Probe Headings

Probe headings are used mainly for the detailed investigation of geology, hydrogeology, presence of gas and rock characteristics. They are generally implemented as a separate contract prior to the award of the main contract in order to provide a better basis for project and tender documents for the later main tunnel. The information learned in the preliminary investigation is also integrated into the contract for the main tunnel, although the full geological risk is transferred to the contractor.

A preliminary investigation of the rock with a probe heading is particularly useful if the overburden is deep, as is the case in the Alps, since tunnel construction holds substantial risks with the correct choice of construction procedure. Good examples for this are the probe headings for the Gotthard [183] and Loetschberg [166] NEAT base tunnels. Especially difficult rock conditions for tunnelling can also make probing necessary. Examples for this are the Seikan tunnel [45], which lies below sea level, or the Freudenstein tunnel shown in Fig. 11-2, where leached Keuper with water inflows tending to swelling had to be tunnelled [63].

In long tunnels, access headings for intermediate starting points can also be used as probe headings. Examples for this are the intermediate starts for the Irlahüll and Euerwang tunnels on the new railway line from Nürnberg to Ingolstadt [78] or also the NEAT basis tunnels Gotthard [183] and Loetschberg [166].

11.2 Construction Options

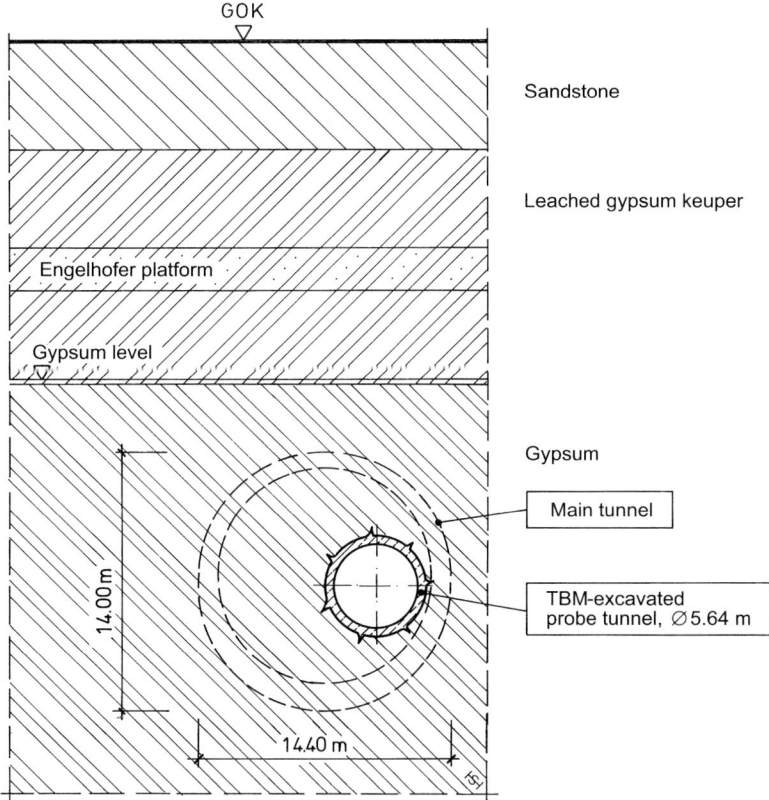

Figure 11-2
Probe heading in difficult geological conditions, Freudenstein tunnel [63]

The implementation of the probe headings in difficult rock conditions can be done in sections, in particular where water ingress is expected, or also along the entire length of the tunnel. The probe tunnel is generally located within the cross section of the later tunnel, but can also be outside the final cross section. The choice of the cross-sectional area to be mechanically driven also depends on the availability of the tunnel boring machine at the time of construction and usually amounts to 8 to 14 m^2 (bore diameter 3.2–4.2 m).

A good example of the application of a probe heading outside the final cross section is the Raimeux tunnel. The probe tunnel was completed with a Robbins gripper TBM 123–133 with bore diameter 3.65 m, already twenty years old, and later cross-connected to the main tunnel when this was bored later. The probe tunnel was used for water drainage and material transport during the construction period.

During the construction period of a road tunnel, it is a part of the entire tunnel scheme and can, for example, be used for maintenance as well as for rescue in emergencies. The TBM advance of the investigation heading can also discover karst cavities and their hydro-geological behaviour.

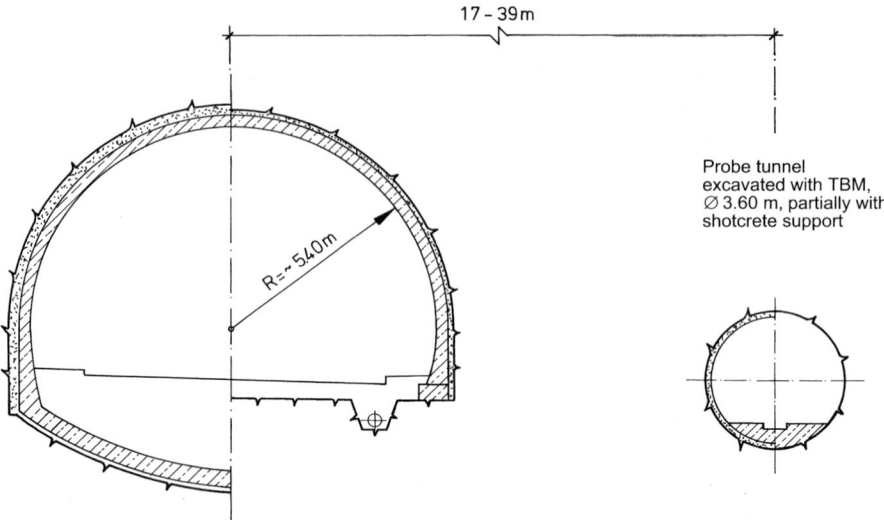

Figure 11-3
Investigation heading outside the final section, Tunnel de Raimeux [60]

11.2.2 Pilot Headings

Pilot tunnels also generally fulfil the purposes of a probe heading. The difference is usually, according to the categorisation here, the time of construction. Pilot tunnels are awarded at the same time as the main tunnel and thus serve mainly operational purposes. As an example, if large amounts of groundwater are expected, the pilot heading can serve as a drainage adit during the excavation of the main tunnel and contribute to the improvement of the drainage. The existing rock is drained and any high water pressure encountered can be relieved.

Due to the size, dimensions and geology, the final cross section is generally excavated by drilling and blasting with shotcrete support; the enlargement of a pilot heading to main tunnel requires less excavation work and less explosive. The pilot heading should then be considered a large cut and relieves the stress at the face [96]. Pilot tunnels near the surface have special advantages, as is the case when tunnelling under buildings with little cover. The Uznaberg tunnel [105] (see Section 11.3.3) serves as an example for traffic tunnels driven by blasting under heavily populated areas, industrial zones or historic buildings.

These machine-driven cross sections also naturally offer all the advantages of a probe heading. In addition to the possibility of investigating the subsoil, preliminary pilot headings can also be used for more effective ventilation during the main excavation. For example, at the Milchbuck tunnel in Zurich, a pilot heading was bored with a Robbins TBM (3.20 m diameter), in order to guarantee efficient ventilation during the construction work with a roadheader (Fig. 11-4).

In conventional tunnels, in particular, effective tunnel ventilation is of real economic advantage, since the working stages drilling, blasting and clearing create a high quantity

11.2 Construction Options

Figure 11-4
Pilot heading with ventilation duct to ensure efficient construction ventilation, Milchbuck tunnel, Zürich

of gas and dust, which contaminates the damp and often warm air in the tunnel and thus represents a considerable danger to the health of the workers. This applies especially to long tunnels, where the sizing of the site ventilation is determined to a large extent by diesel powered engines and is reaching its limit.

11.2.3 Enlargement for Stations, Points or Machine Halls

Particular construction problems are encountered in mechanised tunnelling when the cross section excavated by the TBM is enlarged, not over the entire tunnel length, but at particular locations for later use. Such enlargements are required, for example, for the construction of railway stations, sidings or points locations in railway tunnels or for stopping bays in road tunnels. Examples for enlargements of mechanised excavated tunnel sections are the construction of branches for the Zurich–Thalwil railway tunnel [17], the two-track crossing point in the Vereina tunnel and the stopping bays in the Sachseln road tunnel [133].

The construction process for the creation of an enlargement for a station is shown in Figs. 11-5 and 11-6, using as an example the Mülheim underground railway.

The mixshield with pressurised face there ($\varnothing$ 6.90 m), with a hard rock cutter head, enlarged the 37.5 m^2 cross section after completion of the tunnelling to a 73 m^2 railway station cross section. The segmental lining was partially removed and the cross section enlargement was constructed conventionally with shotcrete support.

Figure 11-5
Enlargement of the segmental llining for the station section, Mülheim underground railway [65]

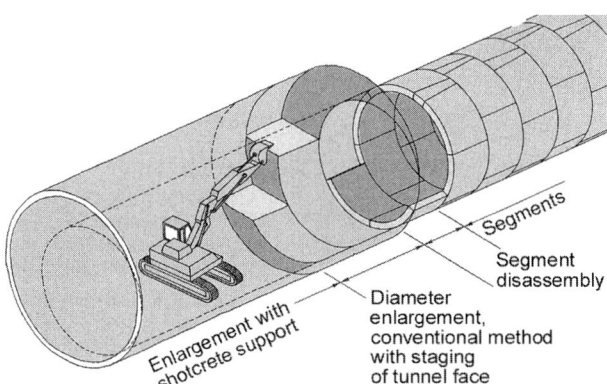

Figure 11-6
Procedure for station enlargement, Mülheim underground railway [106]

Concerning the use of shotcrete in combination with a TBM excavation, its adaptability to the respective geological conditions and variability of possible cross sections for the construction of intersections and branches is a great advantage. This was also demonstrated during the construction of the Channel Tunnel, where a service tunnel bored by a Howden shield TBM ($\varnothing$ 5.76 m) had for logistic reasons to be enlarged over a length of 65 m. The construction process for this enlargement is shown in Fig. 11-7.

Cast iron segments were used for support in the area, which had to be enlarged, which were removed on one side to enable the enlargement. The shotcrete lining of the enlargement was connected to the cast iron segments in the crown with a shoe, and the con-

11.3 Examples

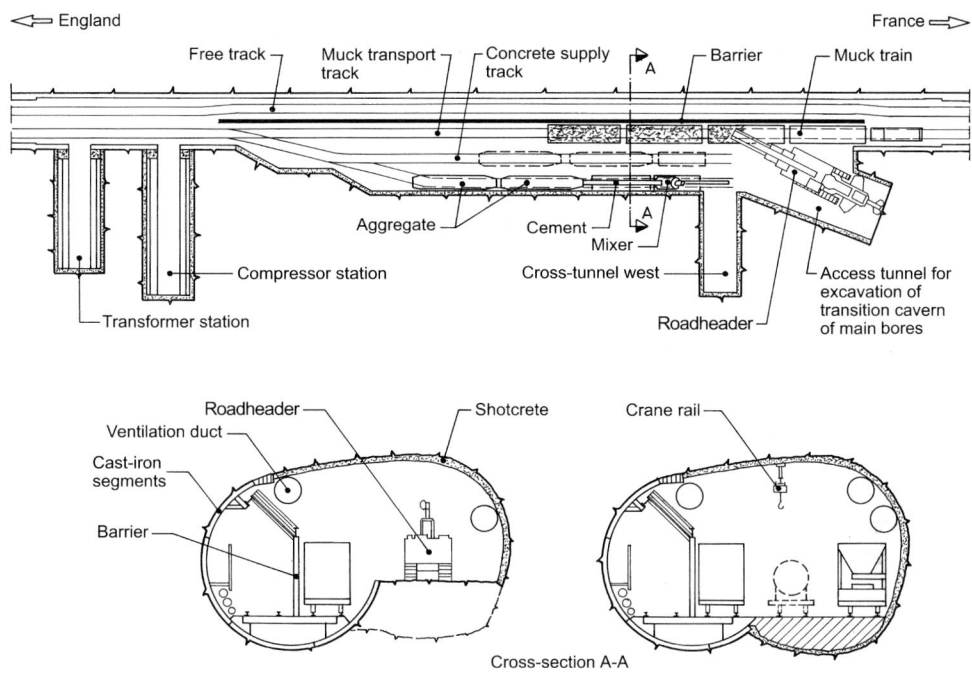

Figure 11-7
Enlargement of the service tunnel at the Channel tunnel [71]

nection in the invert was supported with an invert anchor. The actual enlargement was started manually at first and then completed with a small roadheader, with the cross section being divided into calotte, bench and invert. The completion of the enlargement enabled the access heading to be constructed, which was necessary for the boring of the connection tunnel between the main bores. The enlarged area then provided space for the necessary rail traffic of supply trains carrying aggregates, cement and spoil material.

11.3 Examples

11.3.1 Piora-Mulde Probe Heading

Project

The route of the 57 km long Gotthard base tunnel was intended to cross a geological fault zone called Piora Mulde. According to available geological information, it could not be excluded that the Piora Mulde, shaped like a floating wedge of sugar-shaped dolomite, and additionally under high water pressure, could extend down to the level of the base tunnel. In order to guarantee uninterrupted boring of the base tunnel, it was necessary to obtain certain information about the extent and structure of the Piora Mulde.

Therefore, the mechanised excavation of a 5,552 m probe tunnel was begun in Faido and was driven northwards.

Investigation Scheme

The work on probe heading began in 1993 with a Wirth gripper TBM TB III-450 ($\varnothing$ 5.0 m). In March 1996, the tunnelling was stopped at 5,552 m, approx. 350 m above the level of the base tunnel and approximately 50 m short of the Piora Mulde. Two lateral tubes starting from the probe heading were bored subsequently and connected (Fig. 11-8a).

The front bore served as the entrance to probe chambers. It was intended to continue excavation from one of these chambers with a diameter of 4–5 m through the Piora Mulde with the purpose of testing the machinery, materials and processes, which would be used in the base tunnel. The rear bore served primarily as entrance to a shaft, which could be sunk down to the level of the Gotthard base tunnels, in case the information gained from the probe heading should be insufficient. Additional investigations were performed in a system of headings at the bottom of the shaft. The actual investigation work took place from August 1997 until March 1998. In this period, several probe drillings were performed in a niche of the main heading to investigate the geology in the probe heading as well as at the level of the base tunnel.

Results of the Preliminary Investigation

The results of the probe drillings, which altogether investigated a 200 m wide corridor around the route planned for the Gotthard base tunnel, showed that the Piora Mulde at the base tunnel level mainly consists of dolomite and anhydrite and that the rock is firm and dry (Fig. 11-8b). The thickness of the zone is approximately 125 to 155 m. At the probe heading level, probe drillings had encountered sugar-grain dolomite with over 100 bar water pressure. The difference between these two zones can be explained by the fact that the karst water circulation system bottoms out approx. 250 m above the base tunnel level. A gypsum decke seals and/or stops all the weak points in the rock. This leads to a complete sealing of the underlying dolomite-anhydrite from the water circulation above [184].

Under the circumstances, tunnelling through the Piora zone with known tunnelling methods was regarded as feasible. In addition, as a result of the investigation, further shaft sinking and further probing was no longer held to be necessary.

11.3.2 Kandertal Probe Heading

Project

The 9.5 km long Kandertal probe heading was driven from 1994 to 1997 in order to be able to create project and tender documents for the proposed 34.6 km long Lötschberg base tunnel (Fig. 11-9). The project section Frutigen (north portal) to Kandertal served to investigate geological and hydro-geological conditions as well as the presence of gas.

11.3 Examples

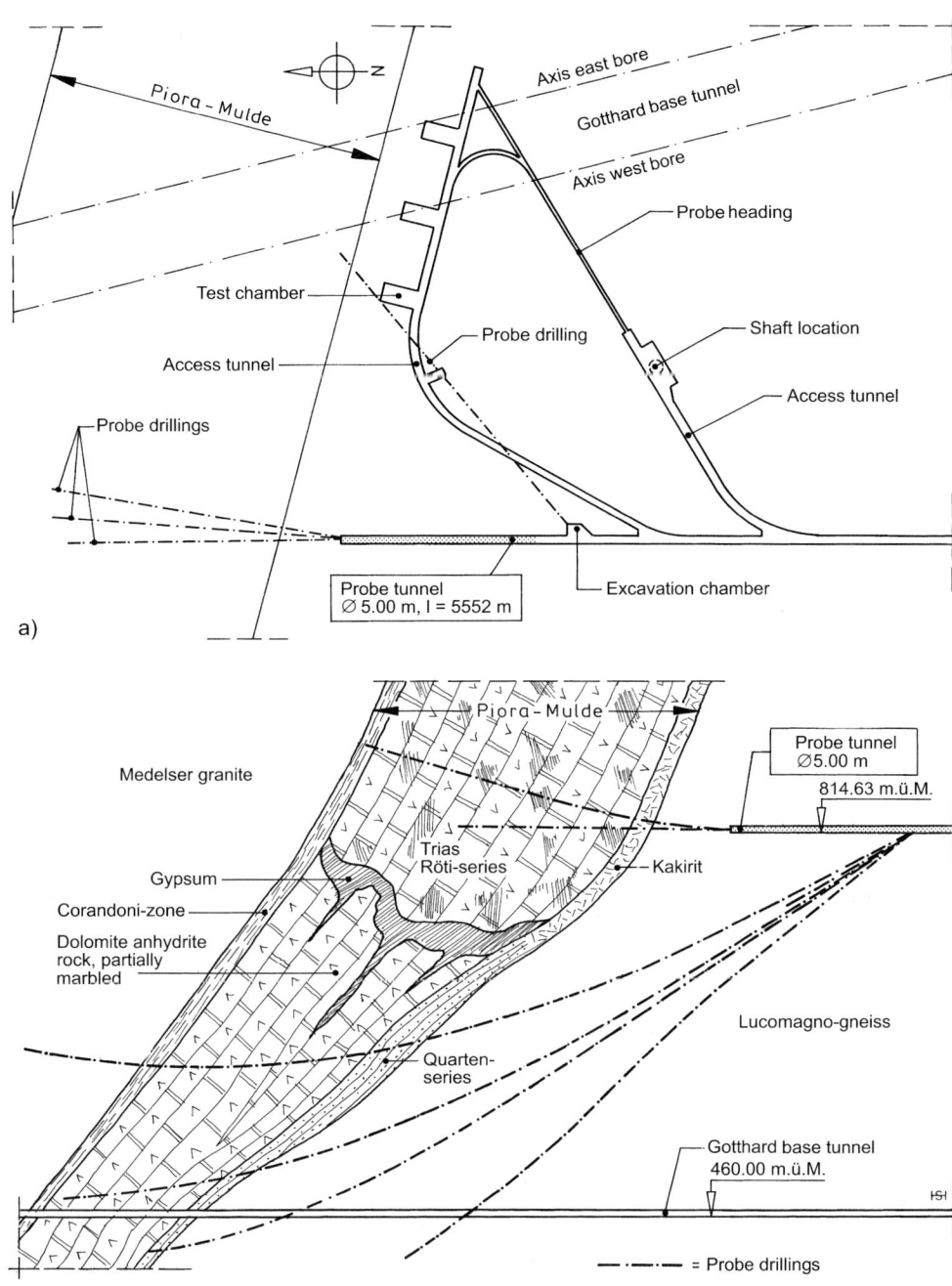

Figure 11-8
Probe system Piora-Mulde, Gotthard base tunnel
a) Overview of probe system [183]
b) Geological longitudinal section [46]

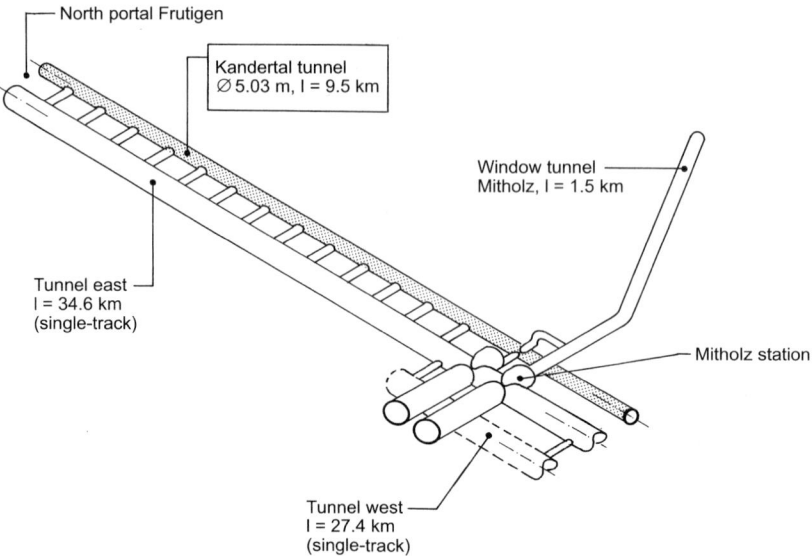

Figure 11-9
Kandertal probe heading, Lötschberg base tunnel [16]

In order to save costs in the construction of the base tunnel, only the east bore between the north portal in Frutigen and the service station planned in Mitholz will be built for the time being. This is possible because the Kandertal probe heading runs parallel and will take over the function as rescue and safety tunnel in the operating phase. Two tunnel tubes are planned south of Mitholz. In the first phase, there is no railway equipment planned in the first section of the western tube. Details of the utilisation of the single-track cross sections of the Loetschberg base tunnel are not discussed further here.

The route of the Kandertal probe tunnel runs from the Frutigen north portal about 30 m to the east of the base tunnel tube on the western side of the Kander valley. After approx. 7.2 km, the Mitholz window heading, which provides ventilation, connects to the side. It was decided to end the probe heading after 9.5 km where the geology changed from the Wildhorn decke in the flysch of the Doldenhorn decke. Starting from this zone by Kandersteg, the geological prognosis southward is considered sufficiently certain.

Tunnelling Scheme

The probe tunnel was driven with a new Robbins MT 1610–279 ($\emptyset$ 5.03 m) (Fig. 11-10) gripper TBM with overcut equipment of 20 cm. The tunnelling advance for the approximately 20 m^2 cross-section varied between 2 and 45 m/working day. The average tunnelling advance was 19.5 m/working day. During the excavation, thrust pressure, tunnelling advance and also the data of various probe drillings were electronically recorded, which was used to evaluate the geology bored through.

Depending on the stability of the rock, it could become necessary to provide support directly behind the cutter head. The following options were available – to install anchors,

11.3 Examples

Figure 11-10
Gripper TBM MB 1610-279 (Robbins) in the starting excavation, Kandertal probe heading, $\emptyset$ 5.03 m [165]

roof arches or additional steel rings. Steel fibre shotcrete support about 8 cm thick was applied about 40 m behind the cutter head. Then, about 55–65 m behind the face, invert segments were installed [165]. The 1.6 m long and 5 t invert segments with integrated drainage were laid into the cleaned invert and grouted with mortar (Fig. 11-11). For additional investigation, independent probe drills installed directly behind the cutter head were capable of probing ahead up to 80 m into the neighbouring rock.

Results of the Preliminary Investigation

For the first 4 km, the Kandertal probe heading crosses the Taveyannaz series and the flysch, which lies under the following Wildhorn decke. The flat base overlap of this decke lies slightly above level of the tunnel and then up towards the brow of the Doldenhorn decke a little under the base tunnel. The geological situation is shown in Fig. 11-12.

Water appears mainly in the form of dripping water or as wells along open fissures. No large karst systems or water-bearing valley fissures were found. Collected together, the annual quantity of water at 15 l/s is very low along approximately 9.5 km heading.

Natural gas was observed from many anchor drill holes, particularly in the slates of the Wildhorn decke. No actual gas blowers were recorded.

The information gained showed that the choice of the location for the route of the base tunnel could be confirmed. Further, the data, supported by rock mechanic measurements on rock samples, provided the information that the section of the Kandertal probe tunnel with its "soft" rock could be tunnelled through either with a TBM drive or by drilling and blasting. Both tunnelling methods were included in the tender documents.

Figure 11-11
Installation of invert segment, Kandertal probe heading [20]

11.3.3 Uznaberg Pilot Heading

Project

The Uznaberg tunnel is, with its two 923 m and 937 m long two-lane tubes, the longest tunnel in the by-pass project Wagen–Eschenbach–Schmerikon (Zurich–Chur). The tunnel has a maximum overburden of 50 m and passes under a heavily built-up area. Due to the very small cover, in some places only 12 m, only low-vibration excavation was possible. In addition to a pure excavation by blasting, another variant was offered – a TBM pilot heading followed by enlargement to the final cross-section by blasting. The client finally decided for the variant with a TBM pilot tunnel (Fig. 11-13), because this possibility was not much more expensive and was better considering the sensitive built-up area.

Geological and Hydro-Geological Conditions

The geology of the Uznaberg range of hills consists predominantly of grey, fine to medium grained sandstones of the lower sweet water molasse. Under these lie marl sandstones and marlstones. The hydro geological conditions showed that because of the partially jointed sandstones, little water was to be expected along the seams, which could have penetrated into the excavated cavity.

Tunnelling Scheme

Both approx. 800 m pilot headings were excavated with a Robbins gripper TBM MK 15 ($\varnothing$ 5.08 m) in the crown area of the final section of the relevant tunnel.

11.3 Examples

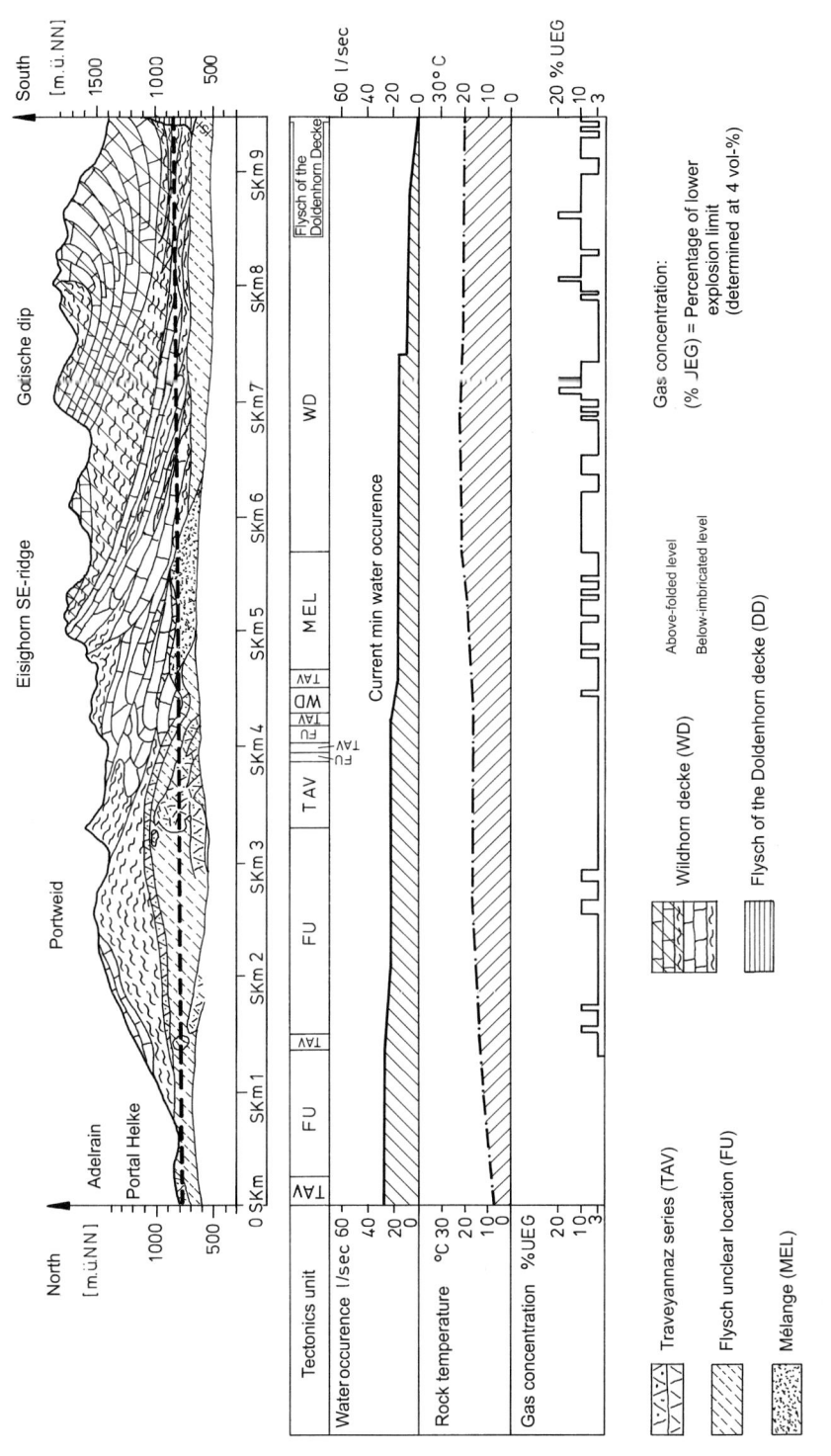

Figure 11-12
Kandertal probe heading – geological results [165]

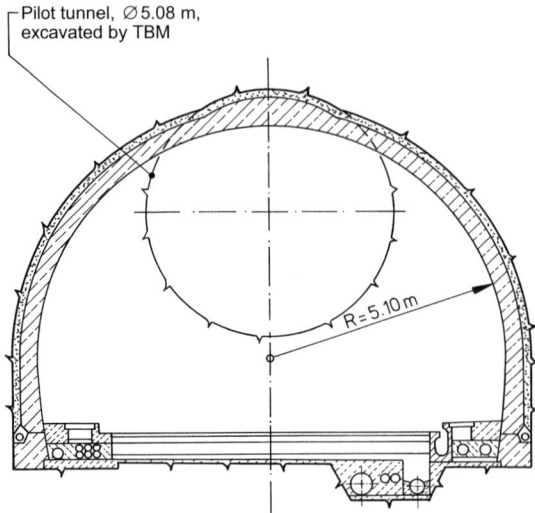

Figure 11-13
Standard section with location of pilot heading, Uznaberg tunnel [105]

The pilot heading of the western tube was driven in 50 working days with an average daily advance of 15.7 m from north to south. After the TBM had been withdrawn, the probe heading for the east bore was started. Due to optimisation of the working processes, the excavation of the second pilot heading was completed in November 1999 after only 28 days with an average daily advance of 27.5 m. After the breakthrough, the TBM was dismantled, so that the enlargement works by drilling and blasting could be begun. The blast pattern and a view of the tunnelling work are shown in Fig. 11-14. The final enlargement to the full size of 80 m^2 was performed in both tunnels at the same time in full face excavation, with the enlargement of the eastern tube being about 100 m behind the western tube.

11.3.4 Enlargement at the Connecting Structure at Nidelbad Zürich–Thalwil Tunnel

Project

The Nidelbad underground connection is located close to Thalwil, at the end of the two-track main tunnel Zurich–Thalwil (compare Chapter 16). At the connection, two single-track tubes separate without crossing in the direction of Thalwil. The separation is staggered in two approximately 155 m long, 22.40 m wide and 12 m branch structures. The construction of the caverns with a complete cross-section of 280 m^2 was done by drilling and blasting using shotcrete. The 1,376 m long tunnel that splits off from the branch structure in the direction of Thalwil at first climbs and then crosses over the double-track tunnel in an arch and drops down to re-join the 703 m long tunnel bore (in another enlargement) in the opposite direction creating a double-track tunnel towards the Thalwil portal.

11.3 Examples

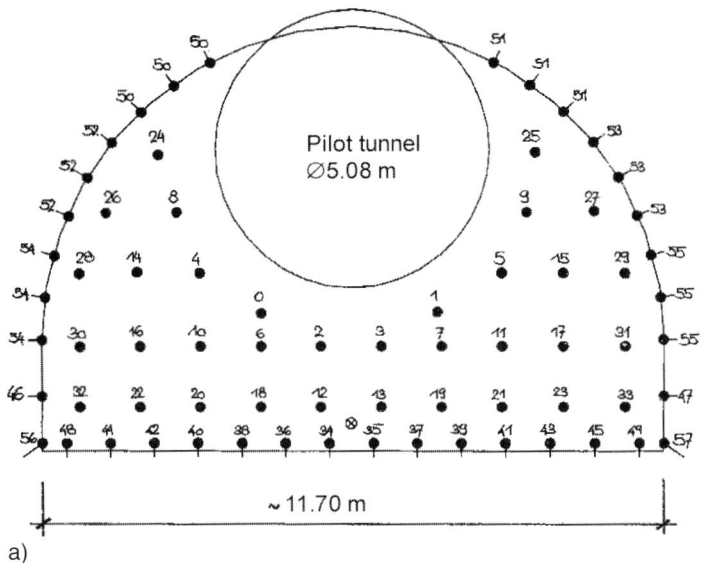

a)

b)

Figure 11-14
Enlargement of the pilot heading to the standard section, Uznaberg tunnel [105]
a) Blasting pattern
b) Tunnelling situation

Enlargement Scheme

The excavation of the two-track main tunnel was done with a Herrenknecht shield TBM S-139 (⌀ 12.28 m) up to the contract section border near Thalwil. There, the dismantling of the TBM began at the start of 2000 simultaneously with the start of conventional excavation of the single-track tunnel towards Thalwil and the actual enlargement of the branching structures. Figure 11-15 shows the plan of the branching structure in the

192   11 Special Processes: Combinations of TBM Drives with Shotcreting

Figure 11-15
Plan of branch structure in the direction of Zürich [116]

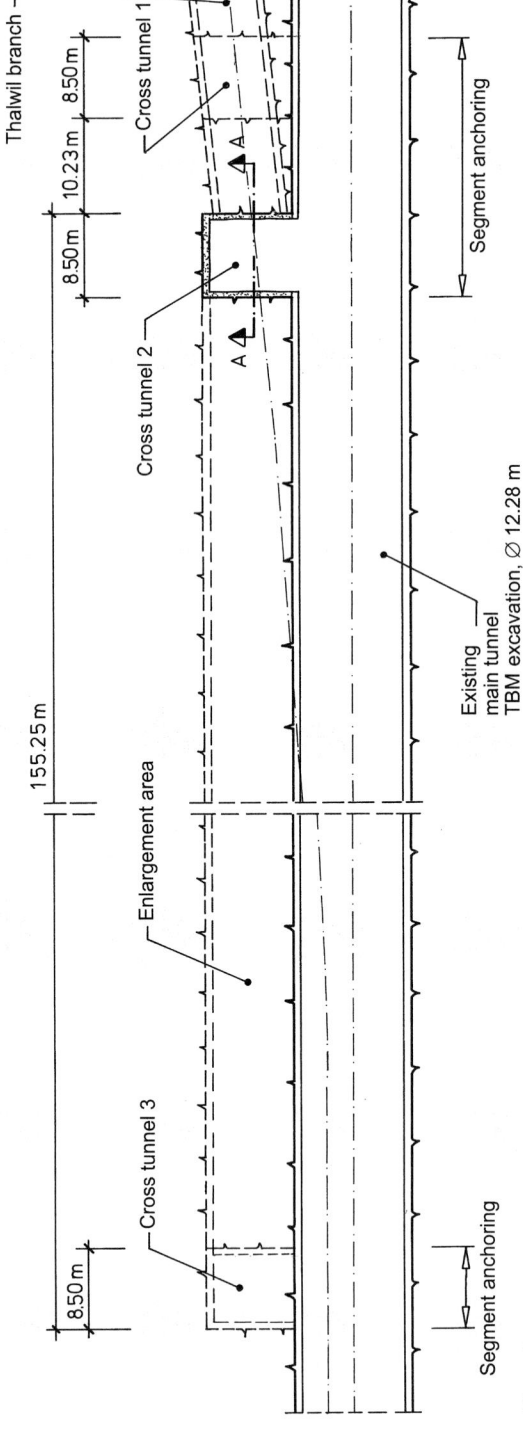

11.3 Examples

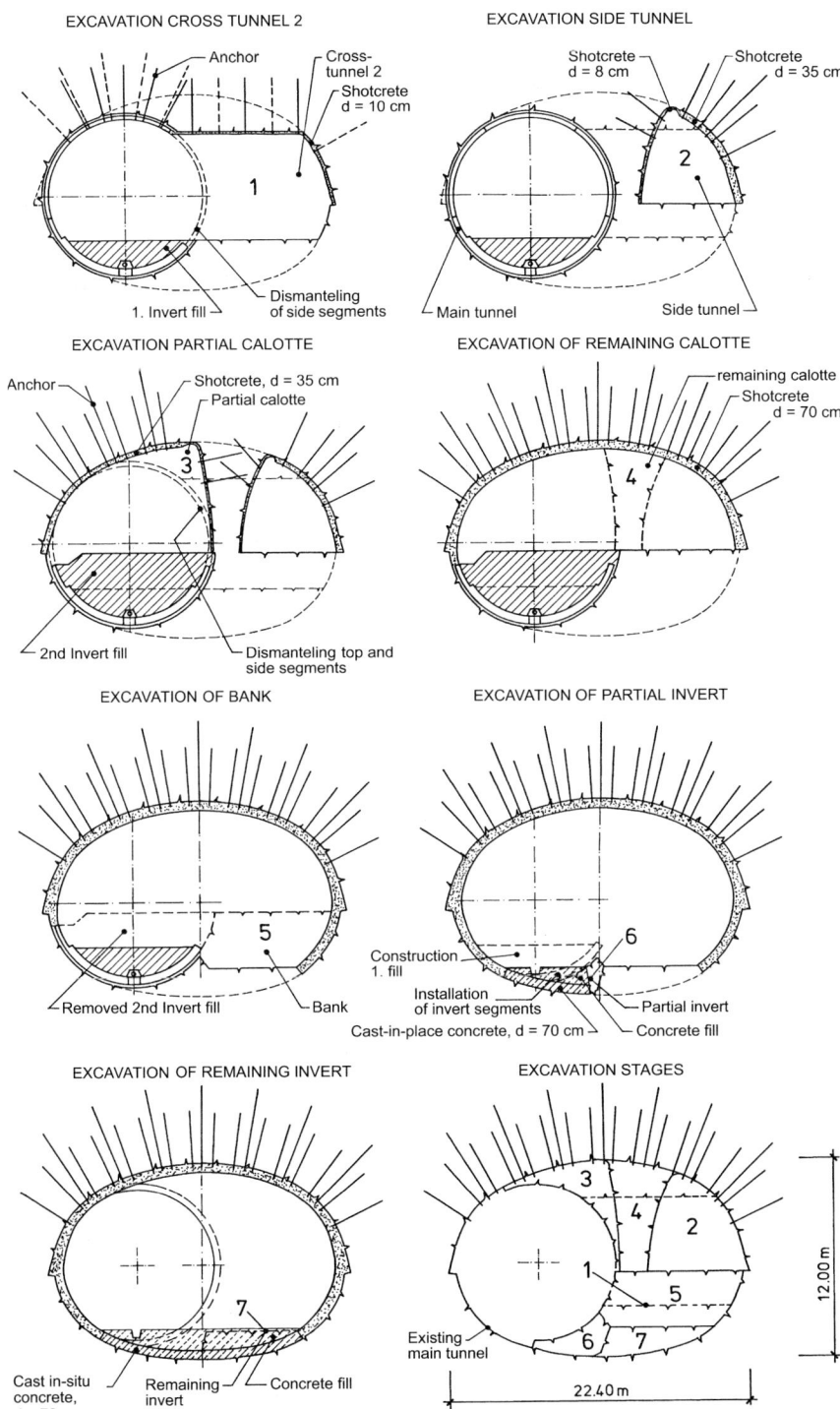

Figure 11-16
Construction stages for the enlargement of the bored section from 118 m^2 to 280 m^2 [116]

Zurich direction with the position of the cross-tunnel 1 for the drill and blast excavation in the Thalwil direction. It also shows the cross-tunnels 2 and 3, which were necessary for the expansion of the segmental lining tunnel section.

The construction schedule used for the enlargement for the branching structure is shown in Fig. 11-16. At first, coarse sand was brought in to fill the invert of the main tunnel in the area of the later cross-tunnel. The top segments were reinforced with anchors. Then the left side segments were broken out and the invert segments were cut over the entire width of the later cross-tunnel. After the partial removal of the lining, excavation of cross tunnel 2 could begin.

Subsequently, as the first step of the actual enlargement, a side heading parallel to the segment tunnel was excavated to cross tunnel 3.

The next step was the further filling of the invert of the main tunnel with coarse sand and the removal in sections of the side and roof segments in the main tunnel, in order to be able to first excavate the partial calotte in the roof area and then the remainder of the calotte.

All partial cross sections were excavated by drilling and blasting, with a uniform section length of 3.00–3.50 m. The support in the calotte area consisted of a reinforced shotcrete vault, $d = 70$ cm, thickened feet.

After the removal of the second invert fill in the main tunnel area, the rest of the bench was taken out. The section length was here also 3.00–3.50 m and reinforced shotcrete was applied for support, $d = 70$ cm.

The remaining fill and then the invert segments were removed from the invert area. Then the invert was excavated in two stages. Cast in-situ concrete was poured to construct an invert vault, $d = 70$ cm, for support. The concrete section length was 10 m with breaking out a maximum of 20 m ahead. Finally concrete backfill was poured.

12 Geological Investigations and Influences

12.1 General

The planning and construction of a tunnelling project is based on knowledge of the geological, hydro-geological and geotechnical conditions. The appropriate investigations of geology and rock mechanics should provide a picture, as realistic as possible, of conditions to be encountered. The goal of these investigations is the description and evaluation of the existing ground conditions as they affect construction work. Based on this, design documents for safe and economic construction and also the planning of measures for temporary and final support of the rock cavity can be produced. They should be adapted to the anticipated construction process to provide specific details and data for design, tendering, estimating, type of contract and invoicing. Therefore this work should be done in close co-operation with specialists in geology, rock mechanics, design, construction management and from machine manufacturers, with the cooperation at best starting at a very early stage of the preliminary investigations.

As part of the preliminary design, using suitable methods of investigation as well as evaluation of general and specialised maps, it can be determined early whether construction is feasible in the selected area considering the technical and economical conditions. It must be investigated, which geological, hydro-geological and geotechnical conditions and rock characteristics will have a particular effect on the structure. As part of the project preparation, geological and geotechnical conditions are to be determined in more detail by in-depth investigations. Furthermore, appropriate data must be available for the contractor for his construction planning, estimating, scheduling of construction work as well as the practical execution of the work.

Since rock conditions are only completely revealed by the actual tunnel excavation, predictions of rock conditions should be continuously checked during the advance and, in case of differences with the conditions encountered, adapted. The responsible specialists should be informed and, if necessary, any planned measures regarding the support of the cavity revised.

The results of the preliminary investigations in engineering geology and rock mechanics, investigations during the advance as well as measurements of the completed structure should be collected in a suitable documentation. This documentation should be in accordance with the national standards [37, 155]. This enables, along with the design and tender documents, calculations and as-built plans, a valuable basis for the evaluation of

- The structural safety
- The fitness for purpose
- The background to later claims and
- Remedying of defects and damage

The type and extent of all investigations must be aligned with size and purpose of the project as well as appropriate statements about rock characteristics (such as stability, working properties, permeability, existing stresses). The conclusion to the investigations must give details of

- The correct choice, construction and technical equipment of the TBM
- The design of the temporary and permanent support and
- The influence of the drive and construction work on the environment

In order that design, tendering and construction documents are as accurate as possible and to avoid as far as possible unpleasant surprises during the excavation, reliable statements about the subsoil, ground water conditions and mechanical behaviour of soil and rock types must be made available.

The design engineer, who has to demonstrate the structural safety, must also be included in the decision process regarding the extent of investigations, since the chosen model influences the assumptions for loading and calculation.

The German committee for underground building (DAUB), the Swiss engineer and architect association (SIA), the Specialist group for underground construction and the Austrian society for geo-mechanics (ÖGG) have, individually and in collaboration, produced and published joint recommendations, which enable a systematic recording of the subsoil [33]. The SIA guideline No. 199, published in 1998, "Recording the rock in underground construction" represents a good basis for the description and the evaluation of the existing ground conditions [155]. However, danger and risks are not defined clearly enough.

Table 12-1 shows an excerpt from the SIA guideline 199, limited to hard rock. It differs only insignificantly from the recommendations of the DAUB and the OGG.

These characteristic values provide a basis for the determination of the stability of the rock, for the planning of construction methods, for investigating structural safety, for the design of support measures, for the determination of any deformations and/or settlement to be expected and for the definition of in-situ measurements.

However, no standards or guidelines replace the intensive discussions required between the geologists and the design engineers. Only this analysis leads to the implementation of geology into civil engineering. It is not sufficient simply to provide all geological documents for the contractor. Also important for the contractor as bidder and as contract partner is the implementation of the geological and hydro-geological information, which the consultant has undertaken on behalf of the client. This influences the geological and geotechnical longitudinal profile and the tender documents.

Knowledge of the course of strata, especially the change from rock to softground and noticeable fault zones, is much more important with TBM operations than with blasting or roadheaders.

In addition to traditional construction site investigations, a supplementary technology available today is reflection seismology, which displays the rock surface and strata in high resolution (Fig. 12-1).

Table 12-1
Description of rock conditions for hard rock (according to [155])

Geometry	• Division into kilometers • Azimuth of runnel axis • Overburden thickness (above crown)	
Geology	• Tectonic unit • Geological unit	
Borders of geological units (strata limits)	• Geological longitudinal profile and further plans • Section lengths	
Description of rock	• Petrography • Anisotropy • Mineral content with scratch resistance <3 • Mineral content with scratch resistance ≥7	• Mineral content subject to swelling • Unfavourable components (quartz, coal, sulphates, mica, clay, asbestos etc.)
Seams	Per seam system • Geometrical position • Geometrical position relative to tunnel axis • Distance • Extent • Opening • Filling/coating • Roughness/undulation • Shear strength (c, φ)	Joint bodies/seam bodies • Basic shape (cubical, plate etc.) • Volume, dimensions
Rock	• Grain cohesion/cementation • Bulk density • Porosity • Water content • Single-axis compression strength $\perp/=$ • Splitting tension strength $\perp/=$	• Elastic modulus $\perp/=$ • Highest/residual strength • Abrasiveness • Potential swelling (swelling coefficient) • Behaviour on water ingress
Geology	• Rhythmic stratification • Primary stress condition (Stress anisotropy, residual stress)	• E-modulus • Rock temperature • Radioactivity
Hydrogeology	• Type of water circulation (pores, joint, karst water) • Porosity • Water pressure in tunnel area, at the invert with variation • Flow direction and speed • Water ingress during tunnelling, permanent water ingress • Water quality and temperature • Existing use of water resources • Springs, Catchments (thermal or mineral water)	
Gas	• Size of gas source, type of gas • Type of source (matrix, reservoir) • Outflow behaviour • Gases of man-made origin (tanks, contaminated ground)	

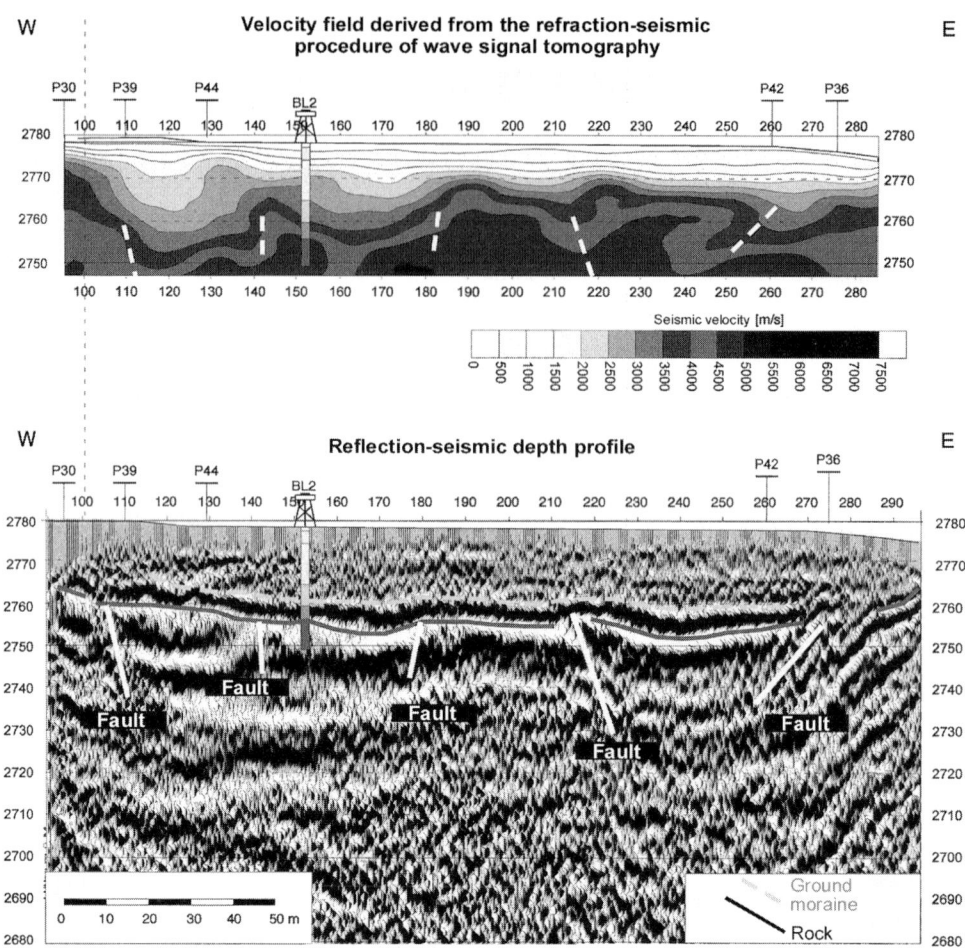

Figure 12-1
Seismic profiles, below as reflection seismology with revealed fault zones and change from rock to softground [49]

For tunnels with little cover, it is significant for the choice of tunnelling method whether any unknown weathered valleys reach into the tunnel profile. Using reflection seismology, depressions were discovered for the Oenzberg and Murgenthal (Rail 2000 Project) tunnels in Switzerland. Lowering the planned gradient enabled a greater overburden at critical locations.

Drill hole photography very quickly enables a complete image of the drilled rock with joints and strata, even without having to take cores (Fig. 12-2). Probe drillings can be sunk in a rapid impact drilling procedure, e.g. with an in-hole-hammer, and then recorded with drill hole photography. The drill hole can even be full of water.

12.2 Influences on the Boring Process

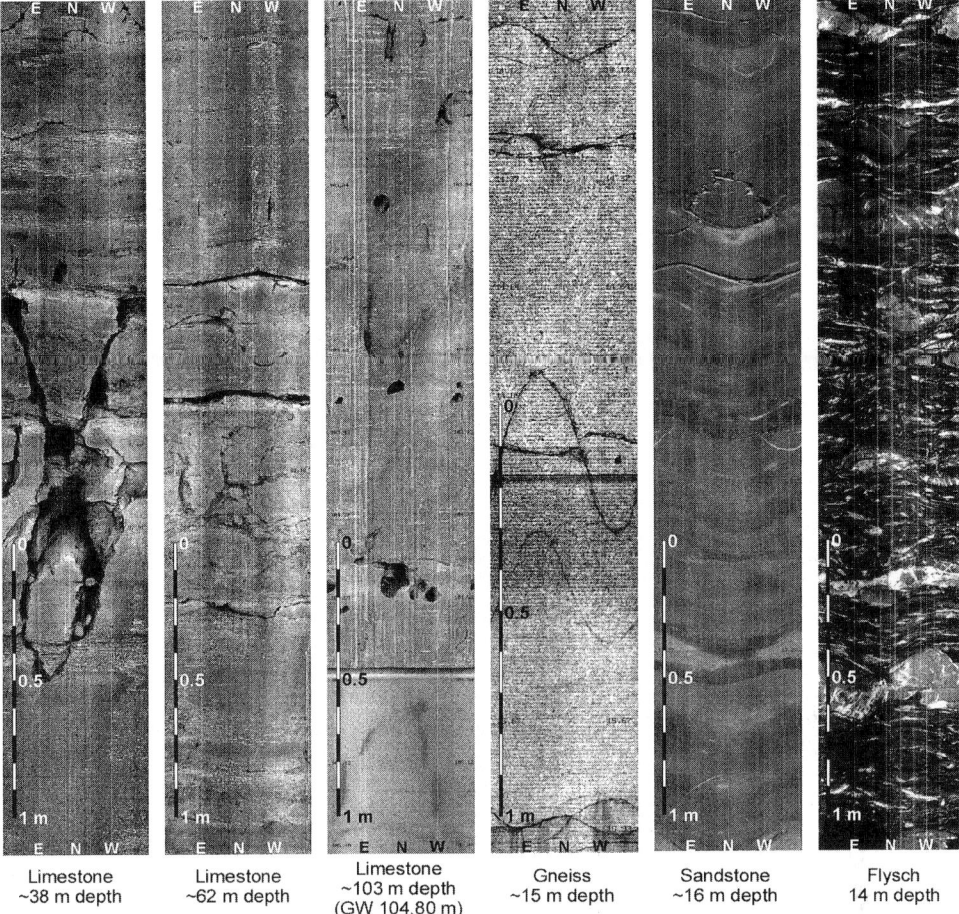

Figure 12-2
Digital drill hole photography with 360° drill hole development, approx. 2.6 m long section of drill hole; left: drill hole in limestone with karst phenomena [49]

This book does not provide a detailed description of the current state of geological and geotechnical investigation technology. We recommend the relevant literature (see Maidl [88]). Further details about construction site investigation can be found, for example, in DIN 4020, "Geotechnical Investigations for Construction Purposes" [37].

12.2 Influences on the Boring Process

Figure 12-3 shows the factors, which determine the penetration during the tunnelling process. The two areas rock and geology are repeated here (see also Chapter 3).

The strength characteristics of the rock, abrasiveness, the joint size and joint consistency are of great significance for the boring process. The latter decides whether the

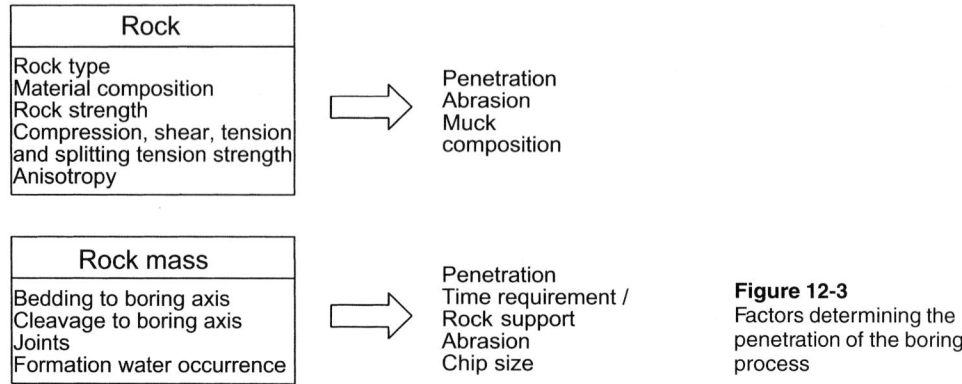

Figure 12-3
Factors determining the penetration of the boring process

use of a TBM with bolted cutter housings, such as are frequently found in small diameter machines, would be feasible at all. Jointed blacks within the tunnel can cause major damage to a TBM cutter head. Adjustments to the cutter head during the advance are expensive and waste time changing cutters. Cutter head modifications like this had to be made, for example, to the Jarva Mk 15 gripper TBM in the pressure tunnel for the Amsteg power station or to the shield TBM of the Colorado-Arizona irrigation system, after cutter housings had been torn off on both machines.

The abrasiveness of a rock can be shown quite well with the CAI test. In less stable rock such as very friable sandstones, abrasiveness can be quite extraordinary. In rock like this, the commonly used methods of wear determination do not supply realistic values. Help is provided in these cases by the LCPC abrasiveness test, as described in Section 3.3.5. If water ingress occurs at the same time in such friable sandstones with high quartz content, as frequently is the case in the lower sweet water molasse, then the sand-water mixture works as a grinding paste.

In the Murgenthal tunnel, sharp grinding sandstone like this caused unusual wear patterns on the cutter head grill bars (Fig. 12-4).

Rock strength is usually determined from cores, which leads to good results. Drill cores from rock with high primary stress, with deep overburden or because of included residual stresses, still deform days after the core has been extracted. Micro-cracks develop in the structure, reducing the strength of the samples. Such false values for strength lead to the wrong conclusions with the choice of the TBM and the cutter. In order to achieve the estimated advance in the tunnel, the contractor is forced to run the TBM with a higher cutter loading, if his machine permits this at all.

Such micro-crack formation in drill cores have been demonstrated microscopically in dense granites or granitic gneiss:

- Amsteg power station in the central Aare granite
- Piora pilot heading in the Leventina granite-gneiss
- Paute project in Ecuador in granodiorites
- Lötschberg tunnel in the Gastern granite

12.2 Influences on the Boring Process

Figure 12-4
Wear on the grill bars of the cutter head, Murgenthal tunnel

In the central Aare granite at the Amsteg power station, the fall-off of the spindle velocity in the drill core through the formation of micro-cracks amounted to an average of approx. 15% four days after core extraction. It is obvious that the compression and shear strengths and also the tensile strength are reduced by such micro-cracks. Within a realistic range of precision, it may be assumed that the strength reduces in relation to the reduction of spindle velocity. In certain parts of the Aare granite around the Amsteg power station, there is a clear, if not very distinct, orientation of the mica minerals. This slight slatiness should lead to compression strengths considerably higher perpendicular to the plane of orientation then in the plane of cleavage. The anisotropy factor in this case is always greater than one. Measurements on drill cores, however, showed an opposite result. This contradiction can be explained only by the formation of micro-cracks. Thin section investigations confirmed these assumptions. Instead of the logical anisotropical values of approx. 1.2, those determined ranged from 0.91 to 0.75.

Rock, which has not been able to relax the stress yet, is excavated during the boring of the face. The strengths that have to be overcome in such cases are substantially greater than material properties determined by conventional methods. If preliminary investigations show differing anisotropic values like this, then supplementary investigations are recommended to clarify discrepancies.

As long as the tensile strength is in a usual relation to the compression strength, the prediction of the compressive strength can be sufficient for approximate sizing. For rock with extremely high toughness – these are to be classified as very difficult to bore – detailed investigations including thin section investigations are unavoidable. If these investigations are not performed, the necessary penetration may not be achieved. The tunnel can then only be completed with a change of cutter type from the disc cutter to tooth or tungsten carbide inserts, accepting the resulting very high costs.

12.3 Influences on the Machine Clamping

The clamping forces extend to about twice the thrust force (see Chapter 4). The large-calibre disc cutters used today require appropriate thrust loads and this to considerable stress, which are transferred into the rock through the gripper units (Fig. 12-5).

The values measured for various TBM while tunnelling show loads of 2–10 MPa onto the side of the cavity. Related to the primary stress condition in the rock, these additional loads are quite substantial.

Through appropriate geological and geotechnical investigations and consideration of sensitivity, it possible to make a better choice of a suitable TBM and to provide measures resulting from reserved decisions, in order not to be taken by surprise in case of occurrence.

Essentially, three different problem areas affect the gripper system of a TBM:

- The rock strength is insufficient for the use of a gripper TBM. Soft ground, like the frequently occurring poorly cemented sediments or in the area of geological fault zones, do not offer sufficient reaction for the gripper units. An access heading for the junction of the Rosenberg tunnel in St. Gallen had to be enlarged with oak baulks in the soft molasse layers to increase the gripper surface (Fig. 12-6).

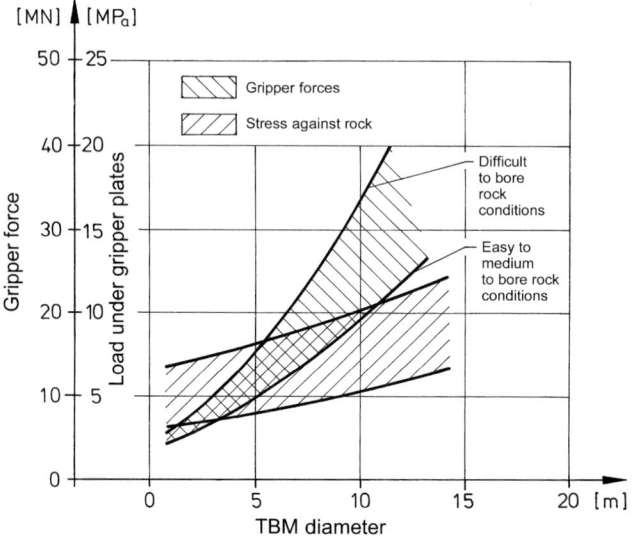

Figure 12-5
Clamping force of an open TBM and resulting loads on the rock as a function of the cuttability and the diameter of the TBM

12.3 Influences on the Machine Clamping

Figure 12-6
Rosenberg access heading with enlarged bearing for the clamping unit

- A Mini Fullfacer was not able to clamp itself at all at the start of tunnelling of the Bochum west tangent, due to a fault zone; abutments had to be concreted (Fig. 12-7) [162]. Bessolow and Makarow report very impressively about the operational problems affecting a TBM in the Baikal Amurmagistrale project in Siberia [12].

Figure 12-7
Abutment for clamping in a fault zone for the Mini Fullfacer, Bochum west tangent

- The high clamping forces applied in very hard rock (classified as difficult to excavate) lead to a redistribution of stress, which can cause or significantly worsen collapses (see also Chapter 4).

- In the Amsteg pressure heading, a rupture in the massive granite similar to a rock burst occurred between the roof of the stator and the front gripper unit (the distance between stator and gripper is approximately 1 m at the start of tunnelling) (Fig. 12-8). This rupture points to redistribution of the stress created by the gripper unit. Myrvang [109] reports rock bursts during the boring of a power station tunnel in the geologically very old granites of Kobbelv in Norway. These happened surprisingly and strongly between the roof of the stator and the gripper unit of the Robbins machines. In some areas, this had the consequence that the gripper system would only work with support, because the outbreaks penetrated too deeply into the walls of the cavity.

- The changing loading of the cavity walls by the gripper units moving forwards stepwise displaces the jointed bodies in jointed rock. The outcome can be collapses around the machine.

Geotechnical investigations and geo-mechanical observations lead to the classification of excavation classes with the relevant support systems. In the higher excavation classes, excavation support will have to be installed as a provisional measure or as part of a system, either in front or within the gripper area of the TBM. Rock support and gripper units then have to be coordinated; otherwise the rock support would be destroyed by the high pressures (see Chapter 9) [131].

Figure 12-8
Tunnel wall of the Amsteg pressure heading with collapses described as rock burst

12.4 Influences on the Rock Support

The ground already deforms to a certain extent in front of the face during tunnelling. Due to the respective geo-mechanical characteristics, deformation of the walls of the cavity occurs more or less shortly after excavation, depending on the size of the cross section. This deformation has to be kept within limits by the rock support. The known method of using curves of rock characteristics best illustrates this phenomenon [85].

The purpose of the rock support should therefore be to select support materials, which allow deformation within a certain range, but nevertheless enable systematic installation with the ensuing high tunnelling advance rates (see Chapter 1).

While simple excavating pressure certainly permits a rigid support, and sometimes even requires rigid support in order to reduce settlement at the surface, genuine rock pressure requires a system allowing deformation after installation.

Proven and available measures, which permit such required deformations, are:

- Steel elements with bell sections, which slide in the connections above a certain loading.

- Deformation slots in the shotcrete support. These can, however, hardly work in a TBM excavation due to the application mostly being done 40–50 m behind the face, because at this location the deformations are hardly increasing any more or have already reached the permitted value.

- Deformable connecting bolts in segment joints or also hydraulic cylinders in the invert segment.

- Deformable ring gap grouting of a segment support with deformable grains in the pneumatic stowing, for example with polystyrene balls coated with cement grout.

It is therefore necessary, when selecting a suitable support system, to investigate the deformation behaviour (see Chapter 15).

13 Classification for Excavation and Support

13.1 General and Objectives for Mechanised Tunnelling

The rock classification for tunnel construction has a long tradition, from both scientific as well as practical viewpoints. The number of the publications is extensive. The reader can find a general summary and specifically details of conventional construction methods in Maidl [88]. The objectives, which are interesting for mechanised tunnelling, are discussed, the basics as well as in practice. Apart from the rock characteristics, the type of ground pressures is of particular interest.

Geologists began early to describe the effect of pressure on rocks. The geo-mechanical engineer today describes rock by measuring its physical characteristics. The contractor describes the rock in terms of balance sheet and performance. He is primarily interested in the advance per unit of time with the components excavation, rock support and delays due to water ingress.

A categorisation of the excavation classes into descriptions preferred by the geologists like stable, liable to rockfall, fractured, squeezing, can only be done by persons in the project, who know the rock from their own experience. Likewise, excavation class categories based on the rock support methods to be used presume good knowledge of the rock behaviour.

The quantitative description of the geo-mechanical engineer permits a good location in the geological longitudinal profile on one hand and an appropriate selection of support methods on the other hand. This is not very useful, in the practice. The contractor estimates a tunnel project, consisting of excavation and support, and he assigns a certain part of the total expenditure to each activity. Correspondingly excavation classes are only of practical use if they take all this consideration into account.

The rock classification, the expression "Vortriebsklassifizierung" *(tunnelling classification)* has become established in Germany, serves during the design phase to help decide the construction and support methods as well as the determination of the most precise possible time and cost plans. It is also an aid for the client in making decisions during the construction phase and can be used for determinations of cost and time as required by the provisions of the contract between client and contractor.

The classification systems known so far were, almost without exception, produced for conventional methods (drilling and blasting). As discussed above, classification provides the basis for accounting and time schedule calculations. In TBM excavation, cuttability also becomes a decisive factor, because in difficult to excavate rock, the boring time for each stroke time increases leaving more time for rock support work.

Apart from the rock classification as a grading of stability, mostly classified according to the extent of support material required, cuttability should be included in an appropriate form, to fulfil the basic objectives. If the contract does not define these matters, then diverging views among the contract parties are inevitable.

13.2 Classification Systems

3.2.1 Classification According to Rock Properties

Terzaghi, Stini and later Lauffer, Packer and Rabcewicz developed rock quality classes, based on rock characteristics or rock behaviour. All grade the rock between "stable" and "squeezing" in a varying number of classes. Using the theoretical findings of rock mechanics, Bieniawski created the RMR system (Rock Mass Rating system) and Barton created the Q System (Quality system) [96]. Both systems are based on quantitative rock parameters without considering the cuttability for a TBM.

13.2.1.1 RMR System (Rock Mass Rating System)

The RMR system [14, 15, 96] was developed by Bieniawski in 1972–1973 and is mainly used in American-influenced areas. Other possible descriptions for this system are Rock Mass Rating system or also "Geo-mechanics Classification". In the last 20 years, the system has proved itself in over 350 practical applications and has been constantly improved by extensions. The RMR System, like the Q System from Barton, is a quantitative classification system, which is calibrated based on completed projects and therefore continuously updated.

In the absence of such systems for TBM excavation with a critical appraisal of conditions, the RMR system can also provide usable results for mechanised excavation.

Process of Classification

The following six parameters are used by the RMR system to classify rock types:

1. Unconfined compressive strength of the rock material
2. Rock Quality Determination (RQD)
3. Separation of joints
4. Condition of joints
5. Water ingress
6. Joint orientation

To use the geo-mechanical classification, the rock is divided into ranges, in which the condition of the rock has almost similar properties. Although the rock in its nature is not homogeneous, individual ranges can be defined on the basis of the above criteria and consulted for investigation (homogeneous ranges). The distinctive properties of a range are recorded on a data sheet and evaluated with the help of Tables 13-1 and 13-2. It is essential here that Table 13-1 is used independent of the orientation of possible fault zones and that the results are corrected using Table 13-2 with regards to orientation and the planned structure

The results from the two tables are added to give a characteristic value, which enables an allocation to a rock class using Table 13-3. The higher this value is, the better is the rock under consideration. The range of added values is between 0 and 100, bad to very good.

13.2 Classification Systems

Table 13-1
Classification parameters and their evaluation [14, 96]

	Parameter		Range of values					
1	Strength of intact rock [MPa]	Point load strength index uniaxial compressive strength	>10	4–10	2–4	1–2	–	
			>250	100–250	50–100	25–50	5–25	1–5 <1
	Evaluation		15	12	7	4	2	1 0
2	Rating of quality of drill core RQD [%]		90–100	75–90	50–75	25–50	<25	
	Evaluation		20	17	13	8	3	
3	Fault zone spacing [m]		>2	0.6–2	0.2–0.6	0.05–0.2	<0.06	
	Evaluation		20	15	10	8	5	
4	Condition of the fault		Very rough surface, not continuous	Slightly rough surface	Slightly rough surface	Slickenside surface or fault displacement <5 mm or	Soft fault displacement >5 mm or	
			No settlement, unweathered rock face	Settlement <1mm, slightly weathered face	Settlement <1mm, very weathered face	Settlement 1–5 mm continuous	Settlement >5 mm continuous	
	Evaluation		30	25	20	10	0	
5	Ground water	Inflow per 10 m length of tunnel [l/min]	None or	<10 or	10–25 or	25–125 or	>125 or	
		Fissure water pressure	0 or	<0.1 or	0.1–0.2 or	0.2–0.5 or	>0.5 or	
		General condition	Completely dry	Moist	Wet	Dripping	Flowing	
	Evaluation		15	10	7	4	0	

Table 13-2
Correction factor for the direction of strike of the fault [14, 96]

Strike and dip direction of fault		Very favourable	Favourable	Acceptable	Unfavourable	Very unfavourable
Evaluations	Tunnels and mines	0	−2	−5	−10	−12
	Foundations	0	−2	−7	−15	−25
	Embankments	0	−2	−25	−50	−60

Table 13-3
Determination of the rock classes from the overall evaluation [14, 96]

Evaluation	100–81	80–61	60–41	40–21	< 20
Rock class	I	II	III	IV	V
Description	Very good rock	Good rock	Acceptable rock	Bad rock	Very bad rock

Table 13-4
Evaluation of rock classes [14, 96]

Rock class	I	II	III	IV	V
Average free stand-up time	20 years at 15 m free-standing	1 year at 10 m free-standing	1 week at 5 m free-standing	10 hours at 2.5 m free-standing	30 minutes at 1 m free-standing
Cohesion of the rock [kPa]	> 400	300–400	200–300	100–200	< 100
Friction angle of the rock [°]	> 45	35–45	25–35	15–25	< 15

Table 13-5
Linkage of RMR system to ÖNORM B 2203

RMR system	Rock type according to ÖNORM B 2203
> 80	A_1
80 to 60	A_2
60 to 50	B_1
50 to 40	B_2
40 to 20	B_3/C_1
< 20	C_2 to C_3

13.2 Classification Systems

In Table 13-4, the practical evaluation of the individual rock classes is described based on examples from engineering practice. Since the geology consists of the different ranges, the most significant range is the one most unfavourable for the proposed construction. Future construction work must be based on this range, independent of other possibly good features of rock stability or other parameters. If two ranges with different parameters dominate the entire cross section, the evaluated numbers are weighted according to the area of occurrence of each, producing an average characteristic value.

Strengths and Limitations

The RMR system is very simple to use, the classification parameters can be obtained from analysis of drill cores or from geo mechanical data. This procedure is applicable and adaptable to many situations in the mining industry, stability of embankments, foundation stability and in tunnelling. The geo-mechanical classification is very suitable for application in expert systems.

Against this, the result produced by the RMR classification method is rather conservative, which mostly leads to an over-dimensioning of support measures. This can be balanced with constant supervision during construction, with the evaluating system being adapted to the local conditions.

In 1996, Bieniawski created a linkage to the rock classification according to support measures required, detailed in ÖNORM B 2203 (*Austrian standard 2203*) (for rock classification according to ÖNORM, see Section 13.2.2).

This rock classification was developed on the basis of the drilling and blasting method. With mechanised tunnelling, the tearing of the rock outside the intended cross-section, caused by blasting and often undirected, is inapplicable. The positive influence of mechanised tunnelling is calculated with the following formula according to Alber et al. [1]:

$$RMR_{TBM} = 0.84 \cdot RMR_{D+B} + 21 \qquad (20 < RMR_{D+B} < 80) \qquad (13\text{-}1)$$

RMR_{TBM} represents the RMR value in TBM excavation and RMR_{D+B}, the value for drilling and blasting. In rock of bad to very bad quality, the difference between conventional and mechanised tunnelling is less significant. The constantly changing loading on the rock due to the gripper clamping has negative effects in TBM operations.

13.2.1.2 Q System (Quality System)

The Q-system of the rock classification was developed in 1974 by Barton, Lien and Lunde [8] in Norway and later further developed by Grimstad and Barton for TBM operations [55]. The system is based on the analysis of more than 200 tunnel construction projects in Scandinavia. Due to this analysis, it is assumed to be a quantitative classification system. It was intended as a system for engineers, to facilitate the design of tunnel support systems.

The Q system is based on the numerical estimation of the following six parameters:

1. Determination of the rock quality (RQD = Rock Quality Designation)
2. Joint set number (J_n)
3. Roughness of the most unfavourable joint as joint roughness number (J_r)
4. Degree of alteration or filling along the weakest joint as joint alteration number (J_a)
5. Water ingress as joint water reduction number (J_w)
6. Stress conditions as stress reduction factor (SRF = Stress Reduction Factor)

These six parameters are combined to three quotients, which result in a weighted mathematical value for rock quality Q as follows:

$$Q = \frac{RQD}{J_n} \cdot \frac{J_r}{J_a} \cdot \frac{J_w}{SRF} \tag{13-2}$$

The value for Q (rock quality), calculated using this formula, ranges from 0.001 to 1000. The presentation is in graphic form, with the value for Q being entered on a logarithmic scale, and rock quality can be read off (Fig. 13-1).

Classification process

The Tables 13-6 to 13-11 provide numeric values for each of the factors stated above. They are interpreted as follows:

The first two factors, RQD and J_n (number of joints), give an overview of the structure of the rock; their quotient is a relative value for the joint body size in the drill core. The RQD index, suggested by Cording, Deere and Hendron [26], shows the relationship L_{10}/L in percent, whereby L_{10} describes the drill core pieces more that 10 cm long in the drill sample length L.

The quotient resulting from the third, J_r (joint roughness number), and the fourth factor, J_a (joint variation number), can be regarded as an indicator of the shear strength at the meeting surfaces between the individual blocks of rock.

The fifth parameter, J_w (reduction factor for the influence of joint water), is a measurement of the water pressure, while the sixth factor, SRF (Stress reduction factor), permits differing interpretations:

a) Loosening pressure in case of shear surfaces and rock containing clay
b) Ground pressure in competent rock and
c) Squeezing and swelling pressures in plastic, not stable rock

This sixth parameter is also regarded as total pressure parameter. The quotient of the fifth and the sixth parameters describes the active stress.

Barton regards the parameters J_n, J_r und J_a as those, which play a special role in the joint orientation. This is implicitly contained in the parameters J_r and J_a, because these are based on the most unfavourable joint.

13.2 Classification Systems

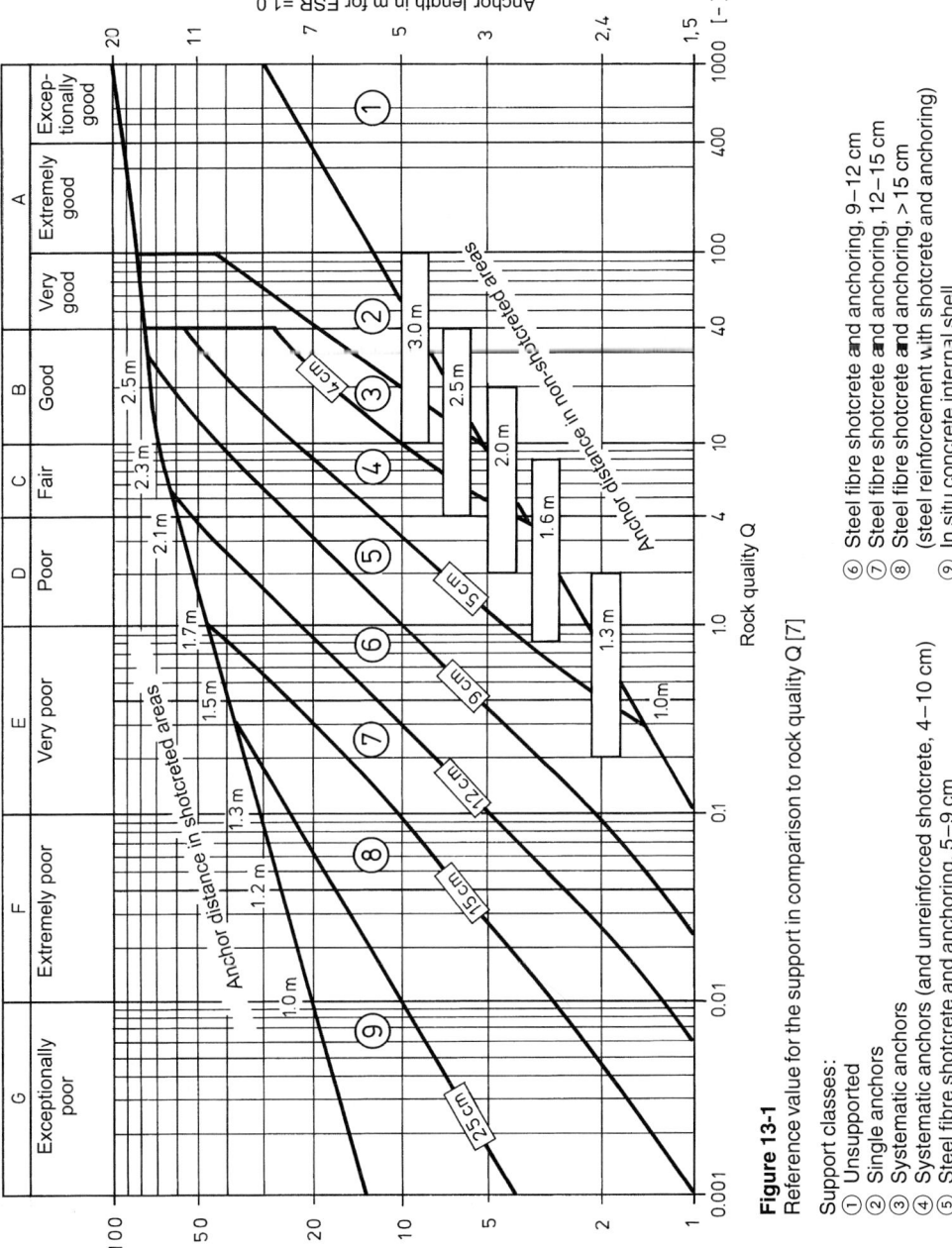

Figure 13-1
Reference value for the support in comparison to rock quality Q [7]

Support classes:
① Unsupported
② Single anchors
③ Systematic anchors
④ Systematic anchors (and unreinforced shotcrete, 4–10 cm)
⑤ Steel fibre shotcrete and anchoring, 5–9 cm
⑥ Steel fibre shotcrete and anchoring, 9–12 cm
⑦ Steel fibre shotcrete and anchoring, 12–15 cm
⑧ Steel fibre shotcrete and anchoring, >15 cm (steel reinforcement with shotcrete and anchoring)
⑨ In situ concrete internal shell

Table 13-6
Dermination of rock quality – RQD [7, 8]

Determination of the rock quality RQD			
Description	Range of values	Remark
Very poor Poor Fair Good Excellent	0–25 25–50 50–75 75–90 90–100	(1) Where RQD is reported or measured as ≤ 10 (including 0), a nominal value of 10 is used to evaluate Q. (2) RQD intervals of 5, i.e. 100, 95, 90 etc. are sufficiently accurate.

Table 13-7
Q system: number of joint systems – J_n [7, 8]

Joint set number J_n		
Description	Range of values	Remark
Massive rock, no joints or very few joints One joint set One joint set plus random joints Two joint sets Two joint sets plus random joints Three joint sets Three joint sets plus random joints Four or more joint sets, random joints Heavily jointed, "sugar cube" etc. Crushed rock, earthlike	0.5–1.0 2 3 4 6 9 12 15 20	(1) For connections of tunnels to cross headings, $(0.3 \cdot J_n)$ is to be used (2) For areas near portals, $(0.2 \cdot J_n)$ is to be used

Table 13-8
Q system: Joint roughness number – J_r [7, 8]

Joint roughness number J_r		
Description	Range of values	Remark

• Rock-wall contact
• Rock-wall contact before 10 cm shear
  – non-continuous joint
  – rough or irregular, wave-shaped
  – smooth, wave-shaped
  – slickenside, wave-shaped
  – rough or irregular, flat-surfaced
  – smooth, flat-surfaced
  – slickenside, flat-surfaced
• No rock-wall contact when sheared
  – The shear zone contains clay minerals sufficiently thick to avoid rock wall contact
  – Sandy, gravely or ground down areas sufficiently thick to avoid rock wall contact
	
	4.0
	3.0
	2.0
	1.5
	1.5
	1.0
	0.5
	
	1.0
	
	1.0
		(1) Add 1.0 in case the mean spacing of the relevant joint sets is more than 3 m.
		(2) $J_r = 0.5$ can be used for planar slickensided joints, which have lineation, as long as the lineation is favourably oriented.

13.2 Classification Systems

Table 13-9
Q system: Joint alteration number – J_a [7, 8]

Joint alteration number J_a		Range of values	$\varnothing_r$ (appr.)
Description			
a)	Rock-wall contact (no mineral filling, only surface)		
	A) Densely filled, solid, not softened, impermeable filling	0.75	
	B) Unvaried sides of joint, only stained surface	1.0	25–35°
	C) Lightly varied sides of joint, not softened, mineral surface, sandy particles, clay-free, loose pieces of rock	2.0	25–30°
	D) Silty or sandy clay surface, no clay fractions (not softened)	3.0	20–15°
	E) Softened or low friction, showing clay mineral surface	4.0	8–16°
b)	Rock-wall contact before 10 cm shear (weak mineral filling)		
	F) Sandy particles, clay-free, loose rock pieces	4.0	25–30°
	G) Strongly over-consolidated, not softened, clay mineral filling (continuous, <5 mm thickness)	6.0	16–24°
	H) Medium or low over-consolidated, softened clay mineral filling (continuous, <5 mm thickness)	8.0	12–16°
	J) Swelling clay filling, e.g. montmorillonite (continuous, <5 mm thick); the value for J_a depends on the percent of swelling clay-size particles, and access to water	8.0–12.0	6–12°
c)	Shear jointing without contact with the rock face (thick mineral filling)	6.0; 8.0 or	
	K,L,M) Zones or bands of disintegrated or crushed rock and clay (see also G, H and J for description) of clay condition	8.0–12.0	6–24°
	N) Zones or bands of silty or sandy clays, small clay fractions, (non softening)	5.0	
	O,P,R) Thick, continuous zones or bands of clay (see also G, H and J for description of clay condition)	10.0;13.0 or 13.0–20.0	6–24°

Remark:
(1) The values for $\varnothing_r$ should be regarded as approximate reference values for the mineral properties of the varied products, should such be present.

Using the Q value, necessary support measures can be determined indirectly, taking into account the actual cross-section size and the future use of the excavated cavity. The ratio for support measures is a function of the parameters stated above. It is determined by dividing the diameter or the height of the excavated cross section by a defined number, ESR = Excavation Support Ratio, that is based on the purpose of the building. This value provides the input value for Fig. 13-1.

The ESR refers to the use of the cavity and the required degree of safety; see Table 13-12 for more detail.

Table 13-10
Q system: Joint water reduction number – J_w [7, 8]

Joint water reduction number J_w			
Description	J_w	Approximate water pressure [kg/cm^2]	Remark
A) Dry excavations or minor water inflow, locally up to 5 l/min	1.0	<1	(1) The values in cases C to F are rough estimates. Increase J_w in case water-lowering measurements are in use. (2) Special problems caused by icing have not been considered.
B) Medium inflow or pressure occasional outwash of joint fillings	0.66	1.0–2.5	
C) Large water inflow or high pressure in competent rock with unfilled joints	0.5	2.5–10.0	
D) Large water yield or high pressure, considerable water loss from individual joints	0.33	2.5–10.0	
E) Exceptionally high water inflow on blasting, which reduces with time	0.2–0.1	>10.0	
F) Exceptionally high water inflow or constant water pressure with no noticeable reduction	0.1–0.05	>10.0	

The relationship between the Q index and the reference value for support determines the appropriate support measures. For temporary support, either the Q-value should be increased to $5 \cdot Q$ or the ESR should be increased to $1.5 \cdot$ ESR, if the preliminary support is intended for a period of more than one year. The ESR can be reduced to 0.5 for long traffic tunnels. The necessary anchor length can be determined from Fig. 13-1. It needs to be taken into consideration that ESR = 1.

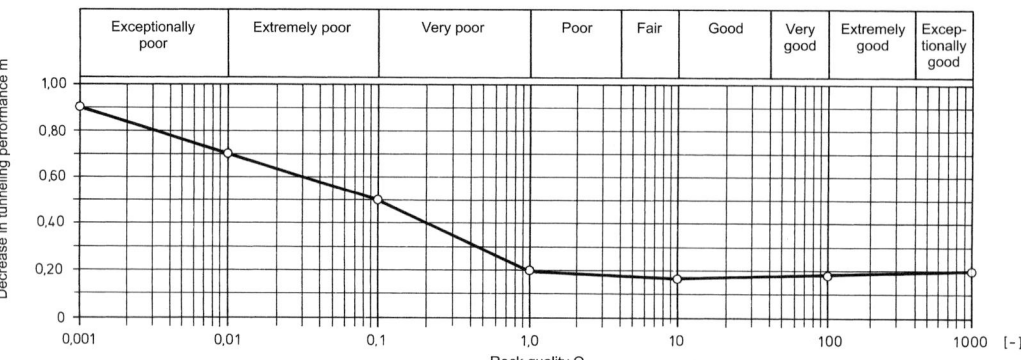

Figure 13-2
Reduction factor m_1 in dependence on Q value [7]

13.2 Classification Systems

Table 13-11
Q-system: Stress reduction factor – SRF [7, 8]

Stress reduction factor – SRF [7, 8]			
Description		Range of values	Remark
a) **Weakness zones intersecting excavation, which could cause a loosening of rock mass on excavation** A) Multiple occurrence of weakness zones with clay or with chemically stabilised rock, very loosened surrounding rock (any overburden) B) Isolated weakness zone with clay or with chemically stabilised rock (depth >50 m) C) Isolated weakness zone with clay or with chemically stabilised rock (depth >50 m) D) Multiple shear zones in competent rock (clay-free), loosened surrounding rock (any depth) E) Isolated shear zones in competent rock (clay-free) (depth 50 m) F) Isolated shear zones in competent rock (clay-free) (depth >50 m) G) Loosened open joints; strongly jointed or "sugar-cube" etc. (any depth)		10.0 5.0 2.5 7.5 5.0 2.5 5.0	Reduce the given value for SRF by 25 to 50%, if the relevant shear zones only influence the excavation, but do not intersect.

Description	σ_c/σ_1	σ_Θ/σ_c	Range of values	Remark
b) **Competent rock, rock stress problems** H) Low stresses, near to surface, open joints J) Medium stresses, favourable stress conditions K) High stress, very dense structure. Normally favourable for stability, possible unfavourable for the stability of the sides L) Light rock slabbing after >1 hour, (massive rock) M) Spalling and rock burst after a few minutes (massive rock) N) Strong rock burst (strain burst) and immediate, large deformations in massive rock	 >200 200–10 10–5 5–3 3–2 <2	 <0.01 0.01–0.3 0.3–0.4 0.5–0.65 0.65–1 >1	 2.5 1.0 0.5–0.2 5–50 50–200 200–400	For strongly anisotropic stress fields (if measured) the following applies: if $5 \leq \sigma_1/\sigma_3 \leq 10$, then σ_c is to be reduced to $0{,}75 \cdot \sigma_c$; if $\sigma_1/\sigma_3 > 10 \, \sigma_c$, then σ_c is to be reduced to $0{,}5 \cdot \sigma_c$; with: σ_1 and σ_3 major and minor stress, σ_c = compression strength and σ_θ = maximum tangential stress (according to elasticity theory). In cases where the overburden of the tunnel is less than the diameter, ESR should be increased from 2,5 to 5 (see H).
c) **Squeezing rock, plastic yielding of not stable rock under the influence of great ground pressures** O) Lightly squeezing ground pressure 1–5 P) Strongly squeezing ground pressure >5			5–10 10–20	Squeezing rock can occur at depths of $H > 350 \, Q^{1/3}$. The rock compression strength can be estimated with $q = \sim 0.7 \, Q^{1/3}$ (MPa) = density in kN/m³
d) **Swelling rock, chemically caused swelling because of reaction with water** R) Lightly swelling ground pressure S) Strongly swelling ground pressure			5–10 10–15	

Table 13-12
Values for ESR depending on intended use [7]

Type of tunnel	ESR
Support only for temporary purposes	1.5 ESR and 5 Q [1]
Pilot tunnel	2.0
Water and sewer heading	1.5
Traffic tunnel	0.5–1.0 [2]

[1] The value is to be reduced to 2.5 Q, if the temporary support will be needed for longer than 1 year.
[2] For long traffic tunnels, ESR should be reduced to 0.5.

The attempt to create a Q_{TMB} index based on a Q index was performed by Barton et al. [7]. This rock classification for TBMs, as a combination of the prognosis for rock support, the prognosis for cuttability and material wear, may serve in individual cases. Generally, however, cuttability and cutter life cannot be reconciled with rock classification. Barton tried, using the values Q, Q_{TBM} and additional values, to estimate the penetration rates PR and advance rate AR of a TBM drive. The Cutter Life Index (CLI) as a wear value (Fig. 13-3) with a capped scale of values is less convincing, considering the life span of cutter rings varies between less than 200 km to over 20,000 km rolling distance.

The penetration rate *PR* and the advance rate *AR* are determined as follows:

$$PR = 5 \cdot Q_{TBM}^{-1.5} \qquad [m/h] \qquad (13\text{-}3)$$

$$AR = 5 \cdot Q_{TBM}^{-1.5} \; T^m \qquad [m/h] \qquad (13\text{-}4)$$

Where the following applies:

$$Q_{TBM} = Q_0 \cdot \frac{SIGMA}{F^{10}/20^9} \cdot \frac{20}{CLI} \cdot \frac{q}{20} \cdot \frac{\sigma_\Theta}{5} \qquad (13\text{-}5)$$

$$SIGMA = SIGMA_{cm} = 5 \cdot \gamma \cdot Q_C^{1/3} \;\; \text{with} \;\; Q_c = Q_0 \cdot \frac{\sigma_c}{100} \;\; \text{for} \; \beta > 60° \quad (13\text{-}6)$$

$$SIGMA = SIGMA_{tm} = 5 \cdot \gamma \cdot Q_t^{1/3} \;\; \text{with} \;\; Q_t = Q_0 \cdot \frac{I_{50}}{100} \;\; \text{for} \; \beta < 30° \quad (13\text{-}7)$$

The value *SIGMA* is intended to take into account the strength of the rock, considering uniaxial compressive strength σ_c and tensile strength, expressed here as the stability index I_{50}, and the angle β between bedding and tunnel axis. Further, the value *m* has to be calculated, which takes the wear patterns of the TBM into consideration:

$$m = m_1 \cdot \left(\frac{D}{5}\right)^{0.20} \cdot \left(\frac{20}{CLI}\right)^{0.15} \cdot \left(\frac{q}{20}\right)^{0.10} \cdot \left(\frac{n}{2}\right)^{0.05} \qquad (13\text{-}8)$$

13.2 Classification Systems

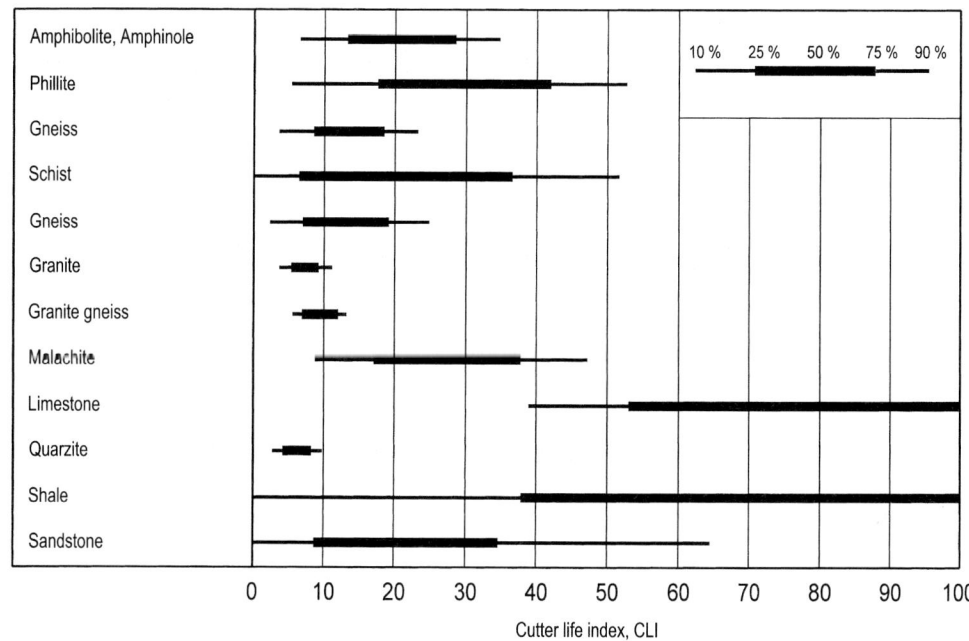

Figure 13-3
Wear index CLI in dependence on the rock type [7]

The time required for a TBM to bore a tunnel section or a specific geological area with the length L, can be estimated with T

$$T = \left(\frac{L}{PR}\right)^{\frac{1}{1+m}} \qquad (13\text{-}9)$$

where:
m = reduction of advance caused by wear [–]
T = time required to excavate a tunnel section
F = thrust force per disc cutter [to]
m_1 = starting value for the reduction of advance caused by wear [–]
D = tunnel diameter [m]
CLI = Cutter Life Index, specific wear number of disc cutters determined in laboratory tests [–]
q = quartz content of the rock [%]
n = rock porosity [%]
γ = rock density [kg/dm^3]

Relationship Between the *Q* and the *RMR* System

Based on over 100 case studies, it was possible to establish an originally not intended empirical relationship between the *RMR* and *Q* systems [14, 15]. For tunnel construction, this gives:

$$RMR \approx 9 \cdot \ln Q + 44 \quad \text{respectively} \quad Q \approx e^{\frac{(RMR-44)}{9}} \tag{13-10}$$

Barton considers the relationship to be expressed in the following formula:

$$RMR \approx 15 \cdot \log Q + 50 \quad \text{respectively} \quad Q \approx 10^{\frac{(RMR-50)}{50}} \tag{13-11}$$

The relationship between *Q* and *RMR* can also be seen in Fig. 13-4.

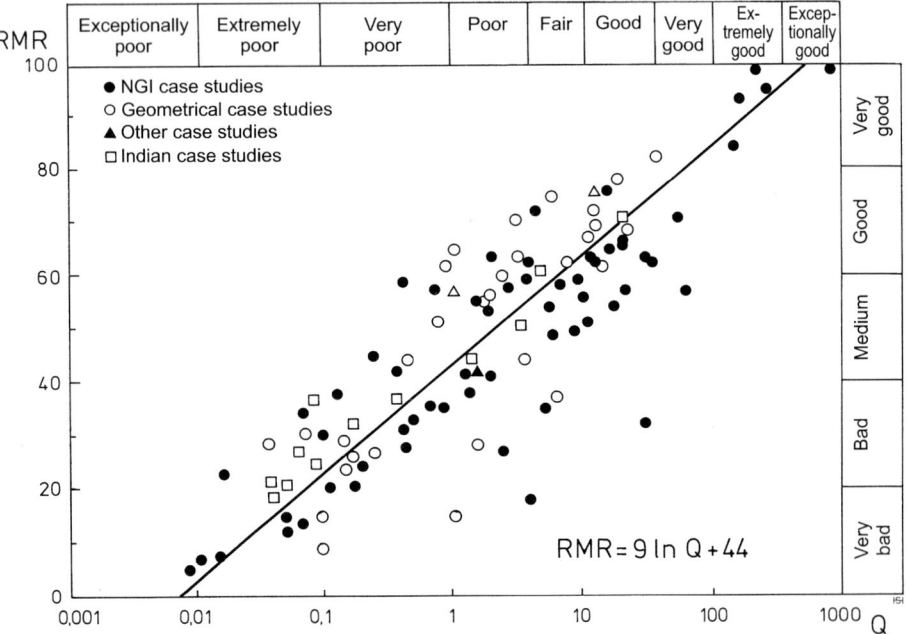

Figure 13-4
Correspondence between RMR and Q system [96]

13.2.2 Classification According to Cuttability and Abrasiveness

A pure excavation classification is not sufficient for TBM operation. Indices, which determine penetration, are therefore necessary for the overall evaluation of the rock to be excavated.

Such a classification can be gained from the characteristics of the geological strata or by direct determination of penetration by means of test excavation, as in SIA 198.

13.2 Classification Systems

Rutschmann [131] determined in 1974 the two significant classifications of rock and cuttability, which determine the performance of TBM operation. He regarded the following four factors as important in determining the cuttability:

- Stress-strain relationship: E modulus at 50% compressive strength E_{t50}
- Uniaxial cylinder compressive strength β_D
- Splitting tension strength β_Z
- Hardness expressions, for example saw hardness SH

The rock classification according to Rutschmann essentially represents clamping ability of the TBM. Both classifications are attempts at a qualitative representation.

Rauscher developed a classification system based on models of the cuttability of the rock. This cuttability is represented as analytic relation between the independent variable F_v (thrust force) and the dependent variable N_B (cutter head drive power) and v_n (net boring speed). Using this relation, he constructed a nomogram, in which a classification of different rock types can be displayed (Fig. 13-5). The different cutting coefficients f_{sp} are entered as a group of hyperbolas and the power relationship κ (the quotient of N_B and the translations power $N_v = F_v \cdot v_n$) as a function of v_n and p (penetration) on one hand and as a variable over F_v on the other. κ can therefore be integrated as the determining value for the classification.

At this point, reference should also be made to the work of Beckmann, who researched influential variables and their effects on tunnelling advances of TBM [11] based on experience and data from over 40 km of tunnelling. Beckmann shows that rock has an influence on the advance performance of tunnel boring machines through both its cuttability as well as its stability. The cuttability, determined by in-situ tests, is divided, just

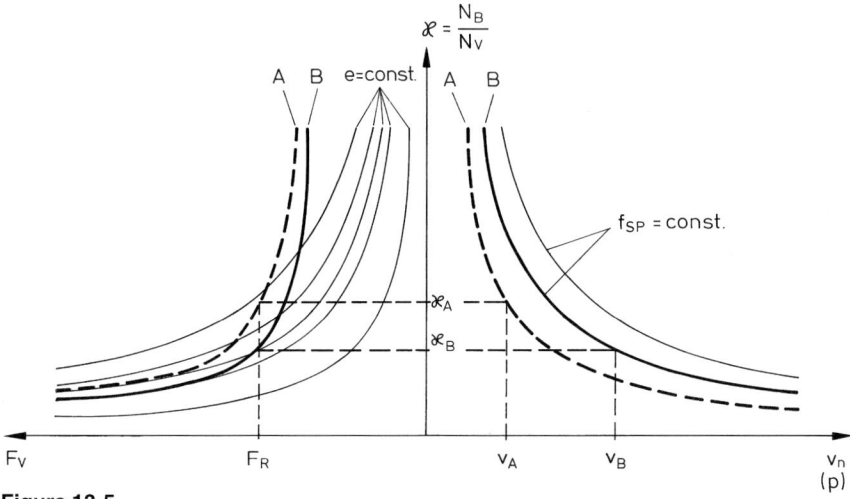

Figure 13-5
Nomogram for the evaluation of different rock types using characteristic curves specific to the rock, according to Rauscher [118]

like the stability, into 6 classes. Both parameters can now be referred to as an additional criterion for a performance payment, with cuttability and stability classes being combined in a price matrix, which can be matched with a performance matrix from the contractor.

Schmid worked out the cuttability according to geological conditions (Fig. 13-6) in 1988 for the Schweizerischer Baumeisterverband *(Swiss master builder's federation)*.

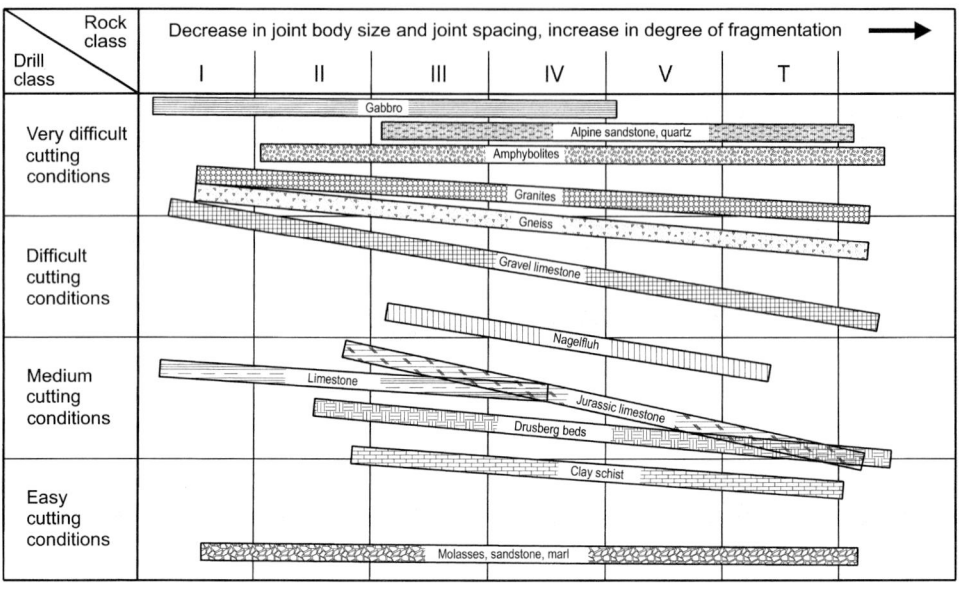

Figure 13-6
Cuttability and geological conditions [143]

The abrasiveness of the rock directly affects the costs of tunnelling. This effect, which can only be determined within limits, arises as a result of the increased cutter changing time and occasional interruptions of tunnelling to exchange wearing parts such as scrapers or buckets. To supplement the cuttability with a wear rate class would make sense for the payments based on wear rates, but hardly for the classification of rock as a basis for advance performance.

13.2.3 Classification According to Type, Extent and Location of the Support Work Required

The experienced tunnel builder is normally capable, in intense collaboration with the geologist, of making a classification of the type, extent and location of support measures. He compares the results of investigations with facts from the completed construction of comparable projects or the knowledge of possible variations from similar types of construction. It is doubtful whether better results can be obtained like this than from the classification of rock characteristics. A better way would be to use his experience in the consideration of the results of the classification according to rock characteristics.

The determination of the necessary extent of support measures, including the obligatory safety margin for the safety of personnel, is clearly only possible on site after the start of the excavation.

For this reason, the national standards developed in Germany, Switzerland and Austria were all produced based on the installed rock support measures.

However, all of them are based on the classification for drilling and blasting, without taking into consideration that in TBM excavation, some support measures cannot be used at any location. Shotcrete sprayed in the machine area is still included in all as a classification parameter.

Without exception, these classifications require no support measures in the lowest classes, corresponding to competent rock. The insurance organisations SUVA, BG Bau, AUVA, however, demand head protection above a cross-section height of 3 m. If, according to the definition, the support installed for the excavation supplies the basis for the classification, then it is unimportant for what reason it is required. It determines the classification in any case.

13.3 Standards, Guidelines and Recommendations for the Classification of Mechanised Tunnelling

13.3.1 Classification in Germany

DIN 18312 „Allgemeine technische Vertragsbedingungen für Bauleistungen – Untertagebauarbeiten", Ausgabe 1998-05 *(German standard 18312, General technical conditions of contract for construction work – underground construction, edition 1998-05)*

The tunnelling classification for tunnel construction projects in Germany is regulated in DIN 18312: 1998-05. The contract classification according to this standard is based on the assumption that the shape and size of the cavity cross section as well as the construction process and measures for excavation and support are given and that the respective categorisation of soil and rock into the individual classes is only undertaken under observation of this default data. In contrast to the term excavation class according to DIN 18312, edition 1992-12, the description tunnelling class has been chosen. The classification according to DIN 18312 has little influence.

The tunnelling classification, which is kept very general, is divided, according to the extent of support work and resulting delay or obstruction of tunnelling, into the classes 1 though 7A in conventional tunnelling, classes SM 1 through SM 3 for shield tunnelling and the classes TBM 1 through TBM 5 for tunnel boring machine excavation (Table 13-13). In TBM excavations, classes are divided according to:

- Support required
- Obstruction of mechanical excavation
- Support installation required in the area of the machine
- Special measures

Table 13-13
Tunnelling classes for tunnel boring machines [38]

Tunnelling class	Type of excavation
TBM 1	Excavation without support
TBM 2	Excavation with support, the installation of which does not hinder the excavation.
TBM 3	Excavation with support immediately behind the machine or already in the machine area, the installation of which does not hinder the excavation.
TBM 4	Excavation with support in the machine area immediately behind the cutter head, the installation of which interrupts the excavation.
TBM 5	Excavation with special measures, the installation of which interrupts the excavation.

While no rock support is necessary in the favourable tunnelling class TBM 1, the excavation is hindered or even interrupted by support measures with increasing instability of the rock and increasing tunnelling class. Tunnelling class TBM 5 requires measures of a special kind, for example for the gripper system of the machine, removal of overbreak in the machine area, for probing of the ground ahead and/or for ground consolidation from the machine.

The classes consider the location of the installation of support, distinguishing between machine area, backup area and rearward area. This classification is only a general division. Sub-classes or detailed information are not offered by this standard. Only in Section 3, "Construction" concerning excavation and support work is it pointed out that all measures to be taken are the responsibility of the client, unless specific contractual agreements regarding type and extent have been agreed.

For shield machines, the classes SM 1 through SM 3 are defined (Table 13-14). A demarcation of the two systems, tunnel boring machine and/or shield machines does not take place.

13.3 Standards, Guidelines and Recommendations

Table 13-14
Tunnelling classes for shield machines [38]

Tunnelling class	Type of excavation
SM 1	Excavation without support of the face
SM 2	Excavation with partial support of the face
SM 3	Excavation with full support of the face

Richtlinie 853 „Eisenbahntunnel planen, bauen und instandhalten" der Deutschen Bahn AG, Ausgabe 10/98 *(Guideline 853 "Design, construction and maintenance of railway tunnels" from German Railways, edition 10/1998)*

The guideline 853 from Deutsche Bahn AG [29], which is to be observed in the design, construction and maintenance of railway tunnels, prescribes that an tunnelling classification according to DIN 18312 is to be included in the tender documents. Existing experience of similar geological conditions is to be taken into account in the categorisation. The safety and support measures required for the relevant tunnelling class are to be given. Further, alternative tunnelling classes are to be tendered for each tunnel section.

The tunnelling class and tunnelling method to be used in each case are determined with the construction based on the suggestion of the contractor in agreement with the client. The agreement of tunnelling classes is done according to the conditions in the "freshly" excavated tunnel. To regulate any disagreements, which arise and cannot be settled, concerning the determination of tunnelling class, a recognised independent expert should be nominated at contract award to act as arbitrator between the client and the contractor [29].

Empfehlungen des Arbeitskreises „Tunnelbau" ETB der Deutschen Gesellschaft für Geotechnik e.V., Ausgabe 1994 *(Recommendations of the working group "Tunnel construction" ETB of the German society for geotechnology, edition 1994)*

The working group "Tunnel Construction" recommends the general classification system according to DIN 18312, which should serve as basis for project related classification of a tunnel in a specific case. Apart from the known classes TBM 1 through TBM 5 and SM1 through SM 3 known from DIN 18312, the additional classes SM-V1 through SM-V5 are defined for shield machines with full face excavation (Table 13-15).

Figure 13-7 shows an example of a classification following the recommendations of the DGGT. This classification is also in accordance with DIN 18312. In this example, support is applied without hindering the excavation

Figure 13-8 shows an example of a classification for shield machines according to the recommendations of the DGGT. The example shows a tunnelling situation with measures of a special kind (drilling) for ground investigations from the machine. Support is a closed reinforced concrete segment ring, which is installed in the protection of the shield. Division into sub-classes is intended to introduce a refinement based on the

Table 13-15
Tunnelling classes for tunnel boring machines and shield machines [30]

Tunnelling method	Tunnelling class	Characteristics
Tunnel boring machines (TBM)		For the categorisation of tunnelling classes for tunnel boring machines, type and extent of the rock support and the location (in the cross section and in tunnel longitudinal direction) as well as installation sequence and resulting obstruction of full-face excavation is decisive.
	TBM 1	Excavation requiring no support.
	TBM 2	Excavation requiring support, which does not hinder the excavation.
	TBM 3	Excavation requiring support immediately behind the machine or directly in the area of the machine, the installation of which hinders the excavation.
	TBM 4	Excavation requiring support in the area of the machine immediately behind the cutter head, for the installation of which the excavation must be interrupted.
	TBM 5	Excavation requiring measures of a special nature, for the installation of which the excavation has to be interrupted (e.g. measures to clamp the machine, to clear falls in the area of the machine, to investigate the subsoil and/or to consolidate in front of the machine).
Shield machines excavating the full face (SM-V)		For the categorisation of tunnelling classes for shield machines, the type of face support, and whether the excavation is hindered or not hindered, are decisive. Installation of the temporary or final support as a closed ring within the protection of the shield. Reorganisation of the technological process is generally not possible.
	SM-V1	Excavation without support of the working face, which does not hinder the excavation.
	SM-V2	Excavation with partial or full mechanical support of the face, which does not hinder the excavation.
	SM-V3	Excavation with the face fully supported by compressed air, which does not hinder the excavation.
	SM-V4	Excavation with the face fully supported by slurry, which does not hinder the excavation.
	SM-V5	Excavation with the working face fully supported by earth slurry, which does not hinder the excavation. Excavation, which requires measures of a special type with the effect of hindering the excavation, is to be divided into sub-classes in the project-related classification (e.g. measures for clamping in SM-V1 → SM-V1.1 or e.g. measures for subsoil investigation from SM-V1 → SM-V1.2).

13.3 Standards, Guidelines and Recommendations

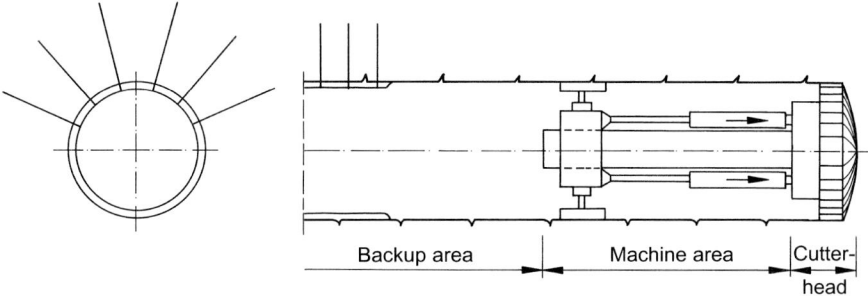

Excavation		Tunnel boring machine
Support, installation without interruption of excavation	Shotcrete	B 25, d = 10 cm
	Reinforcement	Q 257, 1-layer
	Anchors	6-7 SN-anchors, l=4 m, e=2 m
	Support segments	----

Figure 13-7
Example for tunnelling class TBM 2 [30]

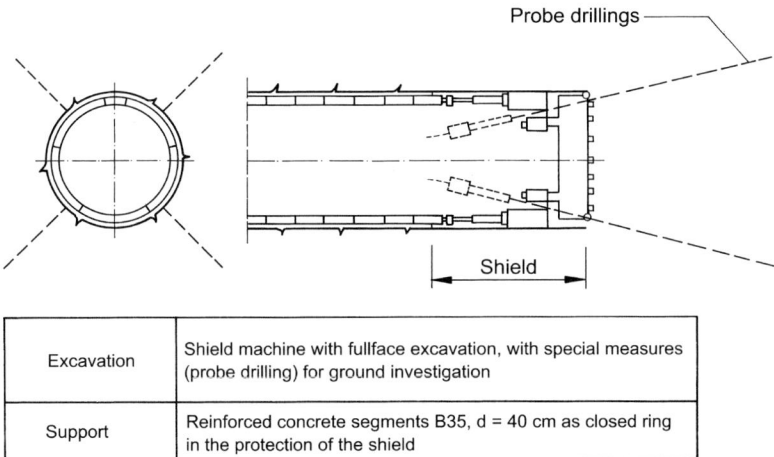

Excavation	Shield machine with fullface excavation, with special measures (probe drilling) for ground investigation
Support	Reinforced concrete segments B35, d = 40 cm as closed ring in the protection of the shield

Figure 13-8
Example for tunnelling class SM-V1.2 [30]

general classification of tunnelling classes, for example by considering worsening difficulties. This should consider project-related particulars.

Special characteristics according to this recommendation are:

- Unusual geological and hydrogeological conditions
- Change of the tunnelling class (which should be taken into account not by the tunnelling class itself, but by an appropriate item in the bill of quantities)
- Effects on the groundwater and, if necessary, appropriate measures

- Indication of tunnelling advance rates for each tunnelling class by the bidder
- Tunnelling under built-up or industrial areas sensitive to settlement
- Quantitative registration of type and extent of support and volume in ranges (shotcrete thickness, number of anchors and lengths, segment types and spacing, type of reinforcement, single and double layer; a certain overlapping of ranges of neighbouring tunnelling classes is appropriate in the opinion of the DGGT)
- Location and installation sequence of the support in the machine area, backup area and machine rear
- Cuttability of the rock (mineral content – in particular quartz content), mineral formation, grain size, compressive strength, tensile strength, hardness, abrasiveness, structure of joints
- Classification by statement of net boring speeds
- Consideration of the particularities of shafts
- Measures to maintain the face support, interventions in front of the cutter head
- Type of annular gap grouting regarding the obstruction of the advance
- Anticipatory measures, like for example the investigation of the subsoil in front of the tunnel boring machine and/or rock consolidation from the machine (injection or freezing)

The classification should be created by the client as part of the design work. The basis for this are:

- The subsoil as recorded in geotechnical investigations and its evaluation for tunnel construction
- The shape and size of the cavity and
- The method of tunnelling, which is determined by the type of excavation and support

After the division into project-related tunnelling classes, a forecast should be made of the proportions of the different tunnelling classes along the tunnel. Clearly different rock conditions should be delineated. The tunnelling classes are determined on the site on the basis of the approved design documents in form of decisions about support agreed between the client and the contractor before the relevant excavation. In case of differences of opinion, the client specifies the tunnelling class. The as-built tunnelling classes are to be documented in an appropriate drawing and compared with the tunnelling class prognosis [30].

13.3.2 Classification in Austria

ÖNORM B 2203 „Untertagebauarbeiten", Ausgabe 1994–10 *(Austrian standard "Underground construction", edition 10/1994)*

In Austria, the classification for the construction of underground structures is currently regulated by the ÖNORM B 2203 "Underground Construction" [111]. As in Germany, the rock conditions during excavation and associated support and safety measures are held in Austria to be the basis for the classification of tunnelling operations and/or the classification of rock into classes.

As part of the revision of the ÖNORM B 2203, 1983 edition, a classification system applicable to continuous tunnelling (TBM) was introduced. The Austrian tunnelling classi-

13.3 Standards, Guidelines and Recommendations

fication is based on the obstruction of the boring work by the installation of support, with the obstruction being quantified by the location of the installation or the latest installation time. This working standard defines three types of rock (A = stable to loose rock, B = fractured rock and C = squeezing rock), which describe the behaviour of the rock for cyclic or continuous tunnelling using an open tunnel boring machine. The 10 sub-types (A1, A2, B1 to B3 and C1 to C5) are again divided into three sub-groups. These types are determined according to the geomechanical description and decided using the Table 1 of the ÖNORM B 2203 (Table 13-16). The table mentioned also gives guidelines, which requirements and measures are normally associated with the individual rock types.

For the forecast rock types, the support and safety measures, which will probably be required, are specified within defined homogeneous ranges and evaluated using the table also indicated in the standard and converted into a characteristic support number (Table 13-17). Knowing the quantity of support measures and the location of installation, an evaluation number can be calculated and a tunnelling classification can be produced.

The ÖNORM B 2203 provides a diagram of the tunnelling classes in a matrix according to Figure 13-9. If necessary, this can be supplemented by project-related data.

A tunnelling class is characterized by two reference numbers. The time difference from starting excavation to the latest start time for support installation serves as criterion for the first reference number. This depends on the stand-up time of the rock type (encountered (Table 13-16). For regulated excavation with a gripper TBM according to this

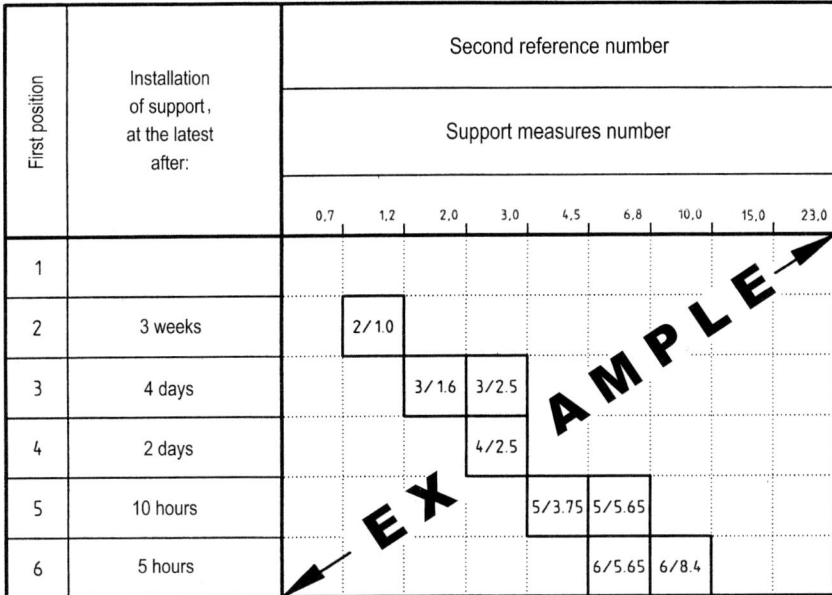

Figure 13-9
Tunnelling classes matrix for continuous tunnelling with a TBM [111]

Table 13-16
Rock types for cyclical and continuous advance [111]

Rock type	Rock behavior	Requirements for excavation and support measures for continuous advance
A Stable rock to rock liable to rockfall. This rock type includes all rock, which can bear loading without signs of rupture.	**A1 Stable** Very quickly subsiding, low deformations, no chips falling after the removal of loose rock pieces.	Support is not necessary. Stand-up time: more than three weeks.
	A2 Liable to rockfall Very quickly subsiding, low deformations, isolated loosening of rock pieces from the crown and upper sides due to jointing.	Support is only necessary in the crown, abutment and upper area of tunnel sides to secure isolated blocks. Installation of support in working area 2 without interruption of the machine advance. Stand-up time: 3 weeks to 4 days.
B Fractured rock. This rock type covers all rock, which due to lack of bonding strength in the joints and/or lack of lack of interlocking, tends to loosen.	**B1 Fractured** Very quickly subsiding, low deformations; low rock strength due to joints and blasting vibration cause loosening and de-bonding of the rock bonding in the crown and upper parts of the tunnel sides.	**B1.1** Systematic installation of support to a low extent, primarily in the crown, abutment and side areas in the work-free area 2, without interruption of the machine advance. Stand-up time: 2 to 4 days. **B1.2** Systematic installation of support in the crown, abutment and side areas in the working areas 1 and 2. The machine advance is affected by the installation of support. Stand-up time: 2 days to 10 hours.
	B2 Strongly fractured Deformations subside quickly; low rock strength determined by the joints, low interlocking, high movement of pieces and blasting leads to quick, deep loosening and collapsing of rock from free unsupported surfaces.	**B2.1** Systematic installation of support measures, beginning immediately behind the cutter head, the duration of support installation generally determines the advance rate. Cutting proceeds only in partial strokes. Stand-up time: 10 to 5 hours. **B2.2** Systematic preliminary support in cutter head area and systematic support installation to the entire working area in working area 1. Stand-up time: without preliminary support, 5 to 2 hours.

13.3 Standards, Guidelines and Recommendations

Table 13-16 (continued)

Rock type	Rock behavior	Requirements for excavation and support measures for continuous advance
	B3 Unstable On opening of only small partial sections, the rock trickles out. Absence of cohesion and missing toothing are the causes of insufficient stability.	Continuous advance with open tunnel boring machines is only possible with special measures. Stand-up time: less than 2 hours.
C Squeezing rock. This rock type includes all rocks, where the rock strength is deeply exceeded. Rock tending to collapse and rock with prominent swelling behaviour are included in this rock type.	**C1** Rock burst Elastic energy is stored in mostly heavy, hard and brittle rocks with high primary stress. Sudden conversion of this energy produces bursts of collapses with pieces of rock falling out. These pieces of rock thrown out of exposed surfaces are mostly chip shaped, the collapse only extends to shallow depths.	Installation of short, narrowly spaced anchors, if required with reinforcing mesh, in working area 1; the machine advance is not essentially hindered.
	C2 squeezing Prominent, long duration and slowly subsiding deformations. Development of ruptures or plastic zones in plastic, strongly cohesive soil.	**C2.1** systematic installation of support measures in crown, abutment and side areas. Stepwise installation of support measures in working areas 1 and 2; the machine advance is hindered by the installation of support; precautions have to be taken to prevent the jamming of the tunnel boring machine. Stand-up time: 2 days to 10 hours **C2.2** systematic installation of support, beginning immediately behind the cutter head; the duration of the installation of support generally determines the advance rate. Cutting is only possible in partial strokes; precautions have to be taken to prevent the jamming of the tunnel boring machine. Stand-up time: 10 to 5 hours.

Table 13-16 (continued)

Rock type	Rock behavior	Requirements for excavation and support measures for continuous advance
	C3 Strongly squeezing Large, long duration and slowly subsiding deformations with high initial deformation speed. Development of deep ruptures or plastic zones.	**C3.1** systematic installation of support in crown, abutment and side areas; stepwise installation of support measures in working areas 1 and 2. The machine advance is hindered by the installation of support; precautions have to be taken to prevent the jamming of the tunnel boring machine. Stand-up time: 10 days to 2 hours. **C3.2** systematic installation of support measures, starting immediately behind the cutter head; the duration of support installation generally determines the advance rate. The cutting can only proceed in part strokes; precautions have to be taken to prevent the jamming of the tunnel boring machine. Stand-up time: 10 to 5 hours.
	C4 Flowing Very low cohesion and friction, weak plastic consistency of the soil leads to the soil flowing in, even with very small, only temporarily exposed and unsupported surfaces.	Advance with open tunnel boring machine is only possible with special measures. Stand-up time: shorter than 2 hours.
	C5 Swelling Soil types with mineral components, which depending on the relaxation through take up of water, experience an increase of volume, e.g. swellable clay minerals, salts, anhydrites.	Support measures effective in the long term accepting the swelling pressure or precautions enable the occurrence of swelling deformations without damage. Advance with an open tunnel boring machine is only possible with special measures. Stand-up time: no indication.

13.3 Standards, Guidelines and Recommendations

Table 13-17
Evaluation of support measures for continuous tunnelling with tunnel boring machines [111]

Support measure		Evaluation factors working areas according to Fig. 14-9		Unit	Remark
		1	2		
Anchors	Swellex and expanding anchor	3.0	2.0	m	
	SN-mortar anchor	4.0	3.0	m	
	Self-drilling anchor	5.0	3.5	m	
	Grout anchor	6.0	4.0	m	
	Prestressed mortar anchor	10.0	6.0	m	
Mesh	First layer	1.5	1.5	m^2	
	Second layer	3.0	1.5	m^2	
Arch segments		3.0	2.0	m	
Closed ring		4.0	3.0	m	
Shotcrete		50.0	15.0	m^3	Theoretical mass without extra thickness and without splashing
Rockbolts	Unmortared rockbolt	5.0		m	
	Mortared rockbolt	7.0		m	
	Self-drill rockbolt	7.0		m	
	Grouted rockbolt	9.0		m	
	Injection rockbolt	12.0		m	
Plates	Flexible plates	10.0		m^2	Installed plates
	Mechanised plates	15.0		m^2	

standard, two hours stand-up time is given as a minimum. If the actual stand-up time is less, a regulation for the compensation of additional costs is provided.

The second reference number is the support measures number. It is derived from the extent and the type of support measures decided upon, which are to be evaluated according to the details continued in the table in the standard, in each case related to the designed cross section of excavation and divided into two working areas (Fig. 13-10).

The intersections of the first and second reference numbers result in matrix fields, in which the contractor has to insert guaranteed unit prices per m^3 excavation and guaranteed tunnelling advance per working day. The horizontal neighbouring fields of the defined matrix fields result in a horizontal and vertical range of tunnelling payments within the fixed limits from the tunnelling classification. If areas are encountered unexpectedly while tunnelling, which have not been considered in the matrix, then new fields in the matrix must be derived linearly from the existing fields. This brings a certain cost security for the client and decreases the potential for disagreements over payments.

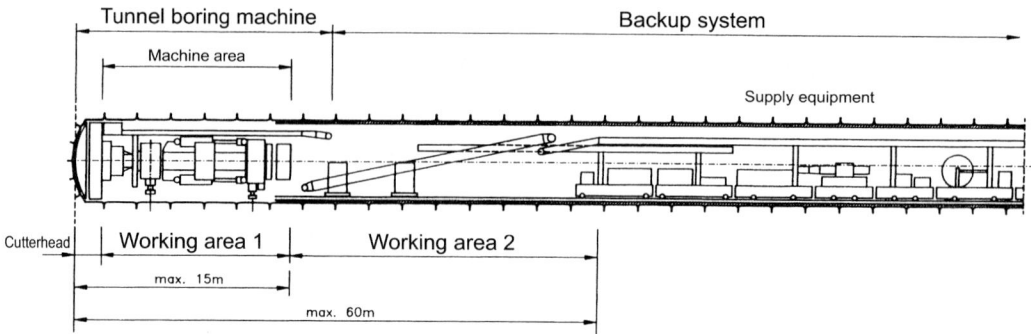

Figure 13-10
Description of continuous tunnelling with a TBM [111]

Working area 1 is defined as the area, which extends from the cutter head to maximum 15 m from the tunnel face. Working area 2 connects to working area 1 and ends at a maximum 60 m from the tunnel face. All support measures required for the geology encountered must be installed by the end of working area 2. Support measures installed in working area 1 are to be evaluated more highly according to the amount of time they obstruct the tunnelling work.

No separate working area is defined for the area in front of the machine. If the installation of support is intended in front of the cutter head, then this requires its own item in the bill of quantities. This is to take into account special safety-related conditions.

Ayaydin summarises the procedure for classifying on site according to Figure 13-11. This procedure is intended to classify only on the basis of overcoming the encountered geological conditions and not on the basis of economic priorities, which are different from the point of view of the client and the contractor [6].

Example: Figure 13-12 shows a tunnelling class according to ÖNORM. The tunnelling class "3/5.4", for example, means that support has to be installed at the latest 10 hours after excavation (1st reference number). The second reference number shows, for example for a tunnel of 5 m diameter, the installation of 5 Swellex anchors with a length of 2 m each and 10 cm of shotcrete in the machine area 1. The support measures number is obtained by dividing the sum of evaluated support measures by the excavated cross section per m^2 tunnel.

13.3.3 Classification in Switzerland

SIA 198 „Untertagbau", Ausgabe 1993, Nachdruck 03/1994 *(SIA 198 "Underground construction", edition 1993, reprint 03/1994)*

The standard SIA 198 Underground Construction was introduced in Switzerland in 1993 and counts as a part of the tender documents and construction contracts. It applies to the tendering and construction of underground construction projects such as tunnels, headings, caverns and shafts, which are constructed using tunnelling methods and in-

13.3 Standards, Guidelines and Recommendations

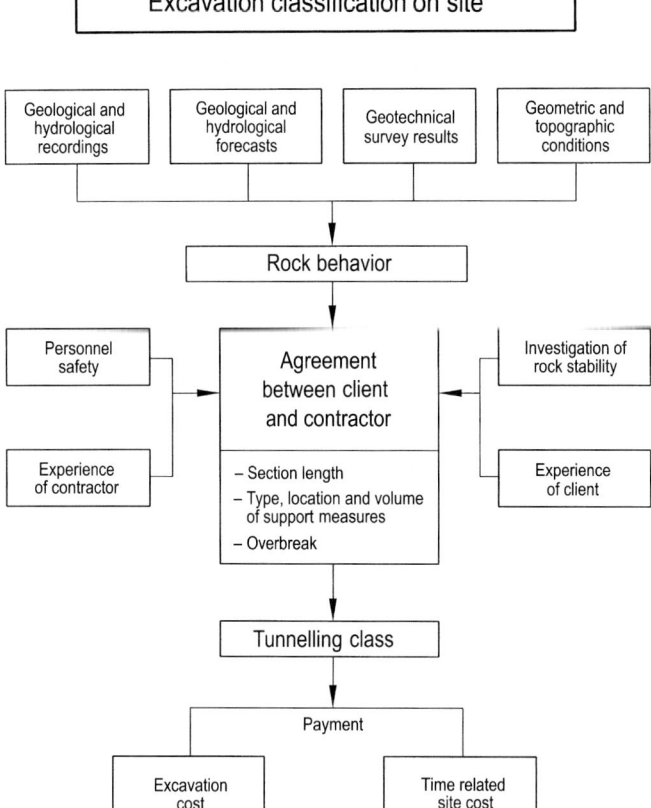

Figure 13-11
Procedure for classification on site [6]

cludes drilling and blasting methods, excavation with full face machines and roadheaders as well as excavations in soft ground.

Section 5.4 of the standard, which deals with the payment for excavation works by tunnelling with a TBM in hard rock, is based on the premise that the costs are connected directly with the type and extent of the support measures required and also with the point in time when these measures have to be effective.

Tunnelling classes in Switzerland are defined according to SIA 198 (1993) on the basis of the expense of the advance. A distinction is made between the excavation class A (full face excavation) and E (excavation in phases with a pilot heading or pilot shaft and subsequent enlargement; both with a TBM advance). The provisions outlined below apply to both these excavation classes.

The tunnelling classes result from a combination of excavation and cuttability classes. The excavation classes are based, as in Germany or Austria, on the required support

First reference number	Installation of support measures at the latest after:	Second reference number				
		Support required				
		Low support expenditure ⟶ High support expenditure				
		0.7	....	....	5.4	....	....	23.0
1	3 weeks							
2	....							
3	10 hours				3/5.4			
4	....							
5	2 hours							

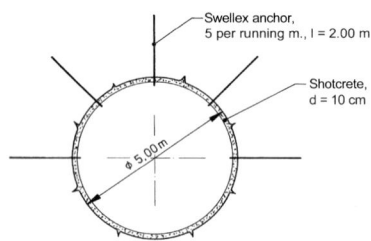

Swellex anchor, 5 per running m., l = 2.00 m
Shotcrete, d = 10 cm
⌀ 5.00 m

Support measures number for machine area 1					
Support measure	per metre tunnel	Length and/or thickness	Amount	Factor area 1 [111]	Total
Swellex anchor	5 St.	2.0 m	10	3	30.00
Shotcrete	15.4 m²	0.1 m	1.54 m³	50	76.97
Sum					106.97
Cross-sectional area: $5.00^2 \cdot \pi/4 = 19.63$ m²					
Support measures number: $106.97 \div 19.63 = 5.4$					

Figure 13-12
Example of the determination of tunnelling class according to ÖNORM
Above: Example of a tunnelling class for TBM advance
Middle: Cross-section with support
Below: Determination of the support measures number

measures and the location of the installation along the TBM. In general, five excavation classes are defined, from class 1 for rock that does not require any support means to class 5 that requires extensive rock support in the TBM area. However, if the immediately and continuously installed rock support consisting of a closed segmental lining, then the subdivision into excavation classes is invalid. Excavation class T is intended in this case.

The consideration of the location where support is installed is more complex for a TBM excavation than for drilling and blasting. The working areas for tunnelling with a TBM are defined in Switzerland as follows:

13.3 Standards, Guidelines and Recommendations

- L1 Machine area
- L2 Backup area
- L3 Rearward area up to 200 m behind the backup

Within the areas L1, L2 and L3, the working zones L1*, L2* and L3* are identified, in which the support is installed because of the requirements of the project and the characteristics of the type of machine (Fig. 13-13).

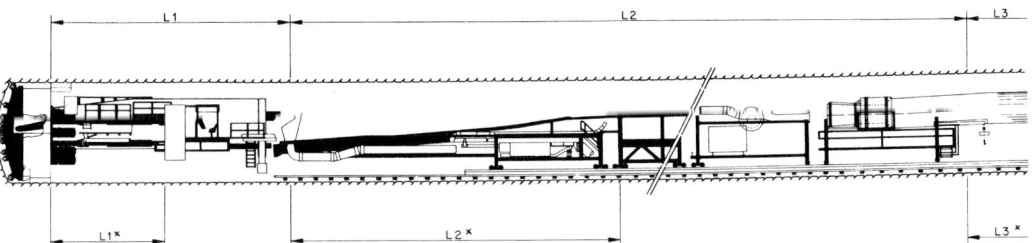

Figure 13-13
Working areas and working zones with a TBM [SIA]

It is the task of the contractor to include with his bid a drawing showing the length of the working areas L1, L2 and L3 and the working zones L1*, L2* and L3*, and to state which support works are possible within these during the tendered work. He must observe the conditions defined in the tender documents regarding the location of the installation of support.

The link between type, extent and location of installation of the support defines the respective excavation class. Support outside the designated working area thus have no influence on the classification of the excavation.

The basis idea behind the definition of excavation classes should be in each case the extent of the obstruction of the tunnelling works caused by support work (Table 13-18). The exact allocation is done using Table 13-21.

The excavation classes apply generally both for drilling and blasting in hard rock (see Chapter 5.2) and also to tunnelling with roadheaders in hard rock (see Chapter 5.3) and also for TBM tunnelling in hard rock (see Chapter 5.4). However, SIA 198 offers differing evaluations.

For the application of the excavation classes, tunnels and inclined shafts are treated equally. For vertical shafts, which are enlarged by a cutter head from the bottom up, there is no subdivision into excavation classes.

The cuttability classes correspond to the effort of excavating the rock with a TBM and consider both the penetration and the wear of tools as determining factors. The client specifies cuttability classes according to characteristic values of rock and geology. In this way, the tunnel route can be sorted into stretches of the same class related to the cuttability comparable geological formations. The number of cuttability classes is de-

Table 13-18
Excavation classes according to Swiss Standard Norm SIA [154]

Excavation class	Obstruction of excavation
AK I	The support causes an insignificant obstruction of the excavation cycle.
AK II	The support causes a light obstruction of the excavation cycle.
AK III	The support causes a considerable obstruction of the excavation cycle.
AK IV	The support causes an interruption of the excavation cycle (immediate support after each excavation stroke).
AK V	The support is installed continuously with the excavation and requires an immediate support of the face or an advance support.
AK T	The support consists of s closed segment ring, which is installed immediately and without delay.

termined by existing conditions and is to be specified for the project in the tender documents. The determining rock characteristic values are to be presented with their scattered values as exactly as possible.

The cuttability classes x, y, z are to be understood as standing for rock with varying cutting characteristics in the relevant tunnel; x could stand for a gneiss defined in the geological longitudinal profile, y for a granite. In another project, x can stand for an eroded limestone, y marl of the Drusberg beds, z a siliceous chalk.

The standard SIA 198 permits also as an alternative a classification of the cuttability classes based on in-situ driven penetration rates. The determination of the net boring rate is done contradictorily on a daily basis along a tunnel route with a test procedure fixed in the contract (test stroke length, thrust force, condition of the cutting tools and their degree of wear). This procedure is only considered suitable in special cases. Too many factors determine the penetration and it is also questionable, which test section is representative.

The division of the excavation into individual classes takes place through the linking of the appropriate tunnelling and cuttability classes. The combination of tunnelling and cuttability classes is accomplished in a matrix and an example is highlighted in Table 13-19. Each entry in the matrix corresponds to a boring advance and support work obstructing the boring progress and is to be provided with a unit price per m tunnelled for the remuneration in this tunnelling class.

With this matrix, it is quite possible that the tunnelling advance is determined solely by the cuttability class. For rock which is difficult to bore it is possible to install support of classes II, III or IV at the same time as the boring of the stroke, as long as it is not shotcrete.

Table 13-19
Matrix of excavation and cuttability classes, system according to standard SIA 198

Excavation classes	Cuttability Classes		
	X	Y	Z
I	I X	I Y	I Z
II	II X	II Y	II Z
III	III X	III Y	III Z
IV	IV X	IV Y	IV Z
V	V X	V Y	V Z
T	T X	T Y	T Z

13.4 Classification Suggestion by the Authors

The classification systems discussed above are based to a substantial extent on support methods, which are not always compatible with mechanised excavation, or they omit the question of the cuttability and clamping process, although these remain important for the tunnelling advance.

These weaknesses lead again and again to major differences of opinion between the contracting parties; it may be because the construction schedule cannot be met or because claims are pending due to changed conditions.

It must be the aim of the classification to create a means of communication, which permits treatment as fair as possible for both contracting parties without, however, believing that every case of changed conditions can be integrated into the classification system.

For tunnelling performance, with the exception of extreme cases, only rock support and the penetration of the TBM to be used are determining factors. An excavation classification has therefore to include both main parts. The wear affects mainly the excavation price.

For TBM tunnelling, the classification must if possible take the following parts into account:

- Support systems as excavation class with the practical possibilities
- Cuttability with geological facts and/or similar, which enable the limitation of the penetration
- Wear with the basics of abrasiveness during excavation, e.g. Cerchar (see Chapter 3); demand: if a test according to Cerchar is not possible, details of the proportion of minerals, which cause sharp wear

Table 13-20
Determination of the excavation classes based on the support for excavation with tunnel boring machines in rock[1]

	Class I	Class II	Class III	Class IV	Class V	Class T
Excavation classes for tunnels						
Machine area L1	Mesh as rock fall safety measure fixed with anchors and bolts	≤ n anchors in the cross-section circumference	> n anchors in the cross-section circumference with mesh and/or isolated flexible metal plates	Anchors with mesh and partial arches Shotcrete as seal on >1/4 of the cross-section circumference Pressure-distributing elements in the area of the grippers	Complete steel ring installation with or without flexibility	Closed segment installation
Backup area L2	Invert segments, if these are provided for all rock classes in the project	**Systematic arrangement** >2; n anchors and partial arch over >1/4 of the cross-section circumference Mesh and shotcrete over <1/2 of the section circumference **Non-systematic arrangement** > n anchors	**Systematic arrangement** Mesh, shotcrete around the entire extent of the section except invert Partial arches with anchors over 3/4 of the circumference of the section **Non-systematic arrangement** Mesh and shotcrete over/ 2 of the section circumference > n anchors with partial arches over < 1/4 of the circumference of the section	**Systematic arrangement** Invert support with shotcrete, if no invert segments are intended Closed steel ring installation, perhaps mounted on invert segments **Non systematic arrangement** Mesh, shotcrete around the entire extent of the section except invert Partial arches with anchors over 3/4 of the circumference of the section Closed steel ring installation with or without flexibility		
Rearward area L3 up to 200 m behind the backup	The support in L3 is unimportant for the excavation classification. In case they are intended, standstill positions are to be provided.					
Values for n (number of anchors per running m tunnel)	max excavation diameter: n	4.00 m 2	6.00 m 3	9.00 m 4	12.00 m 6	

[1] If more than one support measure is in one field of the table, then each measure alone suffices for the classification.

13.4 Classification Suggestion by the Authors

Suggestion for Excavation Classes

The suggested classification of excavation is based on support systems, which can be installed quickly. This enables the interruption of boring to be minimised and increases overall productivity. Table 13-21 shows the suggested excavation classification with guideline values for the support system. The profile types are to be defined for the case in point.

Table 13-21
Excavation classes as suggested by the authors

Gripper TBM with systematic sheeting					Shield TBM
Support system	Very easy (in large section – head protection)	Light	Medium	Hard	Segment
With invert segment					
In machine area	Mesh as rockfall security	Light arch 1–1.5 m spacing with light flexibility, e.g. mesh as a unit	Medium arch 0.8–1.2 m spacing and flexibility as unit	Heavy arch 0.75–1.0 m spacing with heavy flexibility as unit	
In backup area	Shotcrete 30–50 m behind cutter head				
Without invert segment					
In the machine area	Similar, but arch continoues through invert				
In backup area	Shotcrete, also in the invert				
Profile types	Support elements are defined as profile types for tendering				
	Profile type 1	Profile type 2	Profile type 3	Profile type 4	T
Excavation class	I	II	III	IV	T

Suggestion for Cuttability Classes

Tunnel sections with similar cuttability are to be categorised into cuttability classes. This can be in geological part sections (see Fig. 13-5) or, in exceptional cases, also by deciding the penetrations in the construction contract with verification on site using test strokes with a defined condition of the disc cutters and the thrust force.

The procedure with test strokes has, however, grave disadvantages. If the contractor deploys a TBM, which is very powerful considering the conditions, then he is penalised because the payment comes out lower through the higher performance. If he deploys a

TBM with relatively weak performance, then the client is correspondingly disadvantaged.

The categorisation of the geological conditions according to Section 13.2.2 with the determination of the geotechnical properties compressive strength, tension strength for the abrasiveness (Cerchar), produces a good contractual basis.

The contractor tenders the price matrix according to Table 13-19.

Corrections of penetration during construction can be very well regulated in relation to the forecast and measures compression strength according to Gehring. The correction factor leads to a new penetration corresponding to the curve in Fig. 3-18 (see Chapter 3).

$$K_p = \frac{1}{\left(\dfrac{\sigma_{\text{pressure} \cdot \text{measurement}}}{\sigma_{\text{pressure} \cdot \text{prognosis}}}\right)^\lambda} \tag{13-12}$$

K_p = correction factor for penetration
$\sigma_{\text{pressure} \cdot \text{measurement}}$ = measured uniaxial compressive strength
$\sigma_{\text{pressure} \cdot \text{prognosis}}$ = forecast uniaxial compressive strength
λ = exponent 1,0–1,2 (frequently 1.1)

The tension strength would certainly be a decisive factor for the boring process, but the attempt to include this in the evaluation fails again and again. The main reason for this is that, with increasing compressive strength, the tension strength also increases, if not always in the same relationship. As a result of this, the basic value of compressive strength is deemed to incorporate the tension strength as well.

The ratio of compressive strength to tension strength seems to be more essential. This frequently varies between values from 12 to 15 in a range of extremes from 8 to 22. It is therefore important for the construction contract to state the forecast range realistically.

Wear

The increase of wear or also the reduction of wear can, because of the large differences in shape and quality of cutting tools, only be sensibly determined by the formation of a direct quotient of the actual to the forecast values. The increased or decreased cost results from the multiplication of the quotient by the wear costs from the estimation.

14 Tendering, Award, Contract

In tunnel construction projects, which permit mechanised tunnelling, great attention should be paid to matters outside the usual factors specific to tunnelling – these are dealt with in the Handbook of Tunnel and Heading Construction, Vol. II [88]. The tunnel boring machine is described as the leading operation in tunnelling, and all the other parameters like construction time, costs, unreliability, qualification of employees etc. depend on it. All other operations have to be adapted to the requirements of the tunnel boring machine. Not only the technical and personnel viewpoints but also the contractual aspects demand particular consideration and discussion. Above all the influences of the machine technology have to be introduced into the limitation of risks between the client's overall project, the contractor's overall project, which furthermore is the employer of the machine, and the machine employed plus all suppliers of mechanical plant; all has to be discussed and contractually formulated. When it comes to risk distribution, not only questions of stability have to be dealt with, the quality of the project (accident risk) and the settlement risk also influence the overall cost and time risk.

In contrast to drilling and blasting, there is no standardised form of contract, tendering or awarding of contracts in mechanised tunnelling. The details given below are examples of the procedure today in various countries, which are continuously adapted with experience.

In the GATT-WTO agreement concerning public purchasing, the signatory states undertake to award the construction of public works above certain fixed values to the bidder with the most favourable overall price. The agreement clearly states the most favourable tender, not the cheapest. This provision is certainly in the interest of the client, as he is more certain to obtain a qualitatively respectable end product, also from the point of view of cost and time.

It is thus advantageous for the client to include in the tender documents in a suitable fashion the criteria for the selection of the contractor and the criteria for the award, as he intends to apply these in his evaluation of tenders. Receiving such documents, contractors hand in well thought-out tenders specific to the project.

14.1 Procedure Examples

14.1.1 Procedure in Switzerland

14.1.1.1 General

Tendering regulations in Switzerland for public works are largely regulated by the BoeB (BoeB = Federal law regarding public procurement) and the VoeB (VoeB = Public procurement regulations). Both, law and regulations, were required by the GATT agreement. Swiss Railways are not covered by the agreement, but mostly observe it voluntarily.

The statutory regulation provides clear rules for going out to tender, evaluation of tenders and awarding of contracts. Aptitude criteria are a basic first hurdle. Suitable tenders are appraised with acceptability criteria. The evaluation of tenders described below has proved itself in this way many times.

14.1.1.2 Tender Evaluation

The evaluation of the tenders calls for a procedure in good faith observing the submission rules according to the basic principle "one man's right, another man's duty".

- The contractor is obliged to comply with the submission conditions. He can, however, rely on the fact that the client will exclude tenders, which do not comply with the submission conditions.
- The client does not level out the tenders, on the contrary he creates a decision framework that makes the best selection possible within the regulations.

Table 14-1
Award criteria with weighting [139]

Criterion	Weight [%]
Personnel qualifications (with references) • Project manager • Site manager, foremen • Specialist experts	10
Organisation of the bidder • Decision making and technical management • Subcontractors to be used (with references) • Extent %	10
Construction procedure • Intended construction procedure considering all interdependencies	25
Machines to be used • Intended use of machines with details of performance, special devices, age of machinery • Construction ventilation, ventilation scheme, fans, dust removal plant	15
Phase sequence/construction schedule • Phase sequence with details of work in individual activities • Overall schedule with details of average and possible peak performance	20
Quality management • Description of the bidder's QM system • Location of those responsible for QM in an overall organisation chart • Description of the intended control mechanisms to ensure quality	20

14.1 Procedure Examples

Table 14-2
Evaluation of the partial criteria according to Table 14-1

• No details	Mark 0
• Insufficient or cannot be evaluated	Mark 1
• General details, which are not relevant for implementation	Mark 2
• Adequate: details to be evaluated with exceptions	Mark 3
• Good: the details permit an evaluation „suitable" for implementation	Mark 4
• Especially suitable: the bidder is especially suitable in terms of know-how and experience	Mark 5

For large tunnel projects, also those in combination with bridges, overground works and shafts with in the same project, the following system has proved best for evaluating tenders.

Starting from the principle that the most economic tender should be awarded the contract, the first phase is to evaluate the technical criteria in six subsections (Table 14-1) and weight them using a scale of marking that was announced in the tender documents (Table 14-2).

After the technical evaluation is complete, the overall evaluation is performed with the tender price, which is considered in a predetermined relation to the technical criteria. Using this cost/use viewpoint, the most favourable tender can be decided.

14.1.1.3 Quality Management

The client as contract partner could also assume up till now that the contractor takes the care expected of him regarding the quality of his work. If the contractor also offers quality management (QM), then he has a duty of care in line with his QM guidelines.

Such a system may be sufficient in individual cases purely to ensure quality. What is missing, however, are the quality aims defined by the client, which only result from intensive processing of the practical goals during the project phase.

In contrast to engineering or industry generally, the construction industry produces almost exclusively one-off products. QM for construction as a whole – this includes, of course, the client, consultant and all contractors and subcontractors – has to be designed to cope with this speciality. The quality assurance control should not have a purely steering function. The main aim of project-oriented quality management (PQM) must be to prevent mistakes in the design and construction of structures. This occurs in two steps:

- Targeted analysis at the design stage produces risk profiles. These risks are evaluated and measures taken to ensure a fair residual risk within economic bounds. The risk is always to be seen as a danger assessed according to severity and likelihood of occurrence.
- Quality is to be improved by suitable measures in design and construction.

The demonstrable use of such a multi-disciplinary PQM lies in:

- A high level of design with good construction practicality
- Equitable, not excessively overambitious construction quality
- Clear contract provisions
- A relatively high saving of overall costs
  (in the case of the Murgenthal tunnel, the client reckoned this to be 10%)

This sort of PQM is brought together in a multi-disciplinary Q-control plan. Figure 14-1 is a diagram of the procedure of a PQM plan with the most important documents to be produced named. Essential steps in the procedure are:

- Use plan with
  - Formulation of the intended use
  - Connections intended use – serviceability – profitability
  - Risk analysis based on potential danger patterns
  - Safety plan
  - Demonstration of serviceability
  - Profitability with cost relevance considerations
  - Check the achievement of the intended use
- Q requirements for the Q-relevant working areas
- Control plan structure-subsoil with the geotechnical supervision scheme
- Contractor's work instructions: Thus the contractor shows that he has correctly implemented the Q requirements

These work instructions should contain at least:

- Project
- Organisation
- Technical basics
- Working procedures
- Check and control plan
- Traceability
- Accident prevention
- Environmental protection

14.1.1.4 Assignment of Risks in the Contract

The risks, which arise in any construction project, should not be carried solely by one contract partner or the other. The often much vaunted performance specification leads in consequence just as often to a total transfer of risk to the contractor.

The standard SIA 198 "Underground construction" [154] shows in Appendix 5 a possible risk assignment for the main risks in tunnelling, dependant on the condition of the subsoil.

14.1 Procedure Examples

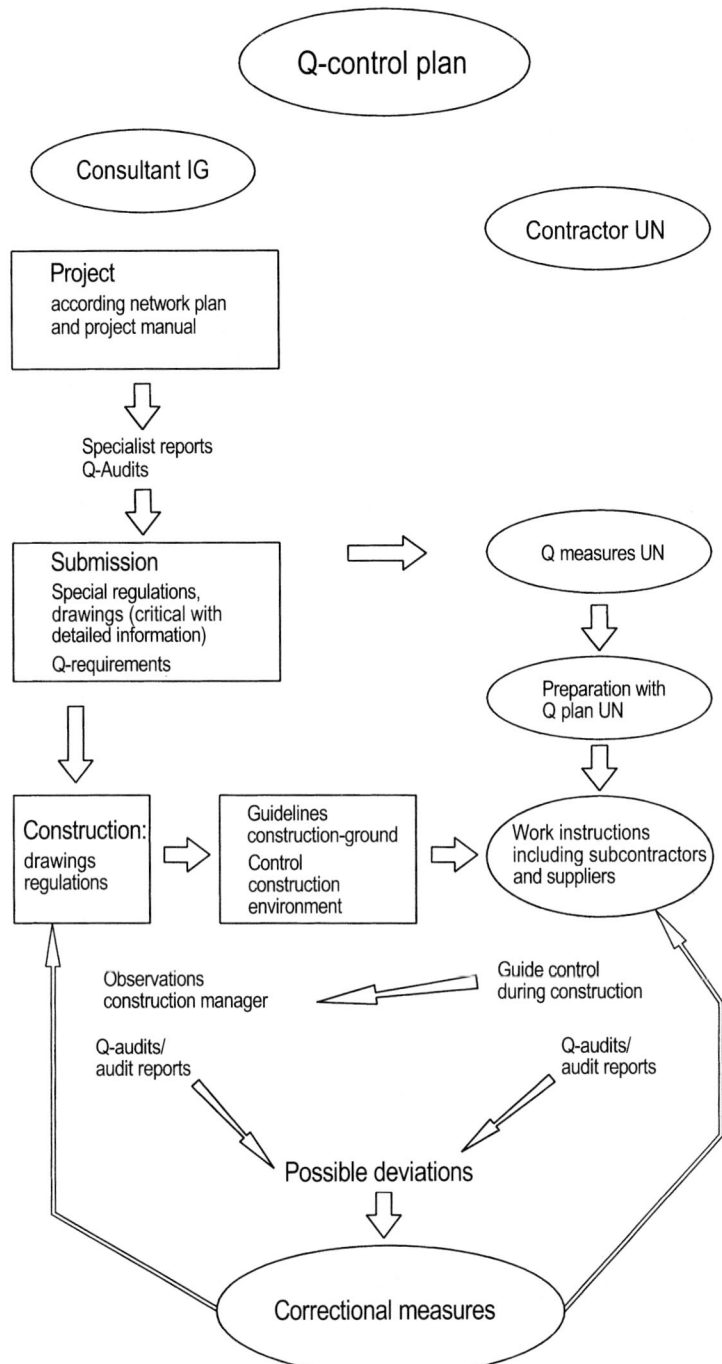

Figure 14-1
Flow diagram of a Q control plan

The following recommendations are made for TBM tunnelling:

Client's Risk

- Gas occurrence
- Geologically caused collapse
- Larger sectional deformations of the cavity than are considered in the contract with
  - Jamming of the tunnelling machine
  - Setting already laid invert segments deeper
  - Enlargement of already bored tunnel sections
  - Rebuilding the tunnelling machine to a larger diameter
  - Rebuilding of formwork to a smaller diameter
- Rock quality lies essentially outside the limit range in the tender documents,
  - Cuttability considerably less easy
  - Gripper plates of the machine can no longer clamp adequately, or they have to be additionally supported due to ruptures
  - the invert is not bearing, The TBM cannot be held at the correct level
  - the rock is so broken that it is equivalent to loose material

Contractor's Risk

- Variation of rock properties within agreed limits
- Advance through rock strata of different hardnesses
- Disruptions to the tunnelling machine through adhesion of the loosened ground
- Qualification of the employed persons

Water inflow represents a special case. Water inflow generally always results in reduced performance. The SIA 198 [154] contains sample contract regulations, which have been found satisfactorily. In principle, a distinction is made between obstructions, which are to be covered by an item in the bill of quantities, and alteration of schedule, because the target performance stated in the contract is no longer achievable.

Table 14-3 deals with the reduction of the target performance. The corresponding reduction is to be made directly to the respective performance in the matrix of excavation/boring class.

Table 14-3
Reduction factors for performance values on water ingress.

	Tunnel with theoretical excavation area $\varnothing \leq 5$ m		Tunnel with theoretical excavation area $\varnothing > 5$ m		Reduction factor
Advance inclination	Rising	Falling	Rising	Falling	
Water ingress quantity [l/s]	10–20 >20–30 >30–40	5–10 >10–20	10–20 >20–40 >40–60	5–10 >10–20 >20–30	0.2 0.4 0.6

14.1.1.5 Geologically or Geotechnically Altered Conditions, Altered Orders, Altered Schedules

With the use of excavation and boring classes and their contractually agreed performance matrix, the adjustment of contractual deadlines results automatically. What remains to be agreed are costs for hire/provision of equipment belonging to the site setup, shorter or longer.

The regulation of altered conditions and exceptional circumstances is much more complex.

Particular subsoil conditions applies if the details in the tender documents are unsatisfactory. However, it makes little sense to avoid these particular conditions with wide-ranging geological data. Authoritative in the contract is the implementation of the geology by the client in the bill of quantities and in the geological-geotechnical longitudinal profile.

The attempt to include all possible and impossible eventualities in the contract has not yet led to success.

4.1.2 Procedure in the Netherlands

The procedure in the Netherlands will be described below using the example of the Botlek tunnel, oner of the largest tunnels on the Betuwe line with the NS Railinfrabeheer (NS RIB) as client. All tendering of other large tunnelling projects in The Netherlands is done relatively similarly.

14.1.2.1 Tendering and Negotiation Procedure with the Botlek Tunnel as an Example [25]

For the Botlek tunnel, the European negotiation procedure was selected, in order to optimally enable bidders to develop their own technical solutions. This consists of a series of fixed steps, which NS RIB worked over to achieve the intended contract tendering process. A speciality of this procedure is that the candidates produce their own technical solutions before submitting a tender. Because the tender is based on a lump sum, the bidders here must produce a considerable amount of design work.

The negotiation procedure consists of the following steps:

Selection

The aim of the selection phase is the select the candidates who are in the position to construct the project in a correct manner. For this purpose, the client creates objective selection criteria in advance.

Call for Tenders

The tenders are based on the draft designs produced by the candidates. The call for tenders consists of a performance specification. In order to be able to ensure that the performance specification leads to a good solution and to be able to evaluate the suggestions from the candidates, the NS RIB produced its own reference draft meeting the specified requirements.

Conditions of Contract

The contract award is made using a clearly formulated Design & Construct (D & C) contract. Because Dutch construction contracts do not provide for draft designs created by the contractor, NS RIB has developed its own D & C contract. This gives contractor the complete responsibility for the buildability of the draft design and the implementation of the construction.

Because of the required type of quality assurance, the bidder group has to possess an ISO 9001 certificate. A quality plan framework created by the candidate is part of the tender.

The liability for the quality of the completed tunnel remains as long as possible with the contractor. The contract stipulates a waiting period of five years and a guarantee of water-tightness of the tunnel of ten years from acceptance.

Consulting Phase

The aim of the consulting phase is that the candidate achieves a technically optimised solution, which then serves as the basis of the tender. Costs are not discussed in the consulting phase. The client consults with each candidate separately and guarantees the confidentiality of the contents of this consulting.

Submission of Tender

The bidder must demonstrate with the submission of a tender that he is informed about all the risks associated with the project and demonstrate with his own risk analyses that the risks are limited and that he can overcome them.

Each candidate submits two tenders. One of these is based on the NS RIB draft design and the other on the alternative developed by the candidate.

As compensation for the expense of the design work, NS RIB paid each bidder about quarter of one million Euro.

The essential points where the bidders' own designs differed from the reference design were:

- Measurement of earthworks and the type of ground improvement
- Construction of the cross-connections between the tubes
- Dimensioning of the segmental lining
- Implementation of the sealing block at the intersection of shaft and tunnel
- Location of shafts
- Type of shield machine
- Diaphragm wall versus combi-wall
- Pile tapes in the ramp areas
- Soil protection in the Oude Maas river

Tender Evaluation

The aim of the evaluation is to select one or more parties to be invited to the negotiation phase. These will be the bidders, who have submitted the economically most favourable tenders.

Because the bidders produce their own draft design, tenders are created, which are not directly comparable. So it is impossible to detail award criteria before the tendering process. This was only described as a framework, with a distinction being made between qualitative factors and the price. The application of this method achieves design freedom. The tender sum is, in contrast to a traditional tender submission, only one of the criteria.

In the evaluation of the tenders, the technical quality proved to be very significant in the selection of the contractors invited for negotiations. These were not the bidders with the lowest price.

The technical solution selected for the Botlek tunnel differs from the reference design mostly in a higher position of the tunnel, which increases the capacity of the tunnel for goods traffic. This is achieved by implementing the construction on the EPB principle. The solution leads to a 5 to 10% capacity increase for the tunnel, which in the future will also effectively be the capacity of the harbour railway. In addition, a more economical solution is being used for the ground improvement and the tunnel lining.

It is the case generally that sharply estimated tenders are submitted. Alternatives offered by candidates always lead to a cost saving.

Negotiation

The aim of the negotiation phase is to achieve agreement in all points affecting the project.

Award

The contract was awarded to the joint venture "BTC Botlek". This consortium is composed of several large Dutch construction companies and the German partner Wayss & Freytag AG.

Arbitration Process

The bidder with the lowest price applied for an arbitration procedure against the decision of the client, after he had been turned down with substantiation. The arbitration court decided, however, that

- NS RIB does not have a duty according to EU rules to issue an explanatory statement and
- NS RIB as client must have a certain design freedom in the use of tendering procedures.

And so the appeal to arbitration of the bidder was turned down.

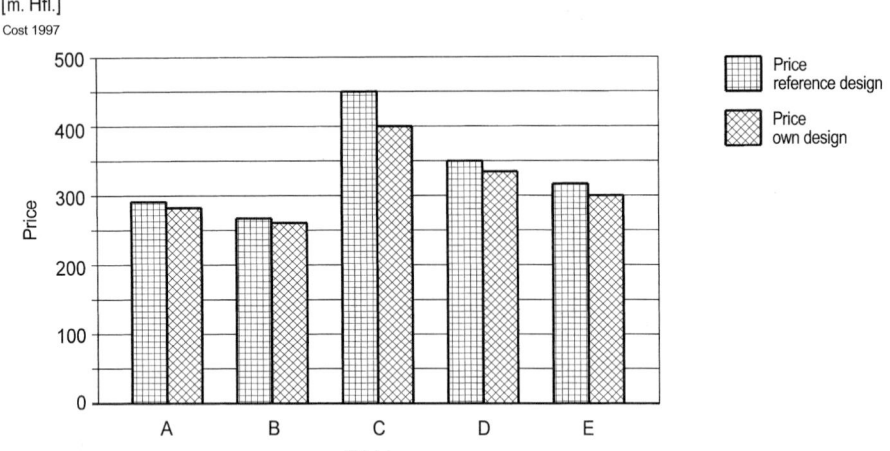

Figure 14-2
Financial evaluation of tenders

The second statement above all is of great importance. If the party issuing tenders did not have this design freedom, the lowest price would always be the deciding factor. In this case, NS RIB would have created no tender documents on a D & C basis in the future, because the quality of the tenders is very important in this case.

The tendering and award process used led to a good result.

It is clear that the candidates took the possibility of delivering a draft design as a challenge in order above all to produce a technical design for the risky boring work in a safe and economic way.

One disadvantage of the process used is the long working time of about ten months. This could have been shortened with a traditional tender method by about three months. A second advantage is the great efforts, which the candidates have to make in order to produce a tender. This is about 10,000 h per candidate for this task with a construction sum of about 150 million Euro. It may be possible to reduce this expense in the future.

The award of the contract for the Botlek railway tunnel has shown that European negotiation procedures are very workable for the tendering of highly complicated building and civil engineering works, and is also beneficial in stimulating the contracting sector to look for innovative solutions.

14.1.3 Procedure in Germany

The procedure in Germany for mechanised tunnelling has been described in detail in [95]. In general, the recommendations of the DAUB [31, 32] provide the basis for the consideration of additional particulars. This has been extended with special conditions; see the publication in the 2001 Tunnelbau-Taschenbuch *(Tunnelling pocket book)* [99].

14.2 Design and Geotechnical Requirements in the Tendering of a Mechanised Tunnelling as Alternative Proposal

14.2.1 Introduction

If the call for tenders for a tunnel project specifies shotcrete, then mechanised tunnelling cannot practically be tendered. Even under favourable geotechnical conditions, the submission of a tender based on an alternative proposal is often impossible, because the scheme decided upon cannot be fulfilled by mechanised tunnelling. This is confirmed by the following examples [91].

Then the geotechnical, design and contract requirements are formulated in the tender documents, which are the precondition for being able to hand in a tender based on mechanised tunnelling.

The costs and the persons active in the phases design, tendering and construction have a decisive influence on the consideration of an alternative proposal for mechanised tunnelling.

Only taking these points into account can a practical and economical tender be produced.

14.2.2 Examples

14.2.2.1 Adler Tunnel

The Adler tunnel is a part of the new railway line from Muttenz to Liestal near Basel as part of the Swiss federal railways scheme Bahn 2000. The underground part of the two-track railway tunnel has a length of 4.3 km and an excavated diameter of 12.58 m. The geology of the Adler tunnel is mostly Jurassic clay, marl and limestone. One third of the tunnel is through gypsum Keuper (anhydride – swellable) with clay formations. Impressive fault zones had to be overcome during the advance. The hydrology of the Adler tunnel is characterised by groundwater containing sulphates and chlorides, which additionally tends to encrustation due to the high calcium content. The water pressure was max 1.0 bar and the water inflow 10–20 l/s.

Comparative design and tendering was carried out in two variants, shield TBM and the shotcrete method.

After a detailed risk analysis, the TBM variant was preferred with the initial stretch being constructed using shotcrete. A major reason for this decision was the swelling behaviour of the anhydrite.

The TBM advance of the Adler tunnel was completed at the start of 1998. At the start of the preparatory works to construct the starting area by shotcreting, a collapse occurred fairly early. There was also a collapse during the following excavation by the tunnelling machine. During the shield tunnelling, there was yet another collapse after 1.720 m, which had the effect of a permanent blockage of the machine.

With the aid of additional measures like dewatering from above, advance injections through the shield and further upgrading of the machine, the shield excavation con-

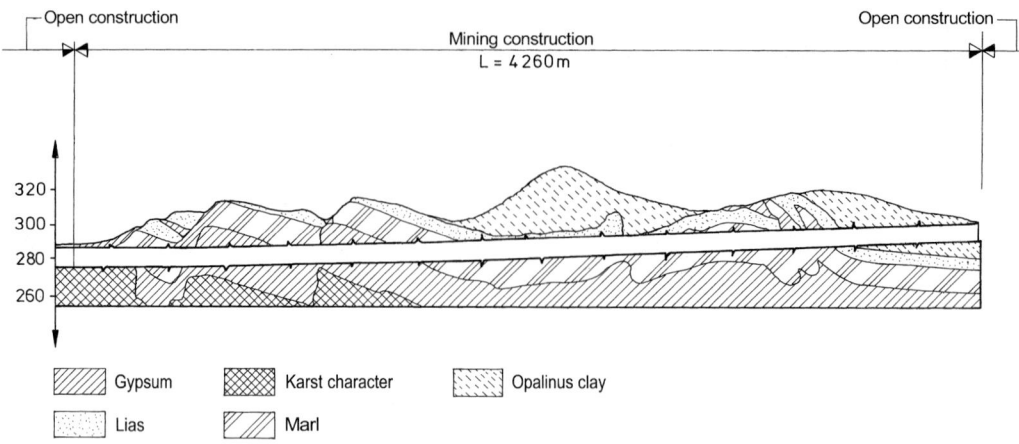

Figure 14-3
Geological longitudinal profile of Adler tunnel

tinued without problem and achieved an average advance 15.5 m/day. Even in a fault zone, 5.4 m/day was achieved.

The extra cost, which arose, led to an arbitration process, at which all problems leading to disruptions and expense were investigated with the parties involved. Despite the stated additional technical measures and the alteration of the organisation structure on the site, the choice of a shield machine was proved correct.

14.2.2.2 Sieberg Tunnel

The Sieberg tunnel is part of the the high-speed railway (HLS) Wien–Salzburg. The two-track tunnel has a length of 6.5 km and an excavated diameter of 12.5 m. The contract award was at the end of 1996. The mining underground section of the Sieberg tunnel lies completely in strata of the molasse zone with layers of siltstone and fine sandstone. These layers are described as Miocene schlier and Oligocene schlier. There were prominently weathered zones at the four valley crossings,. The hydrology of the Sieberg tunnel mostly has two floors of groundwater. The maximum water pressure is approx. 2.9 bar. The water ingress during the advance was very low and seldom reached values of 5 l/s. The overburden was max. 55 m. At the four valleys, this reduced to less than 10 m in places. There were no buildings to tunnel under.

The call for tenders with the shotcrete method included a tunnelling classification according to Ö-Norm *(Austrian standard)* in seven classes for the calotte and five classes for the invert. Of these classes six and seven required advance support work (rockbolts, IBO anchors to form a pipe umbrella) aver approx. 3.5 km. A 1.26 km long section of the tunnel stretch was to be built in cut and cover. The tender documents contained no details or conditions for a shield tunnelling.

14.2 Design and Geotechnical Requirements in the Tendering

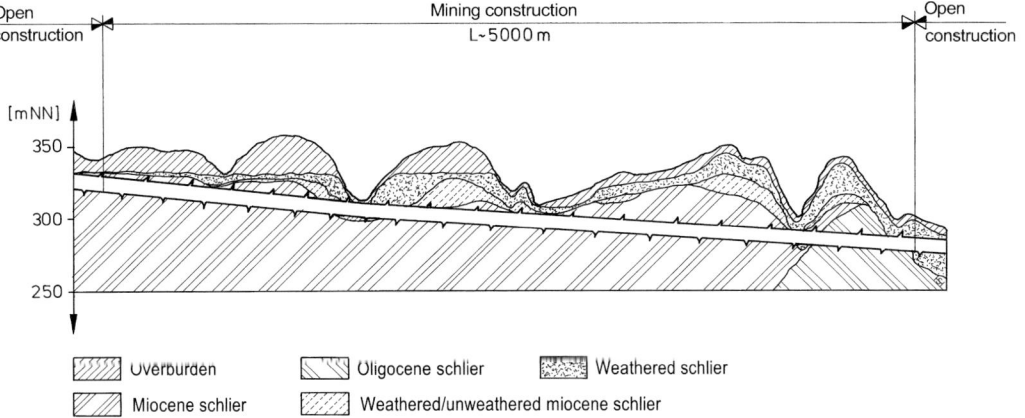

Figure 14-4
Geological longitudinal profile of Sieberg tunnel

Although shotcrete method had been put out to tender, an alternative proposal using the shield method for the entire length was submitted. This was intended to reuse the shield machine from the Adler tunnel. The flat rock cutter head was fitted with disc cutters and picks and moveable to all sides, although to withdraw it into the shield was not possible. The permanent support of the tunnel was intended to be a segmental lining. No additional measures in the areas with shallow overburden were given in the alternative proposal.

Mechanised tunnelling was rejected because the tender was incomplete. Even the assurance of the bidder to bear all risks and costs was considered as impractical by the client.

The difficult geology and the initial experiences from the Adler tunnel made all parties involved cautious.

The experience with the call for tenders and the award of the Sieberg tunnel shows clearly that the client should have included the criteria and the specification for a tender as shield tunnelling in the tender documents. At least a preliminary design would have been necessary to enable competition.

14.2.2.3 Stuttgart Airport Tunnel

The airport tunnel is part of the extension of the S-Bahn *(urban railway)* network in the Stuttgart metropolitan area. Two single-track railway tunnels with a length of 2.2 km each and an excavated diameter of approx. 8.5 m were constructed. The S-Bahn tunnel lies in Jurassic clay-silt-stone with intercalated limestone and sandstone beds of varying frequency. Furthermore, springs and raised horizontal stress can be expected along the route.

The clay-silt-stones are generally weakly permeable, so that only a very slight water ingress is expected in these strata. The limestone and sandstone beds have a greater per-

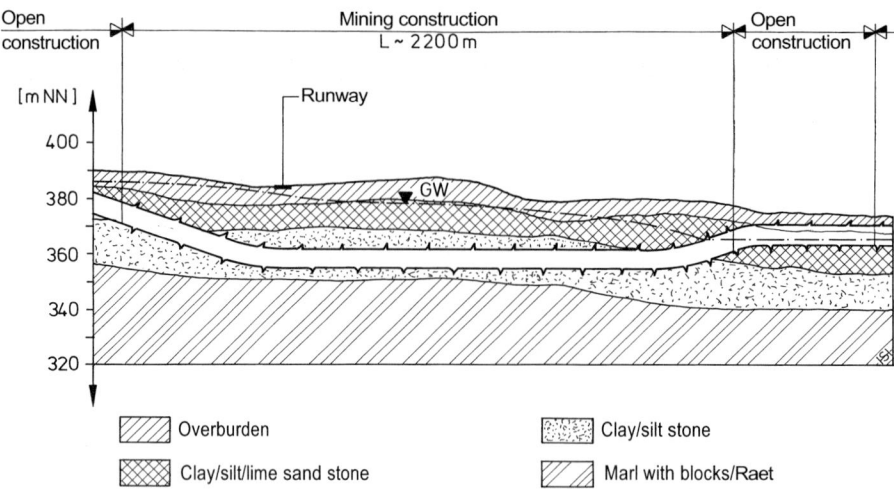

Figure 14-5
Geological longitudinal profile of Stuttgart airport tunnel

meability, with the result that more water ingress is to be expected here. The forecast maximum water pressure along the tunnel route is about 2.5 bar. It is possible to inject both the permeable limestone/sandstone and the clay-silt-stone. The environmental requirements result from having to tunnel under many streets and buildings as well as the take-off and landing runways of the airport with a strict specification regarding settlement. The minimum radius of the tunnel route is R = 300 m.

The draft call for tenders called for shotcrete method to be used. This was to be carried out without groundwater lowering as a low-settlement drive. The tunnel cross-section was to be circular.

Nonetheless, an alternative proposal was tendered using shield tunnelling with water control where necessary by compressed air. The closed cutter head of the full-face machine would be equipped with discs and scraper blades, and on account of the low radii along the route, an articulated shield was intended. All required equipment was of the newest technology to minimise settlement. The lining of the tunnel was, as is usual with shield tunnelling, to be constructed with segments.

The alternative proposal was rejected, despite the price being lower.

The grounds stated for this were:
- The possibility of excavating the limestone beds was questioned, although there was already positive experience from the Adler tunnel and the cutter head was equipped with discs and scraper blades.
- The possible jamming of the shield was also seen as a problem; this was solved with overcutters.
- The water-tightness and the durability of the segmental lining was also doubted, this although single-shell segmental lining is the standard technology today.

- It was feared that the use of shield tunnelling would not be able to satisfy the permissible levels of settlement specified in the tender documents. Against this must be said that no blasting is necessary for a shield tunnel and synchronous grouting with the newest control methods was tendered.

Furthermore, controlled overbreak and quick closure of the rings also spoke for the use of a shield machine. Both offer advantages compared with the shotcrete method.

Shield tunnelling was rejected although there was a wealth of experience under similar conditions (see Adler tunnel) and the solution would have been less expensive and technically better.

14.2.2.4 Rennsteig Tunnel

The Rennsteig tunnel will be, once completed, with a length of 7.9 km the longest road tunnel in Germany. The four lanes of the motorway A71/A73 will be led though two tunnel tubes. Connecting to the crossing of the ridge of the Thueringer Wald, the tunnels Alte Burg, Hochwald and Berg Block are to be built, making an overall length of 12.6 km. The excavated diameter of the tunnel is 11 m. The geological conditions at the Rennsteig tunnel are characterised by Permian porphyry, which is mostly very compacted and has a great hardness. Intercalated and below this layer are further strata of sandstone, claystone and Keuper clay. The groundwater level lies variable above and below the gradient. The maximum water pressure is approx. 8.5 bar and the water inflow is estimated at up to 5 l/s per 100 m tunnel. The maximum overburden of the Rennsteig tunnel is 200 m. The tunnel also crosses the operational Brandleite two-track railway tunnel. As a constructional safety measure, emergency bays will be built in both tubes at 700 m spacing, and the two tubes will also be connected every 350 m with escape headings.

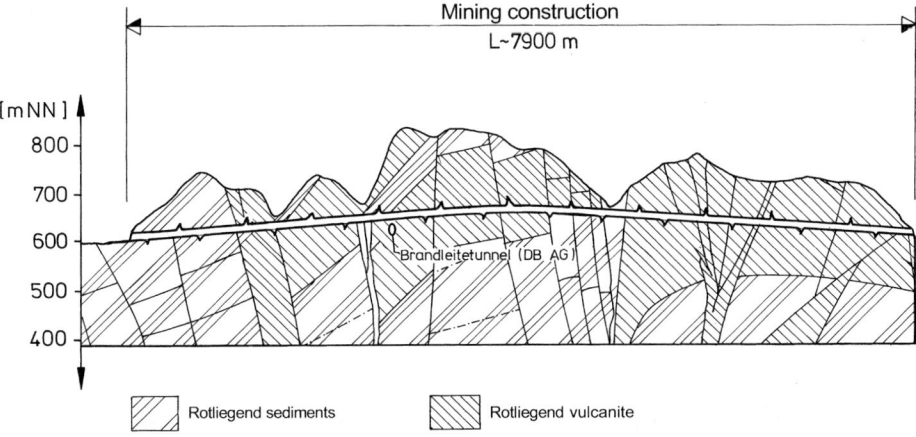

Figure 14-6
Geological longitudinal profile of Rennsteig tunnel

The draft call for tenders called for the use of shotcrete construction. Firstly shortly before the completion of the tender documents, a few conditions for an alternative proposal using a shield machine were included in the tender documents.

Quotation from the tender documents:

"Under the existing technical conditions, the use of a tunnel boring machine with shield (TBM-S) is basically possible. The construction process can be alternatively tendered by the bidder, taking the bulk material handling into account."

The tender documents referring to the alternative proposal contain short statements about construction method, machine concept, segmental lining, bulk distribution and transport.

The Rennsteig tunnel was awarded as a shotcrete project, because this seemed to be less expensive than a mechanised tunnel excavation, which was tendered about 10% more expensively.

Meanwhile, the first tunnelling experience for this tunnel is available. It was clear after half of the route that the geological conditions would have been suitable for mechanised tunnelling. The rock classes are often more favourable than forecast and the water inflow is considerable less. According to experience in Switzerland, better advance rates would have been possible with a TBM.

14.2.2.5 Lainz Tunnel

The Lainz tunnel connects the Westbahn, Suedbahn and Donaulaendebahn railways around Vienna with an overall length of about 5.5 km. The project can be divided into the Hetzendorf tunnel with a length of about 2 km and the Lainz tunnel of about 3.5 km, of which 0.3 km is shotcrete construction. The diameter of the two-track tunnel is 13.8 m.

The Hetzendorf tunnel runs over its entire length through the sediments of the Vienna basin. These are characterised by very changeable and extremely heterogeneous beds of gravel and silt-clay. The water pressure is max. 1.8 bar and high water inflow has to be expected. Only pore water pressure has to be expected in the Tegel layers, owing to the low permeability. The local conditions are marked by tunnelling under inhabited areas and the resulting limited access and working area. Furthermore, the cover is shallow, so that increased standards of permissible settlement are imposed.

The main feature of the route of the Lainz tunnel is the moderately hard rock consisting of 80% claystone and silt and marlstone. Along 20% of the stretch, there are nappe outliers, sandstones and limestones. Water is only a problem locally, but can reach inflows of up to 10 l/s. The water pressure is max. 6.5 bar.

The soft ground area (Hetzendorf tunnel) was specified as a cheek drift heading with shotcrete. The water-bearing strata are drained from the surface using deep wells. Ground injections are intended to minimise settlement.

14.2 Design and Geotechnical Requirements in the Tendering

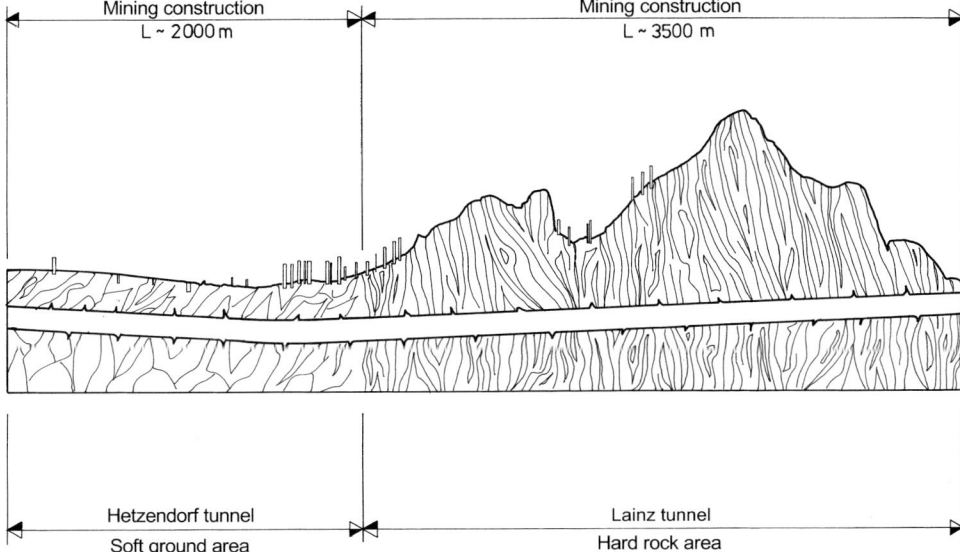

Figure 14-7
Geological longitudinal profile of Hetzendorf and Lainz tunnels

The stretch in hard rock (Lainz tunnel) was also specified with shotcrete. This section is to be excavated by a roadheader or a tunnel excavator in the partial sections calotte, bench and invert. In some areas, anchors, rockbolts and pipe screens will be used.

As an alternative to shotcrete work, investigations were carried out for both stretches for shield tunnelling as one-track or two-track solution. The requirements on the tunnelling process will be best met in the soft ground by a shield machine with liquid-supported working face. The final support is to be single-shell segment lining.

The application of a shield machine in the hard rock area is possible. It should, however, also be possible to seal the excavation chamber because of the water pressure. According to the Swiss method, a two-shell lining with umbrella waterproofing can be used as the final support.

The example of the Lainz tunnel shows clearly the influence of the local conditions. This means that the shield starting area and thus the excavation cannot be adapted optimally to the local conditions.

The construction of points bays after the shield tunnelling worsens the basic premises for a shield tunnel. Emergency niches could according to the project handed in be replaced by box bays inside the shield cross-section.

The use of a tunnelling machine is basically possible under the existing geological and hydrological conditions. The implementation of the requirement for external emergency bays and points bays is time and cost consuming, so that shotcrete work is at the moment held to be more suitable.

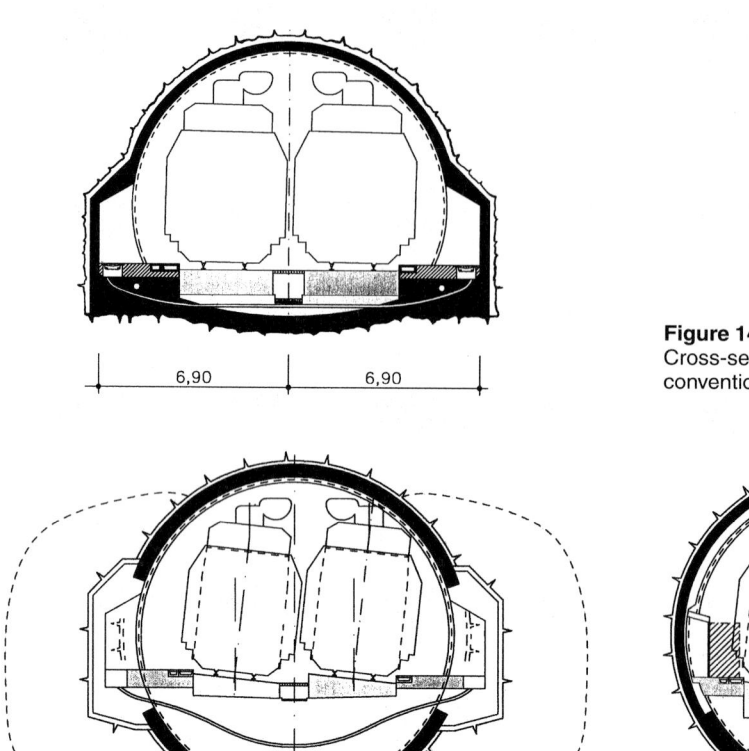
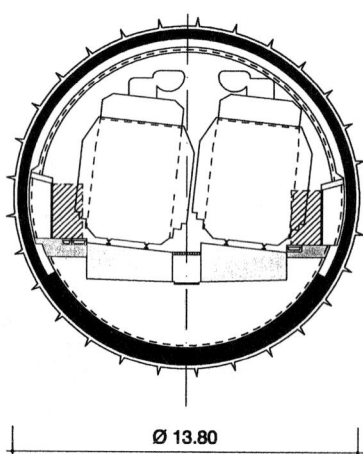

Figure 14-8
Cross-section with safety niches –
conventional construction

Figure 14-9
Cross-section with safety niches –
shield tunnelling

Figure 14-10
Cross-section with box niches –
shield tunnelling

The following figures show examples of the method of construction of the safety bays. Figure 14-8 shows a cross-section with safety bays in the conventional method.

The construction of box and safety bays for a shield tunnel is shown in Figs. 14-9 and 14-10, this requires an opening of the segmental lining as shown in Fig. 14-9. The box bays can be integrated into the existing opening of the segment lining.

14.2.3 Additional Requirements for Mechanised Tunnelling Concept in the Tender Documents

The necessary geotechnical details, which should be given in the specification in order to enable alternative proposals for machine tunnelling, are discussed below. The basic design concepts for mechanised tunnelling and the arrangement of the construction are also brought together. Finally, the requirements for the tender documentation and con-

14.2 Design and Geotechnical Requirements in the Tendering

tract are given. Genuine competition between the two construction methods is only possible on this basis.

14.2.3.1 Geology and Hydrology

For an alternative proposal for mechanised tunnelling, authoritative details are necessary about the type and extent of the obstructions to be found in the subsoil. The geological data should be evaluated regarding the overcoming of fault zones.

The soil mechanical parameters to enable the evaluation of any danger of jamming and the parameters describing wear characteristics of excavation tools and machine parts (seals) should be determined and evaluated.

The suitability of the subsoil for injecting and conditioning should also be evaluated. Decisions should be made about the environmental acceptability of the materials used.

The knowledge of geological conditions should also be used to evaluate the suitability of the excavated material for tipping, recycling and separation.

Finally, when the use of a tunnel boring machine is being considered, it is important to know whether a gripper machine will be able to clamp into the excavated walls of the tunnel.

14.2.3.2 Design and Construction Process

The design concept should be optimised for mechanised tunnelling regarding the arrangement of niches, emergency stopping bays and cross-cuts. The number of niches should be reduced to a minimum, the tunnel cross-section should be enlarged to integrate escape routes (niches, emergency bays, cross cuts etc.) and the segmental lining scheme adapted for the required emergency measures.

Additional measures should be implemented affecting the delivery and assembly times, segment production and logistics, that is, all demands placed on the construction company. Further aspects affecting the construction company are the access and exit routes, the construction of niches and connecting headings, the analysis of breakdowns and risks, measures against fire and explosion and a geometrical measurement programme.

The list of requirements for a mechanised excavation should be specially widened to include the categories site investigation, preliminary drainage, advance ground water lowering, overcoming of obstructions, bulk materials management, recording of operational data, shield steering with safety measures and the depiction of faults. A rebuilding of the tunnelling machine and the change of spoil removal to another type of operation should also be included in the list of requirements.

The design work should also take into account the specification of the lining like, for example, measures against fire and explosion, waterproofing and collision loads. Requirements resulting from geometrical criteria (curve radius, tolerances), influences from shield operation and openings for grouting should also be considered in the design of the lining.

14.2.3.3 Specification and Contract

The specification of the construction project for mechanised tunnelling should also include the regulation of risk arising from subsoil conditions. It is the task of the client to describe the subsoil in detail and make geological forecasts. The contractor is responsible for the expert processing of the subsoil and should, if required, carry out his own soil investigations.

The contract should also contain provisions describing the implementation of additional measures, which could become necessary when overcoming fault zones or when boring under buildings at risk of settlement. The possibility of breakdowns should also be considered in the contract and payment regulated in case of shorter or longer stoppages. The quality criteria should also be stated for the tunnel lining, in order to be able to compare the tenders fairly, detailed requirements should be given of the payment system, the level of waterproofing to be achieved, the tolerances and the necessary fire protection for the tunnel lining.

There should be special requirements in the specification of mechanised tunnelling regarding settlement, calculation data for the stability of the face, technical requirements on the shield construction and additional measures from the shield. Control and regulation programmes including quality assurance system as well as loading on the segmental lining should be determined.

14.2.4 Decisions Based on Cost

The decision phases and the contract parties must be considered in the costs (cost control). The decision phases should thus be divided into design phase, tender phase and construction phase.

14.2.4.1 Design Phase and Preparation for Tendering

Three cases should explain the influence, which the consultant can exert in the preparation of tenders:

Case 1

The consultant estimates the cost of a shield tunnelling approximately without the basis of a completed design and comes to the conclusion that the costs are higher. He recommends the client not to include the possibility of a shield tunnelling in the tender documents, and also to reject this as an alternative proposal.

14.2 Design and Geotechnical Requirements in the Tendering

Case 2

A shield tunnelling is basically permitted as an alternative proposal, but the design and geological information required for the preparation of a tender are, however, not included. The alternative proposals handed in are then incomplete, too expensive or perhaps also too cheap, if manipulated and not estimated (for example, necessary additional measures have not been included).

Case 3

A shield tunnelling is considered and specified alongside the shotcrete method and the two methods compete. This procedure has often been implemented effectively in Switzerland.

The consultant has a decisive influence on the acceptance and implementation of a shield drive.

14.2.4.2 Tendering Phase

A mechanised tunnelling tendered on the basis of a design intended for the shotcrete method is generally more expensive (approx. 10–15%). This means that mechanised tunnelling frequently has no chance of an award. If shield tunnelling is subjected to genuine competition with a suitable design, then it has good chances of being able to compete on price with shotcrete work.

The conclusion could be drawn that the consideration of which process to use should be decided in competition and not according to the opinion of the consultant.

But the question of costs is not finally dealt with at his stage. The construction and final payment stages also have to be considered.

14.2.4.3 Construction Phase and Final Payment

There is a noticeable tendency for shotcrete projects to generally cost much more than the tender price. The experience from Switzerland is different, that apart from a few exceptions (e.g. Adler tunnel), no meaningful cost increases have been noticed for the invoiced shield construction projects.

The theme of cost certainty is only one criteria at the moment, but further investigation could lead to more detailed information.

14.2.5 Forecast

Shotcrete construction and mechanised tunnelling methods will still be able to define and maintain their scope of application in the future. If, however, the local project conditions like tunnel length, cross-section, geology and hydro-geology offer suitable conditions for the use of a shield machine, then the considerations of machine technology should already be included in the design phase.

To include these details of geotechnical, design and contract requirements in the tender documents provides a clear and defined basis for the evaluation of mechanised tunnelling tenders. This is practised successfully in Switzerland. Without cost advantages and lower risks, this would nor have been possible.

Mechanised tunnelling techniques are innovative and highly technical, providing a greater measure of work safety and mechanisation and so provide a decisive contribution to future-oriented tunnel construction.

15 Tunnel Lining

15.1 General

The tunnel lining, as final support for the rock, has to guarantee stability, durability and serviceability for the entire period of use of the tunnel. It secures the interior against the surrounding rock, forms a seal against water coming in or going out, transfers the interior loads from installations and traffic and serves, depending upon the type of TBM, as reaction for the thrust cylinders. The design and the constructional layout of the lining depend on the requirements for the tunnel use, the loading and construction conditions.

The construction process normally requires a circular cross-section. The internal radius is determined by the requirements resulting from the intended purpose, e.g. the required loading gauge in traffic tunnels or the required flow cross section in water or ventilation tunnels.

The dimensions of the lining are determined from the loadings, which are mainly ground and water pressure. Due to the circular shape of the tunnel cross section and the loading, which is usually not symmetrical, system and support line are rarely the same. This means that the tunnel lining is not only subjected to external pressure but also bending in a transverse direction, which for a concrete lining can require reinforcement. Water pressure acting on the lining is influenced by natural conditions on one hand and by the effect of dewatering on the other. With tunnel systems subjected to water pressure, natural water conditions take over after completion of the tunnel construction. In this case, the lining must be designed to resist all-round water pressure. Other tunnel systems experience no water pressure because the water is led away in drains, so water pressure cannot build up.

The machine technology used exerts a great influence on the types of lining, which can be used. With a gripper TBM, the temporary support can be provided by shotcrete with in-situ lining cast later. If the TBM is equipped with a shield and thrust cylinders, it is necessary to install a lining of reinforced concrete segments capable of bearing immediately.

15.2 Design Principles for Tunnel Linings

15.2.1 Single-Shell and Double-Shell Construction

The lining of tunnels can be either single-shell or double-shell [89, 92, 95].

With double-shell lining, there is a constructive and functional separation of the individual shells. The outer shell, which is designed to resist the ground pressures to be expected and installed during the excavation, provides immediate support for the excavated cavity. The outer shell is not normally required to be serviceable and is not designed to be waterproof. This is provided by the second shell, which is installed in the

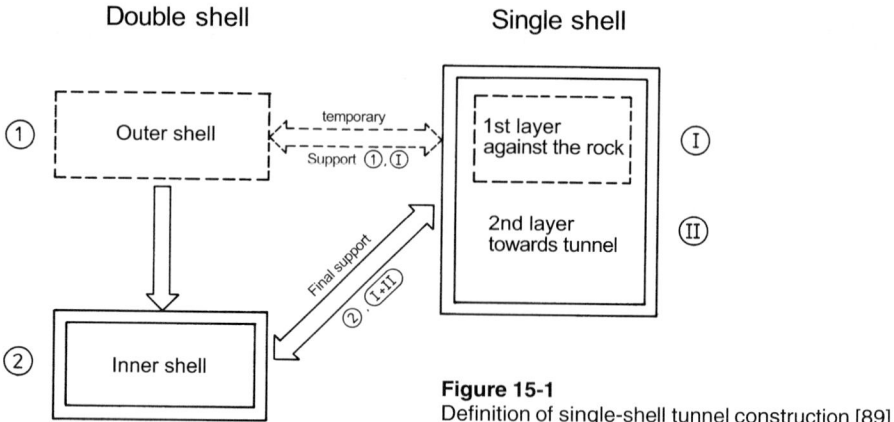

Figure 15-1
Definition of single-shell tunnel construction [89]

usable tunnel and forms the permanent lining. If the tunnel is to be subject to water pressure, then the inner shell is designed to resist it. The inner shell additionally has to support the ground pressure in the long term, if the stability of the outer shell cannot be guaranteed over the entire period of use of the tunnel. This is the case, for example, when aggressive water, which can attack concrete, rots the external shell. The inner and outer shells are generally separated structurally by plastic sheeting, in order to preserve the inner shell from unnecessary loading. In waterproofed tunnels, this function is provided by the waterproofing layer.

Single shell constructions are to be differentiated into the genuine single-shell solutions, which are implemented by a construction system, and the bonded solutions, which fulfil the lining requirements with two or more layers, with the individual layers all being involved in bearing load as a composite. The first group includes, for example, tunnels with a single-shell segmental lining or extruded concrete shell and also tunnels in stable rock, which need no support for structural reasons, but for aesthetic or operational reasons receive a flat surface, perhaps by casting a concrete inner shell. To the second group belong, for example, tunnels whose lining consists of a composite construction of shotcrete for immediate support to enable safe driving and a subsequently cast bonded concrete lining to complete the tunnel support.

For both variants, the single-shell construction or the composite section must guarantee the structural stability and the serviceability and here above all the water tightness of the tunnel over the entire life of the tunnel.

Because both the use of segment of linings and shotcrete are possible as part of a TBM tunnelling, the whole spectrum of single and double shelled solutions in tunnel construction is discussed here. A new development coming into use is the application of steel fibre reinforced concrete. Table 15-1 gives a schematic overview of the construction types already in use.

The consideration whether it is better to construct one or two shells is mainly driven by economic considerations. The construction risks for a waterproof tunnel also have to be

15.2 Design Principles for Tunnel Linings

Table 15-1
Matrix of the construction possibilities for tunnel lining

		Inner shell or single-shell construction			
		Shotcrete	In-situ concrete	Segments	Extruded concrete
Outer shell	Shotcrete	x	x		
	Segments		x		
	Extruded concrete		x	x	
Single-shell construction		x	x	x	x

taken into account. It is sensible to construct segmental lining as the final internal lining of the tunnel, as is already standard practice worldwide for tunnelling with liquid-supported or earth pressure shields in unstable ground. With a TBM, however, there are still a few points against this. Single-shell lining is only cost effective under certain conditions.

Comparison of tenders for many larger traffic projects in Switzerland have shown that double-shell construction is still generally 5–10% cheaper, because the production costs are normally lower due to the segments being thinner, containing less reinforcement and the lower precision required; higher advance rates are possible because of the shorter time required for installation, and the grouting of the annular gap can be done by blowing fine gravel instead of expensive mortar grouting, which is required, at least, to compress the sealing gaskets in the longitudinal joints of the segments.

15.2.2 Watertight and Water-Draining Forms of Construction

There are two main methods of dealing with the ground water affecting the tunnel in service. The groundwater can be collected in drains and led away, or the tunnel can be waterproofed all round, designed so as to be able to resist water under pressure. Figure 15-2 shows the functional principle of both systems using the example of a single-shell segmental lining (left) and a double-shell cross-section with in-situ concrete inner shell (right).

In the case of tunnels with drainage, the tunnel arch is constructed waterproof to protect the interior against water ingress. This can be achieved using plastic sealing membrane outside the inner shell or by constructing the arch out of waterproof concrete: The groundwater flows round the technically waterproof tunnel lining to the drainage located in the sides, where it is collected and led away. This system, known as drainage with umbrella waterproofing, is suitable for both percolating water and water under pressure. Because no hydrostatic pressure can build up around a drained tunnel, the design of the segment only has to consider ground pressure.

One disadvantage of systems, which drain off the water, is the sometimes considerable maintenance of the drainage necessitated by sediments and scale formation with the en-

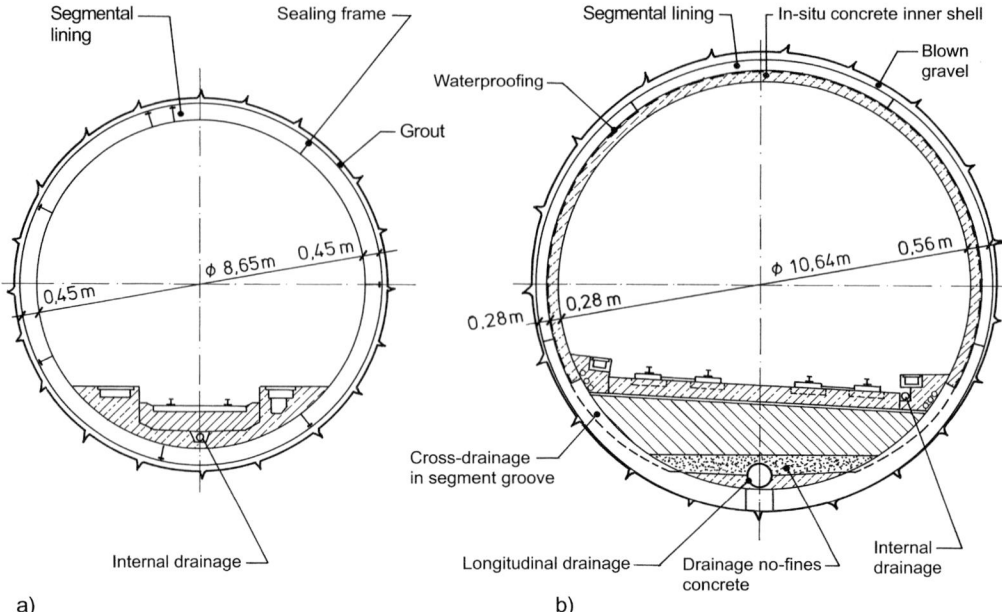

Figure 15-2
Constructional principle of waterproof and drained tunnels with segment linings
a) Botlek tunnel, internal diameter 8.65 m, single-shell, water pressure resistant
b) Murgenthal tunnel, internal diameter 10.64 m, double shelled, drained

suing operating costs. These have to be flushed regularly to preserve the function. Recent investigations show that the use of shotcrete low in eluates, alterations of the detailing of the individual drainage elements and alterations to the overall system like, for example, exclusion of air by backing up the water, can reduce the formation of scale. Further results of running research projects at the Ruhr-University, Bochum are to be awaited.

Water-draining tunnel systems represent, through the lowering of the water table, an influence on the water cycle and thus of the ecological system [88]. It can therefore be necessary to construct a watertight tunnel for environmental protection reasons, to avoid any effects on the natural water cycle after the completion of the construction work.

Since the 1960s, the use of improved materials (plastic sealing membranes, joint bands made of artificial materials, waterproof concrete, new injection compounds, etc.) has enabled the construction of tunnels with a durable all-round seal against water under pressure [147].

Although the construction of single-shell, waterproof segmental linings has become standard practice for shield tunnels, the proportion of tunnels with in-situ cast linings detailed to resist water pressure is very low. For example, an investigation of experience with tunnel waterproofing in Switzerland examined 239 projects, of which 233 were waterproofed to resist percolating water and only 6 to resist water under pressure [147, 185].

In a study of the experience with waterproofing systems in German road and rail tunnels, 9 tunnels altogether with waterproofing designed to resist water under pressure were analysed. It was concluded in the summary that the waterproofing classes required in the specification were not achieved at first without subsequent additional measures. Systems with plastic sealing membranes were often damaged during the fixing of reinforcement and this caused leaks. This was, however, noticed in most cases only after the water pressure was active through visible damp patches or water ingress; this resulted in extensive repair measures, which were mad more difficult by the continuing water pressure. With a single-shell segmental lining, the waterproofing can be impaired by faulty installation of the segments, leading to damage in the form of serious cracking or spalling.

Experience shows that it is never possible in practice to achieve the required standards of waterproofing without additional measures, even with two-layer waterproofing systems. A concept should be developed already in the design phase to enable the repair and refurbishment of the waterproofing system even under active water pressure.

15.3 Lining with Concrete Segments

15.3.1 General

Segments are precast concrete elements, which are built together to form a ring and serve as the tunnel lining. A particular characteristic of segmental lining is therefore, in addition to the segment blocks, the high proportion of the construction formed by the joints. These can be differentiated into longitudinal joints between the segments of a ring and ring joints between the individual rings.

The use of segments in TBM tunnelling is essential whenever it is impossible to achieve the thrust force required to drive the machine because clamping into the rock is ruled out by the rock properties. In such cases the thrust forces are resisted by the already installed lining, which act as an abutment in the longitudinal direction of the tunnel. This necessitates immediate load-bearing capacity, which cannot be provided by a shotcrete or cast in-situ concrete lining ring.

Figure 15-3 shows the spectrum of construction types of segments for single and double shelled construction in tunnelling.

The segments are usually installed in the protection of the tail shield of the TBM with the help of an erector, or behind the shield braced directly against the rock. In a subsequent working stage, the annular gap between the segment ring and the rock is filled with suitable material through appropriate openings in the segments and/or through the tail shield or grouted. This helps to limit any loosening of the rock, enables continuous transfer of the external ground pressure loading into the lining and provides the bedding necessary for stability and strength of the segment ring.

Concrete segments are standard practice today and have largely replaced steel or cast iron tubbings for cost reasons. For the application of steel and cast iron in tunnel lining, see also [39, 68, 87, 95].

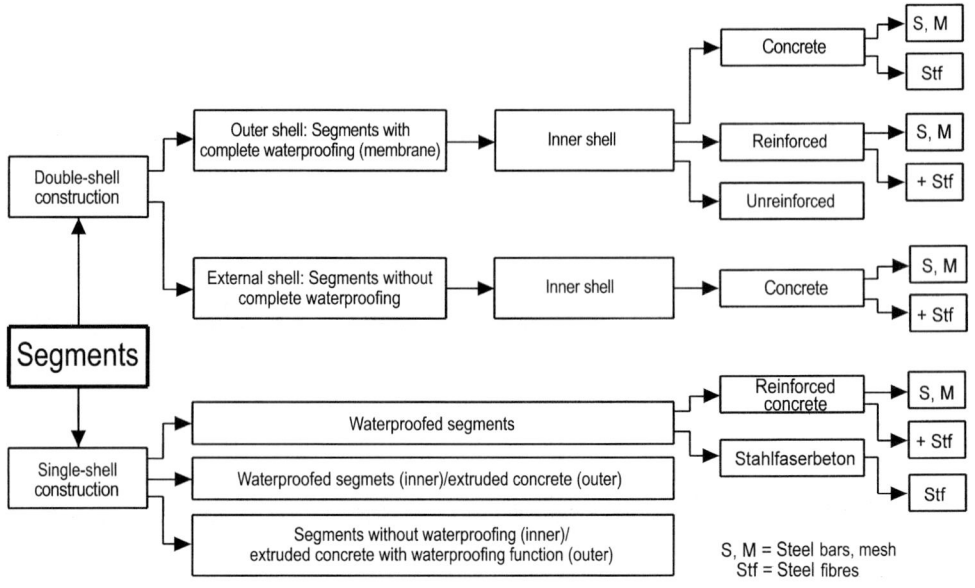

Figure 15-3
Segments for single and double-shelled construction

Typical project requirements for the segment lining, resulting from the geological and hydrological conditions on one hand and engineering and economic aspects on the other, have led to a multiplicity of different detail variants of concrete segments.

The thickness of segments is determined according to structural and detail considerations. The minimum thickness is mostly determined by the requirement to transfer the thrust loading and the resulting contact surface of the thrust cylinder shoes. The thickness used is usually between 20 and 50 cm. With increasing cross section size, the segment thickness also has to be increased; segments of 60 cm thickness were required for the 4th tube of the Elbe tunnel.

The width of the concrete segments varies between 1.00 and approx. 2.00 m, with a tendency to larger segment widths being recognisable due to technical developments in formwork technology as well as in the technology of transporting and installing the segments. This shortens tunnelling times and also reduces the length of joints. With increasing segment widths, manufacturing and installation inaccuracies lead to more concentrated loading of the joint areas and therefore to increased problems with spalling and cracks. Further, the use of larger width segments reduces the clearance for driving curves, and the required stroke length of the thrust cylinders is increased.

The segments need to be reinforced in ring direction to resist bending moments from external loading. To guarantee serviceability, the installation of a nominal minimum reinforcement in ring and longitudinal directions is recommended. The splitting tension loading in the ring joints created by the thrust cylinder load must be covered by suffi-

15.3 Lining with Concrete Segments

cient reinforcement. This applies likewise to the transfer of the eccentrically acting pressure forces into the longitudinal joint.

15.3.2 Construction Types

15.3.2.1 Block Segments with Right-Angled Plan

This variant is the type used most often. A ring is built with five to eight single segments and one key stone. Due to the rectangular geometry, the result is flat ring joints and each ring is alone stable and load-bearing. The key stone, which is the last segment of the ring to be installed, is generally smaller then the other segments and has a wedge shape. This enables the closing/completion of the ring by inserting the keystone along the direction of the tunnel. The spreading effect can cause a pre-loading of the segmental lining in ring direction.

It is advantageous for structural reasons to design all ring segments the same size, which means that the keystone has about the same opening angle as the other segments. A smaller easily manageable keystone is better for the installation of the rings. It must be considered, which system is more effective for each project. Large keystones are being used more and more.

The use of block segments with rectangular plan has prevailed for single-shell waterproof tunnels. The sealing is achieved through the provision of sealing gaskets all round, which are set into the groove provided in the ring and longitudinal joints. In addition to the provision of gaskets, high ring quality is necessary to guarantee the watertightness, along with minimisation of restraint loading and high stiffness of the segment tube. The latter is usually accomplished by staggering the longitudinal joints of neighbouring rings (Fig. 15-4). On one hand, this reduces the deformation of the pipe, in combination with the additional ring joint toothing coupling neighbouring rings (see Section 15.2.3.2) and on the other hand the problem of waterproofing the crossing joints.

A completely different procedure for key stone installation is usual in Switzerland, the five element segmental lining and an invert key (Fig. 15-2b). The segment elements themselves are not sealed, but are supplemented by a sealing foil between the segments and the inner shell. The rings are installed using the process shown in Fig. 15-5: First, the two invert segments are set and then the side wall segments. These have to be supported temporarily on a carrier roller, since the segments are not bolted to each other and because there are no retaining forces available during installation from the thrust cylinders. Then the crown segment is raised into position with the erector. The erector remains in this position to support the segments. Then, the invert joint is spread apart with bolts located in recesses in the invert segments and the straight keystone lowered into the gap created. Finally, the force spreading the invert segments is removed and the erector and the carrier rollers, which are now no longer required, are withdrawn from the segments, upon which the crown segment sinks slightly. The segment ring is now in its final position and the pressure ring of the thrust cylinders is brought forward again.

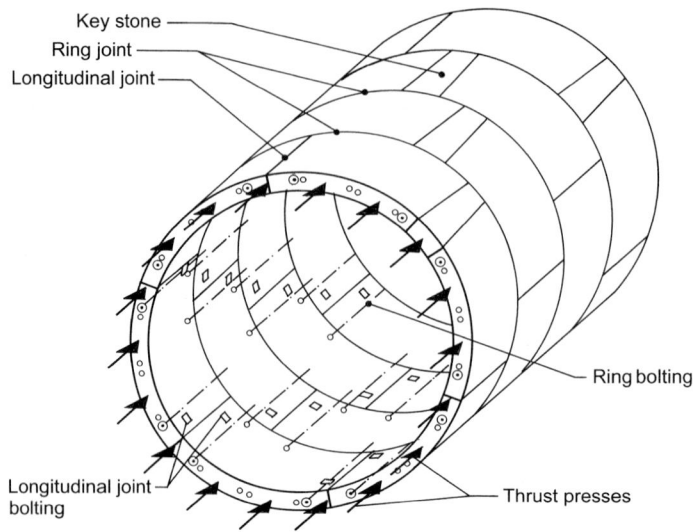

Figure 15-4
Block segment with keystone and staggered longitudinal joints

The advantages of this method are the avoidance of excessive restraint forces caused by violent forcing of the wedge-shaped key segment into position and the short installation time per ring (15–20 minutes). The disadvantages are less precision and the resulting spalling of the edges of the segment.

In general, the route of a tunnel is a curve in three dimensions, which the TBM has to maintain as precisely as possible. The installed ring must therefore follow the direction taken by the TBM without causing damage to the lining by contacting it with the shield [2]. This would create an opening of the ring joints on one side in curves. This is not acceptable for single shell tunnels specified with waterproofing, since the sealing gaskets could lose their effectiveness. To enable driving round curves without needing an opening of the ring joints, the rings are narrower one side. In principle, two different systems can be used:

With the use of a tapered universal ring, it is possible to set any direction by appropriate turning. The advantage is in the low expenditure on formwork and logistics when using the same ring, which has a decisive economic advantage for the quantity production of segments. Bearing in mind the requirement that the longitudinal joints need to be staggered, it should be noted, however, that in turning the ring to steer the shield correctly, a compromise is often necessary. Another disadvantage of the universal ring is the arbitrary position of the keystone. If the keystone ends up in the invert, then the installation of rings has to start with the segment in the crown, which in this case is only held in position by the pressure of the cylinders. This must be taken into consideration in the TBM concept, with additional support measures being provided for crew protection in the erector area.

15.3 Lining with Concrete Segments

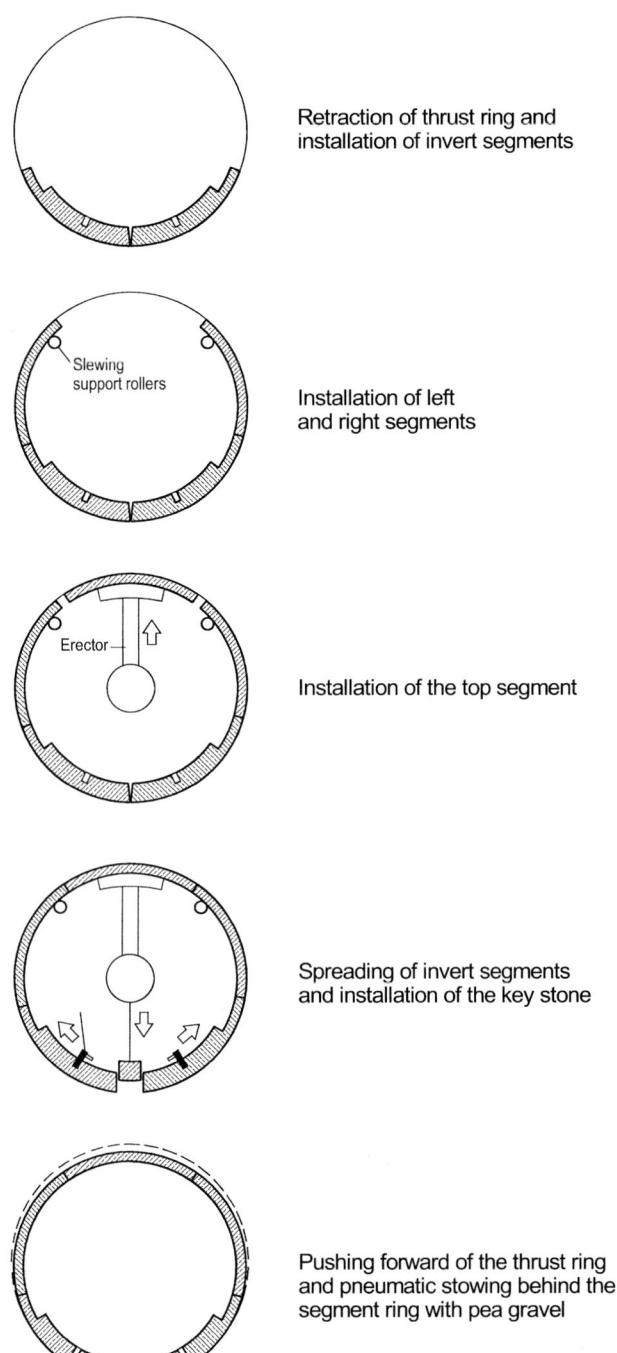

Figure 15-5
Ring assembly process for a five-part segmental lining with keystone at the bottom

Production of several ring types such as right-turning, left-turning and straight-ahead rings can preserve the geometry of the longitudinal joints and the defined position of the keystone. The disadvantages are the relatively high logistical expense and the resulting higher cost of segment production.

The use of block segments requires, with traditional shields, the synchronisation of excavation and ring installation That means that boring can only resume after the ring installation is complete. This restriction results primarily from the limitation of the stroke of the thrust cylinders.

The desire for a continuous installation sequence led to the development of spiral segments, which were used several times in Berlin and Stuttgart (Fig. 15-6). However, this solution did not catch on.

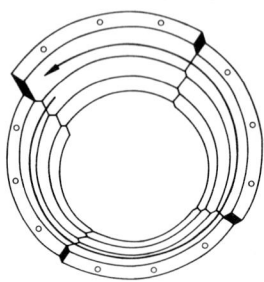

Figure 15-6
Spiral segments, Berlin underground railway, 1965/66 [95]

15.3.2.2 Hexagonal or Honeycomb Segments

Hexagonal or honeycomb segments were used during the construction of the Happurg water tunnel as long ago as 1961 [95]. The type is used more often today as the outer shell of a double-walled tunnel cross section, but also as single-shell solution, particularly for water tunnels when using a double shield TBM [54, 84, 175, 176]. Because of the hexagonal form of the individual segments in this system, there is no continuous ring joint, since it is stepped by a half of the segment width between radially neighbouring stones. This results, with simultaneous locking of the ring joints, in an altogether more rigid tube than could be accomplished with a rectangular block segment design without stepped longitudinal joints. An installed "ring" consisting of hexagonal segments may, in the smaller diameters, only need four segments, with the invert and roof segments facing each and also the side segments facing each other (Fig. 15-7).

A great economic advantage of this segment system is that the ring installation only requires one type of segment. This leads to substantial cost savings in segment production compared with the production and use of rectangular block segments. Only the invert segment, for operational reasons, is often different from the standard design (see Section 15.2.2.3).

Disadvantages of this type of segment result with increasing cross-sectional diameter in the size of the segments and the resulting difficulties in transport and assembly. Hexa-

15.3 Lining with Concrete Segments

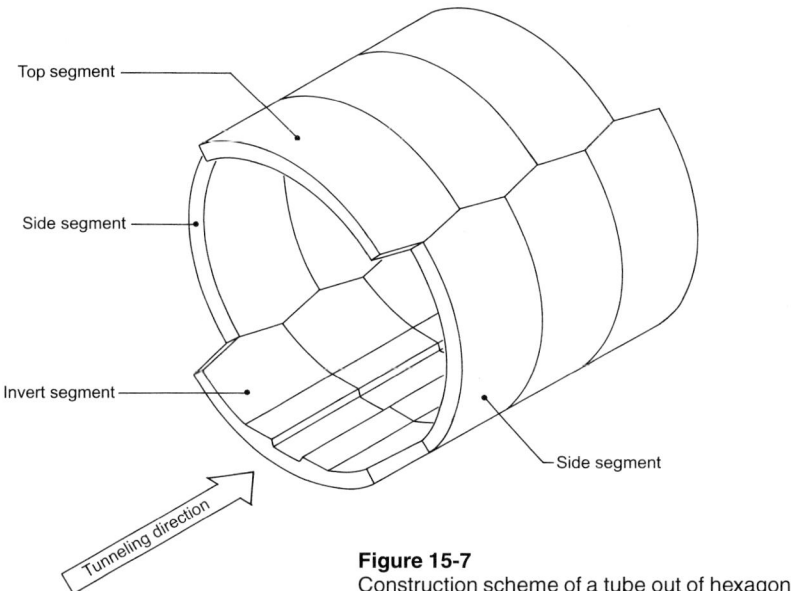

Figure 15-7
Construction scheme of a tube out of hexagonal segments [176]

gonal segments are therefore more often used for smaller cross sections with diameters up to 4.50 m. The fact that larger diameters can also be constructed is show by the examples of the pressure tunnels Plave II and Doblar II in Slovenia, which both have a tunnel diameter of 6.98 m [175].

15.3.2.3 Rhomboidal and Trapezoidal Segment Systems

The desire for an almost completely automated process of segment installation led to the development of segment systems with dowels and guiding rods [177]. Optimisation of the segment geometry is supposed to reduce the need for forcing during the installation and the provision of a tongue and groove in the longitudinal joints as well as centring dowels in the ring joints to achieve high assembly accuracy and quality.

Figure 15-8 shows the segment system used in the construction of a single-track railway tunnel in Paris [177]. The segment ring consists of a trapezoidal invert segment, four rhomboid segments and a trapezoidal keystone. Due to the sloping longitudinal joints of the segments, the sealing frames of the segments only come into contact in the last few centimetres of installation. However, if the joint is twisted, the slanted longitudinal joints cause misfitting and forcing of the deformed ring.

The connection in the ring joint is made by three assembly Conex-dowels per element (Fig. 15-9), or one dowel for the keystone. For precise installation, 5 cm thick guiding rods are used in the longitudinal joints. This system was successfully used in the construction of the railway tunnel in Milan.

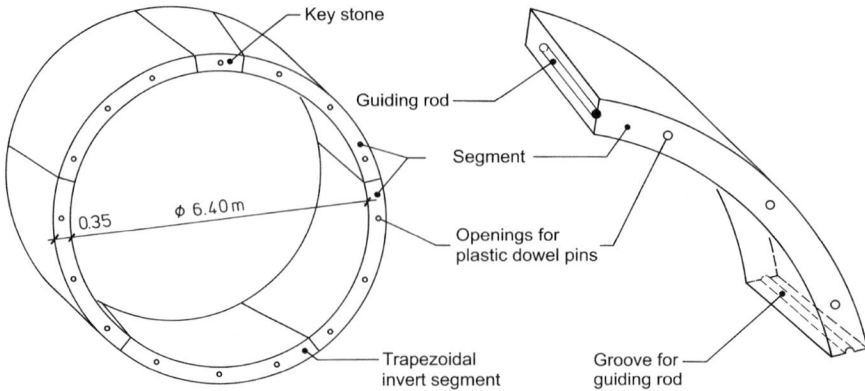

Figure 15-8
Tunnel cross-section EOLE Paris, internal diameter 6.40 m

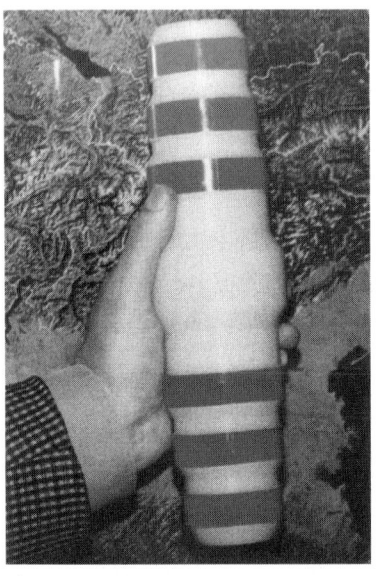

Figure 15-9
Conex Lining System, Socket pin segments,
Passante Milano (Mayreder) [95]
a) Plastic socket pin
b) Pre-assembled socket pins in segment

15.3.2.4 Expanding Segments

In stable rock bearing little water, the rings can also be installed behind the shield. By expanding of the segment, the ring is braced against the solid rock and adopts a stable position. Grouting or filling behind the segments is not necessary if a regular circular tunnel section has been cut during the excavation.

Expanding segments originated in the construction of underground railways in London in clay soil with a long stand-up time [73]. Figure 15-10 shows a typical cross section, which was used to a large extent in the construction of the Channel tunnel on the British

15.3 Lining with Concrete Segments

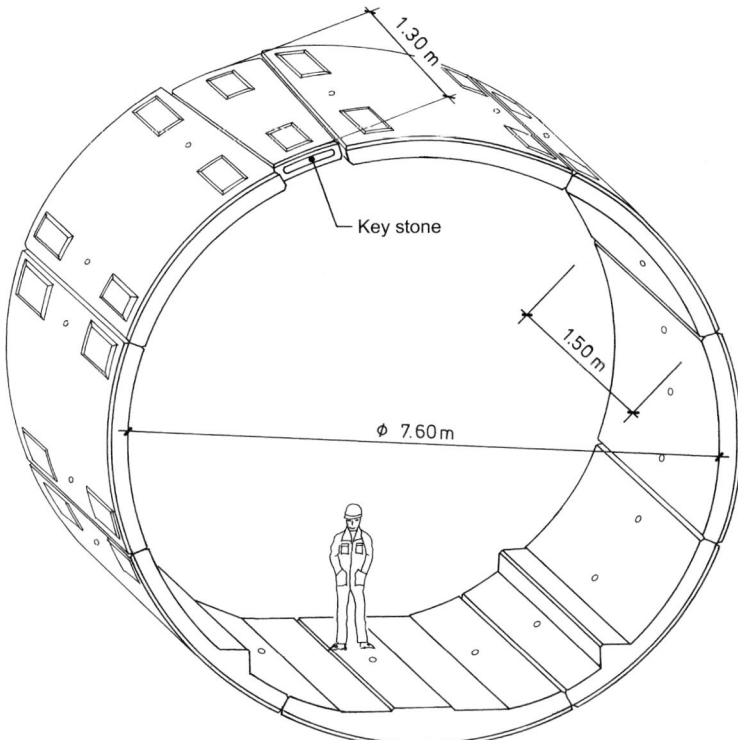

Figure 15-10
Typical cross-section of the Eurotunnel with expanding wedge-block lining on the British side [42]

side. The expansion of the segment ring is accomplished by pushing in the wedge-shaped keystone, which is shorter in comparison to the other segments [10].

This lining system enables short ring installation times with resulting higher tunnelling advance rates. A big disadvantage is in the fact that each ring has a different geometry and this may result in large joint offsets. Due to the relatively large joint twisting in the longitudinal joints, spreading segments are usually unsuitable for single-shell, water proof linings.

15.3.2.5 Yielding Lining Systems

According to the models connected with the use of the shotcrete method, squeezing of the excavated cavity causes the development of a so-called rock bearing ring, which participates in the relaxation of the ground pressure. This makes the installation time for the support specially significant, since this to some extent determines the share of the loading bearing on the lining. If a decision is made to install the lining earlier, this means that less time is allowed for the rock to relax though deformation, thus increasing the load applied to the support. It should, however, be noted that, depending on the geological conditions, the ground pressure may rise again as the rock progressively deforms and destabilises.

The immediately load-bearing and relatively stiff segmental lining resists the ground pressure immediately once the shield has advanced and the gap between the lining and the rock has been filled. This can lead, in the case of deep overburden or or in swelling ground, to over-proportionally high loadings, which in extreme cases cannot be carried by the lining.

For this reason, various parties have attempted to develop flexible lining systems, which permit controlled convergence of the tunnel wall. These measures can be differentiated into measures concerning filling the annular gap behind the segments and concerning the longitudinal joints.

Flexible support can also be implemented with concrete segments by using a special mixture for filling the annular gap. A mixture of polystyrene balls and sand can be used for this purpose. Particularly suitable would also be polystyrene balls surrounded with cement slurry, which would solve the problem of separation. Figure 15-11 shows the potential deformation of a polystyrene – sand mixture.

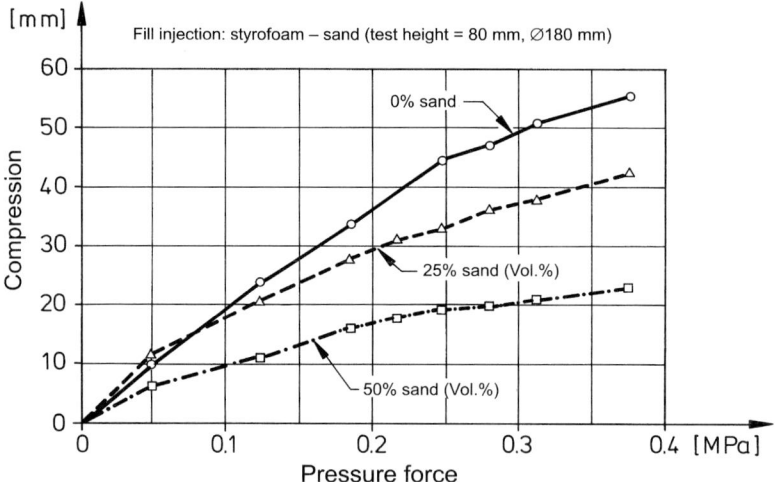

Figure 15-11
Test on yielding backfill material with high potential deformation [142]

The resilience of the segmental lining can also be achieved by the use of compressible inserts in form of steel or plastic profiles in the longitudinal joints. Figure 15-12 shows schematically how convergence is enabled by the plastic deformation of a longitudinally inserted steel tube. The compression force in the lining is limited by the transverse compression taken by the steel tube in a plastic condition.

The design of flexible joint construction shown in Fig. 15-13 is a continuation and refinement of this principle. A plastic hose filled with water under pressure serves as a

15.3 Lining with Concrete Segments

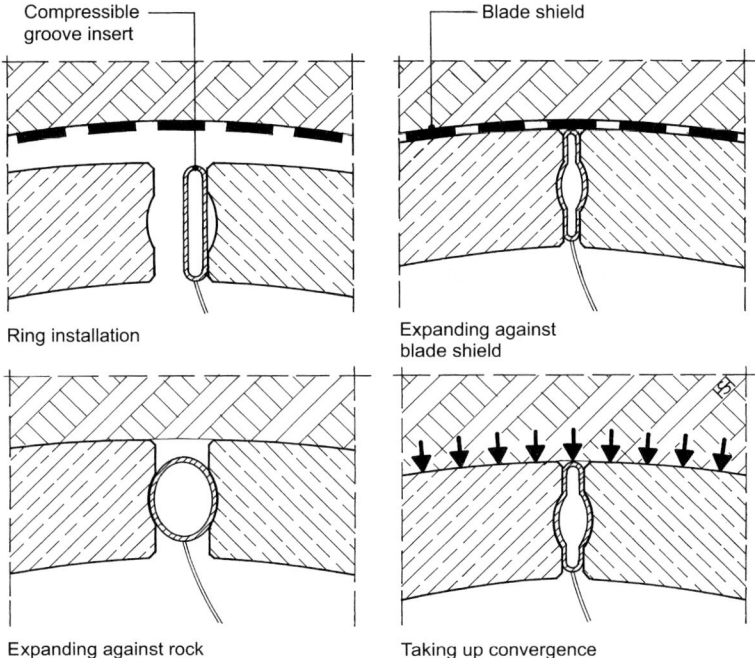

Figure 15-12
Yielding construction of longitudinal joints with yielding steel pipe [123]

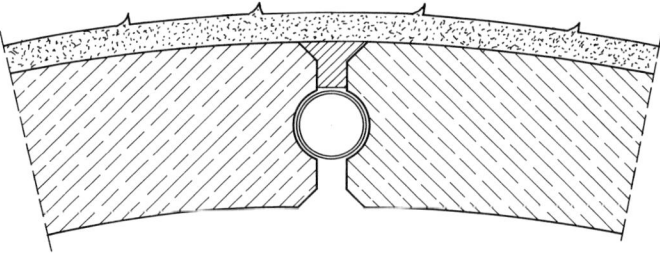

Figure 15-13
Proposal for a yielding joint construction by Thompson [167]

flexible element, which is grouted with mortar after the ground deformation has subsided. Valves in the water outlet pipe, which open at a defined water pressure, prevent overloading of the segments and enable the compression of the segment ring through the reduction in volume of the hoses as water is let out.

Another suggestion proposes the use of closed, ductile plastic elements (Fig. 15-14), which are simultaneously intended to fulfil waterproofing requirements. The element, which takes up almost the entire meeting surface, forms a system of chambers with a compressible filling of aerated concrete. The design of the chamber system permits the setting of the load-deformation behaviour of the system.

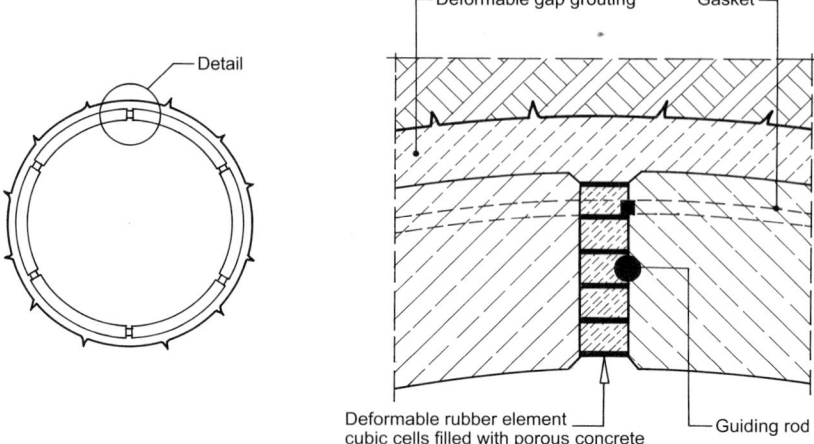

Figure 15-14
Proposal for a yielding element out of plastic [164]

Using a compressible porous concrete for the annular gap behind the ring is another method of providing flexibility for a certain deformation.

A segmental lining with Meypo compression elements in the longitudinal joints was used in a backfill station at the Ibbenbueren mine. During the construction of the filling station, the effective ground pressure corresponded to a depth of 1650 m, which increased with further coal mining to values which equated to a depth of approximately 2000 m.

The tunnel lining shown in Fig. 15-15 consists of eight segments with eight longitudinal joints, of which four are formed as compressible joints. The cross-section has a clear diameter of 8.5 to 9.5 m and a lining resistance of over 1000 kN/m^2. The flexibility over the circumference amounts to 6 · 30 cm, which corresponds to about 6% of the total circumference.

The mechanism for allowing flexibility of the Meypo compression elements is shown in Fig. 1516. The essential components are the shear ring with its hardened shear pin to ensure the shear force bearing capacity and the compression plunger, which folds together in waves under compression.

The working curve of the compression element shows nearly ideally elastic plastic behaviour. Up to 85% of the maximum load, the compression element behaves stiffly, until after only a few millimetres insertion the loading, which can be taken, is held by further plastic deformation.

The effectiveness of the flexible lining system has been demonstrated with the insertion of the plunger up to 17 cm [9]. The initial application, however, showed that the costs of this system are too high for ordinary tunnel construction. Another suggestion therefore proposes the use of demountable compression elements, which are removed after the convergence has subsided and the longitudinal joints have been concreted, and

15.3 Lining with Concrete Segments

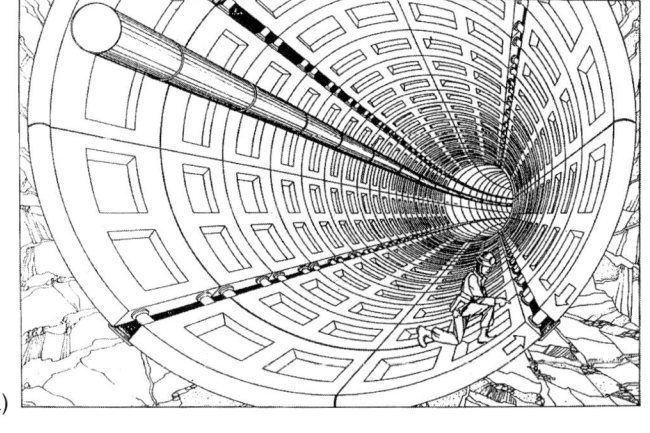

Figure 15-15
Yielding segmental lining at the Ibbenbüren backfilling station [19]
a) Schematic diagram
b) Completed segment lining

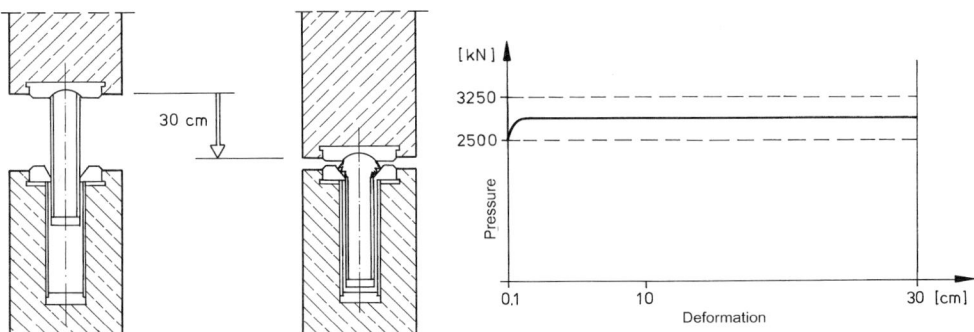

Figure 15-16
Construction and working curve of Meypo compression elements [19]

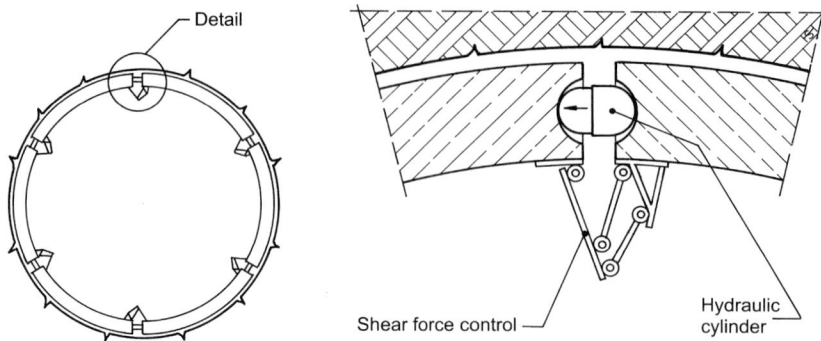

Figure 15-17
Suggestion for a reusable yielding lining [9]

re-used for ring installation behind the tunnel face (Fig. 15-17). Hydraulic cylinders are used as flexible elements. The transfer of shear forces is provided by removable framed steering systems. This system has not yet been proven in practice.

15.3.3 Joint Detailing

The proportion of joints in the tunnel tube is relatively high due to the segmental building of the individual rings and the ring-wise production of the lining. These are the longitudinal joints running parallel to the tunnel axis and the radial ring joints, which differ in function and design. The suitability of the selected joint design regarding load-bearing capacity, risk of spalling and water-tightness should be demonstrated, if possible, by the implementation of an appropriate series of tests [13, 28, 149, 150].

15.3.3.1 Longitudinal Joints

The longitudinal joints transfer axial forces in the ring, bending moments due to eccentric axial forces and shear forces from external and sometimes also internal loading. This is mostly performed by the contact of the contact surfaces, but in some cases also by the bolting of the longitudinal segment joints. With the usual segment systems, the longitudinal joints act structurally as joints or effectively as joints (concrete joints) with a limited capacity to resist bending moments.

The design engineer responsible for the detailing essentially has three groups of different joint types available. These can be differentiated into longitudinal joints with:

- Two flat surfaces
- Two convex surfaces and
- Convex/concave surfaces

As a further possibility, the formation of a tongue and groove could be mentioned, which is however designed in most cases with flat contact surfaces.

15.3 Lining with Concrete Segments

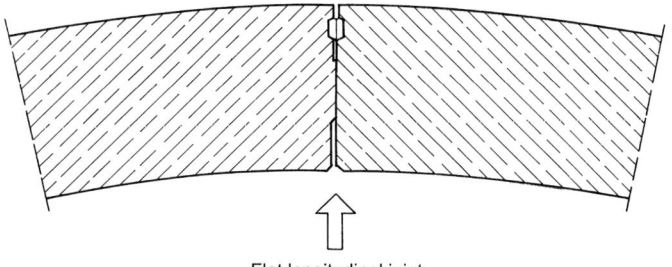

Figure 15-18
Flat longitudinal joint detail

Longitudinal Joints with Flat Contact Surfaces

With a design with flat joint contact surfaces according to Fig. 15-18, the free rotation of the segments is hindered by the geometry. This can mean that, in addition to the axial compression load in the longitudinal joint, bending moments can also be transferred, which reduces the bending loading on the segment.

The turning of the longitudinal joints happens, because of the elastic and plastic compression, at the joint surfaces. Figure 15-19 makes clear the factors affecting the resistance to turning. Because exclusively compression forces are transferred, a stable condition is reached if the resultant of the external forces acts within the cross-section.

In order to avoid the compression stresses at the outer edge of the section acting outside the core, the area surrounded with reinforcement, the contact surface is normally reduced. The width of the neck of the joint is then usually to of the segment thickness. At this point the splitting tension loading resulting from action of concentrated compression stresses should be considered; corresponding reinforcement should be designed to cope with this.

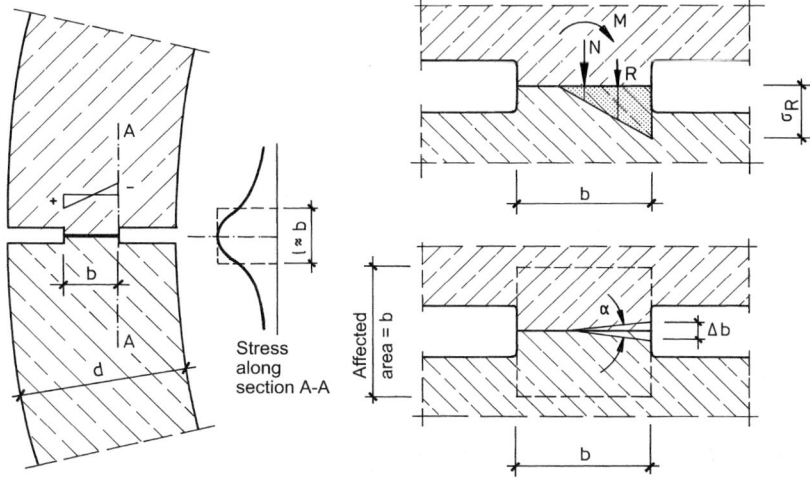

Figure 15-19
Diagram showing the determination of resistance to rotation of flat longitudinal joints [70]

The reduction of the joint neck width leads to a higher rotation capability of the longitudinal joints. This should especially be taken into account with single-shell waterproofed segments, where this rotation capability should be limited in order to keep the sealing profile compressed.

The detailing of longitudinal joints with flat contact surfaces offers advantages during segment erection in that the offsets resulting from unavoidable inaccuracies in building the ring do not lead to concrete spalling.

Longitudinal Joints with Two Convex Contact Surfaces

With flat longitudinal joints, the splitting tension load increases with continuing rotation because of the restriction of the contact surfaces. If there is a large axial force, there is a danger that the concrete spalls off at the outer corners, which can extend into the area of the sealing gaskets in the case of single-shell waterproofed linings. If both the compression forces and the rotation angle become very large, then the convex form of joint shown schematically in Fig. 15-20 is recommended, whose contact surface is not dependent on the rotation angle [10].

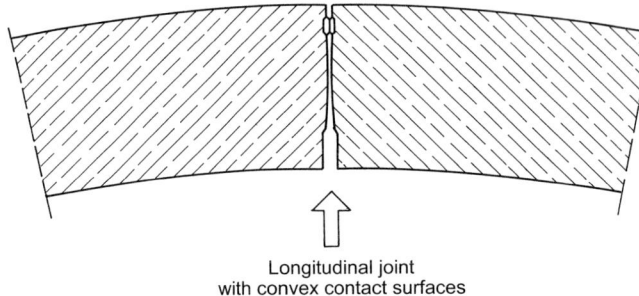

Figure 15-20
Longitudinal joint with convex contact surfaces, Great Belt tunnel [10]

Longitudinal joint with convex contact surfaces

Figure 15-21 shows the scope of application for flat and convex longitudinal joints depending on the compression force and the rotation angle resulting from evaluation of tests.

The radius of curvature of the convex surfaces depends on the segment thickness, the size of the loading and the permissible rotation. The design of the radius of curvature has to meet various requirements. If too small a radius is selected, the surface through which the loads are transferred is limited, resulting in an increase of the splitting tension loading. If the radius is selected too large, this limits the ability of the segment to rotate about the joint.

This system is not stable while the segments are being installed, because there is not sufficient ring compression due to the rotation finding no resistance. Suitable measures (e.g. temporary bolting) should be provided to ensure that the ring does not fall apart during installation.

15.3 Lining with Concrete Segments

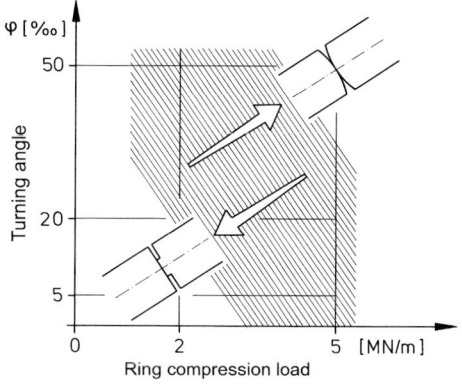

Figure 15-21
Scope of application of flat end convex longitudinal joints [10]

Longitudinal Joints with Convex-Concave Contact Surfaces

Longitudinal joints with convex-concave contact surfaces according to Fig. 15-22, which are also described as articulated segments, normally have a high rotation capability. To reduce the friction limitation and to provide large amount of play during assembly, the radius of curvature of the concave surfaces is generally chosen to be large.

Great care should be taken with the detailing on account of the large angle of rotation, that unwanted contact surfaces at the edges of the joints with resulting spalling from the concentrated loading of compression stresses are to be avoided. The edge areas of the concave side of the joint are particularly endangered, because it is impossible to provide enough reinforcement order to avoid damage to the edges, only the central part of the longitudinal contact surface is normally formed as a joint.

The convex-concave arrangement of the contact surfaces leads to a better stability of the ring during assembly, which is why this is preferred for expanding segments [10]. The centring effect provides another aid to assembly.

Articulated segments are mainly used for double-shelled tunnels, because for single-shell, waterproof applications the problem of waterproofing the joints has not yet been satisfactorily solved [75].

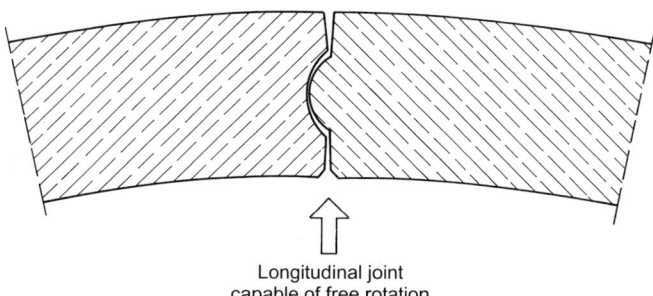

Figure 15-22
Construction of longitudinal joint capable of rotating freely, Quarten tunnel [75]

Longitudinal Joints with Tongue and Groove

Joints with tongue and groove (see Fig. 15-25) offer a good guide during assembly and transfer, if the contact surfaces are straight, axial forces, moments and shear loads. Because the tongue cannot be effectively reinforced, the concrete spalls off if the play in the joint is only slightly exceeded.

A special form of joints with tongue and groove is the one-sided groove with insert. This is mostly only used for smaller keystones, in order to avoid them falling out.

15.3.3.2 Ring Joints

The ring joint level is positioned orthogonally to the tunnel's longitudinal axis. Its loading results primarily from the introduction and transfer of the thrust cylinder forces during construction. With neighbouring rings having different deformation patterns, coupling forces (transverse forces) are created in the ring joints when the deformation is hindered.

The load from the thrust cylinders is applied through shoes (in order to increase the support surface) into the facing side of the ring joint plane and transferred through the segment to the next ring joint. Since unavoidable inaccuracies in the production of the segments mean that a flat contact surface in the ring joint cannot be assumed, there are normally localised contact zones, which load the segment like a wall-type beam. To avoid unacceptably high stress and the resulting formation of longitudinal cracks, intermediate loading plates are used to distribute the load. These are bonded in the centre of the ring joint and are positioned longitudinally to ensure axial application of forces, ideally exactly one behind the other (Fig. 15-23). The materials used are Kaubit or hard fibre board. Deviations in the position of the load distribution plates have to be designed for by additional radial reinforcement in the ring joint area.

Since the height of the surface where the load acts does not extend over the entire thickness of the segment, the loading from the thrust cylinders results in splitting tension, which must be allowed for with appropriate reinforcement.

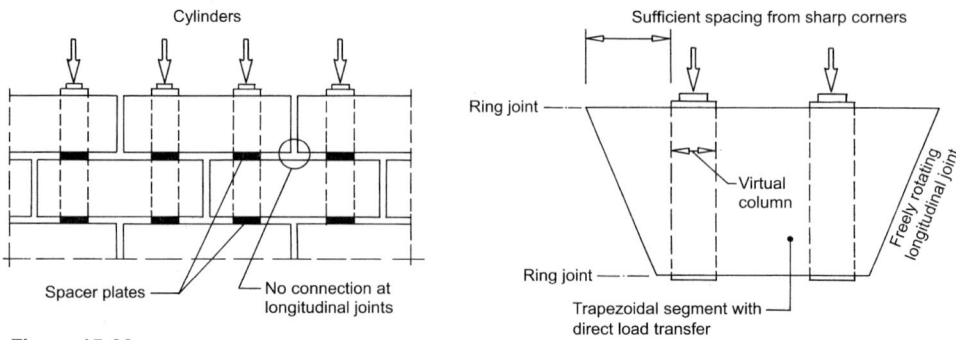

Figure 15-23
Transfer of the jacking forces in the ring joints [115]

15.3 Lining with Concrete Segments

Flat Ring Joints

The simplest form of ring joint design is the flat, level joint as shown in Fig. 15-21. Each ring of this type supports itself without interacting with neighbouring rings, or at least not intentionally. The coupling is only through friction, although the time-dependent release of stored elastic pre-tension from the thrust cylinder forces, for example due to concrete creep, needs to be considered.

The flat ring joint does not offer any assistance with the assembly of segments. Support can be provided, for example with the use of plastic pegs as centring dowels (see Fig. 15-9). Transverse load transfer through joint is not, however, intended. This can be implemented for small coupling forces with a durable bolted connection, e. g. Great Belt tunnel.

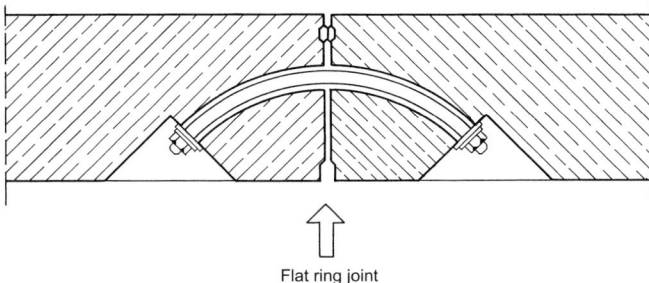

Figure 15-24
Flat ring joint at the Great Belt tunnel [10]

Tongue and Groove Systems

To simplify the installation of rings on one hand and for load-bearing coupling of neighbouring rings on the other, there are various possibilities for designing the ring joint. For single wall, waterproof segment constructions, in particular, the deformation of the joints has to be kept as small as possible.

One possibility is to design the ring joint as tongue and groove system. The tongue usually takes up half of the segment thickness and has a height of 10–25 mm. To enable a defined transfer of the coupling forces, bituminous packing strips can be installed in addition to the load distribution plates, as in Fig. 15-25.

Reinforcement of the tongue and groove elements to cope with the coupling forces is difficult to design while simultaneously providing the necessary concrete cover. Assembly inaccuracies and resulting forcing loads can therefore lead to damage in the ring joint area. To minimise this damage, the groove is made larger then the tongue. The available play usually only amounts to a few millimetres and is quickly eaten up by manufacturing and installation tolerances.

In addition to tongue and groove, convex-concave designs of the ring joint are also known, as shown in Fig. 15-26. Here, also, it is the edge surfaces of the concave surfaces, which are at risk of damage during assembly.

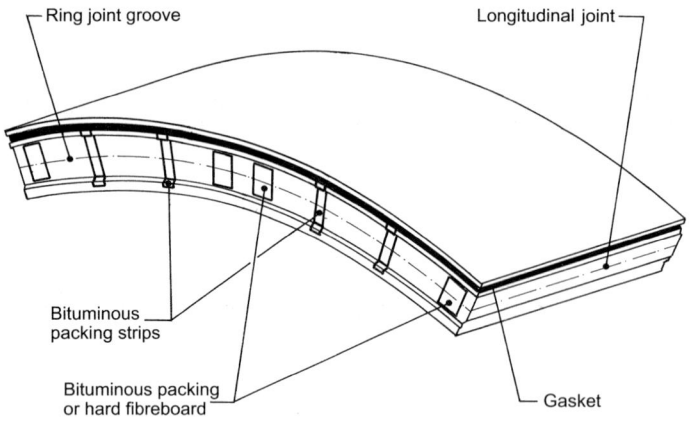

Figure 15-25
Segment equipment in the ring joint as tongue and groove system [28]

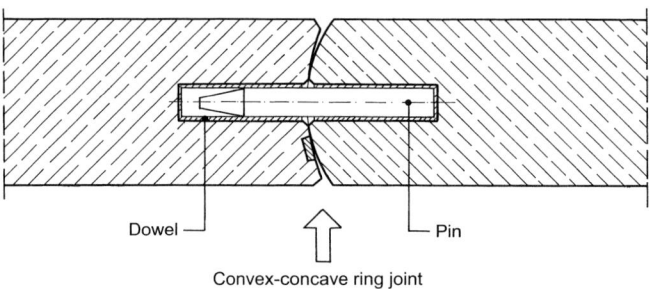

Figure 15-26
Convex-concave ring joint detail [156]

Cam and Pocket System

In contrast to the tongue and groove design, the cam and pocket system represent a point coupling, for example in the quarter points of the segments (Fig. 15-27). The system restricts forcing strains caused by assembly inaccuracies to the cam and pocket. The coupling loads, however, represent locally restricted loads, which with the tongue and groove system are spread more into the ring joint. Depending upon the prevailing conditions, the coupling locations can be highly stressed in operation.

Because the cam is normally thicker then a tongue, it can be reinforced to a certain extent. The height of the cam should be selected so that a failure would occur at the cam and not shear off the edge of the pocket, so that the waterproofing is preserved.

15.3 Lining with Concrete Segments

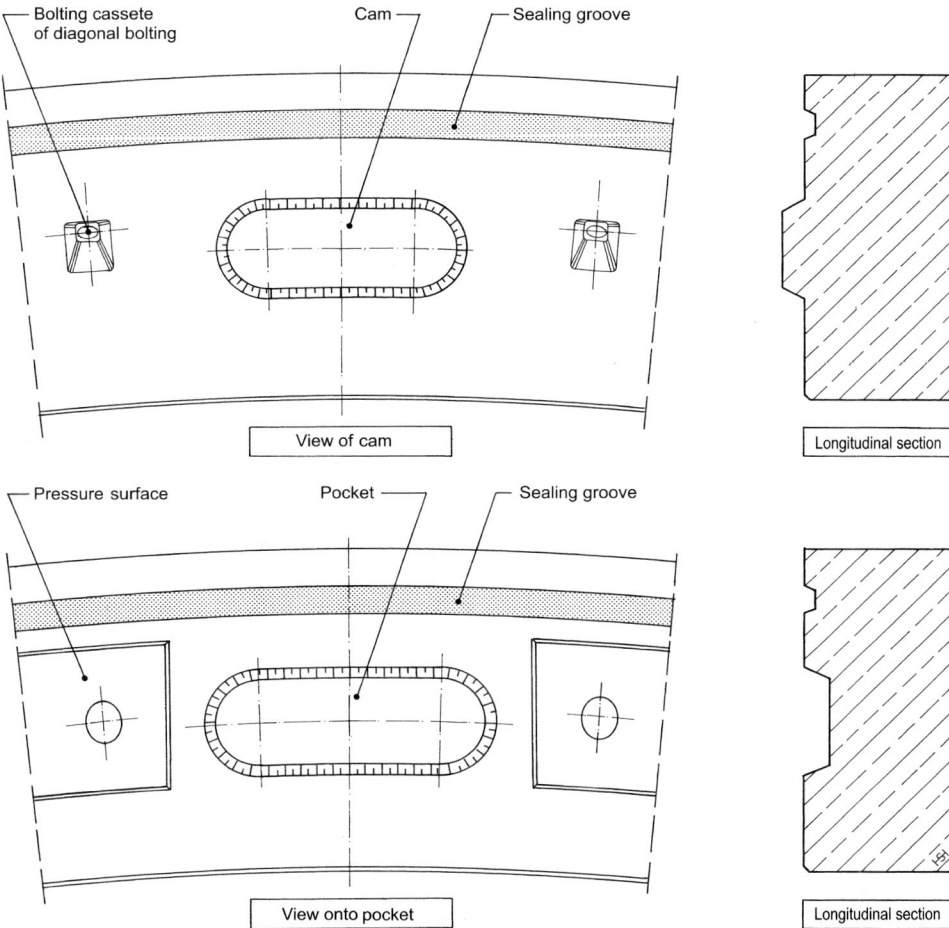

Figure 15-27
Ring joint with cam and pocket system

15.3.4 Steel Fibre Concrete Segments

External loading of the tunnel tube often causes compression loading with little eccentricity, which does not require any reinforcement for structural reasons. However, the provision of nominal reinforcement is usually required and necessary. In such cases, steel fibre concrete can be considered as a practical and economic alternative to conventional reinforcement [95].

Steel fibre concrete is considered to be a relatively ductile building material with high working capacity. The crack distribution effect of the steel fibres in the concrete meets the requirements for water tightness in single wall segment linings. The positive influence on crack behaviour through the typical compression and bending loadings in tunnel linings should be pointed out particularly [36, 44, 110, 112].

A further advantage from the use of steel fibre concrete is particularly the strengthening of the sensitive corners and edges of the segments, which due to the required concrete cover is only insufficiently done by conventional reinforcement [27]. The extent of damage here can be reduced substantially.

For segment production, higher contents of steel fibre can be included than with cast in-situ concrete. As a rough figure, 60–80 kg/m^3 can be stated, with a fibre length of up to 60 mm.

15.3.5 Grouting Annular Gap

As a normal part of using segments in shield TBM tunnelling, a gap is created between the rock and the lining, the so called annular gap. This gap must be filled with suitable material to provide the necessary bedding of the segment tube, to evenly distribute loads from rock pressure and to counteract any possible loosening of the rock.

15.3.5.1 Filling with Gravel

With a TBM tunnelling in hard rock, the annular gap is usually filled with a fine-grained and closely graded gravel (Fig. 15-28), which is blown into the gap with, for example, an ordinary dry shotcrete machine [174]. The ring gap should be filled as soon

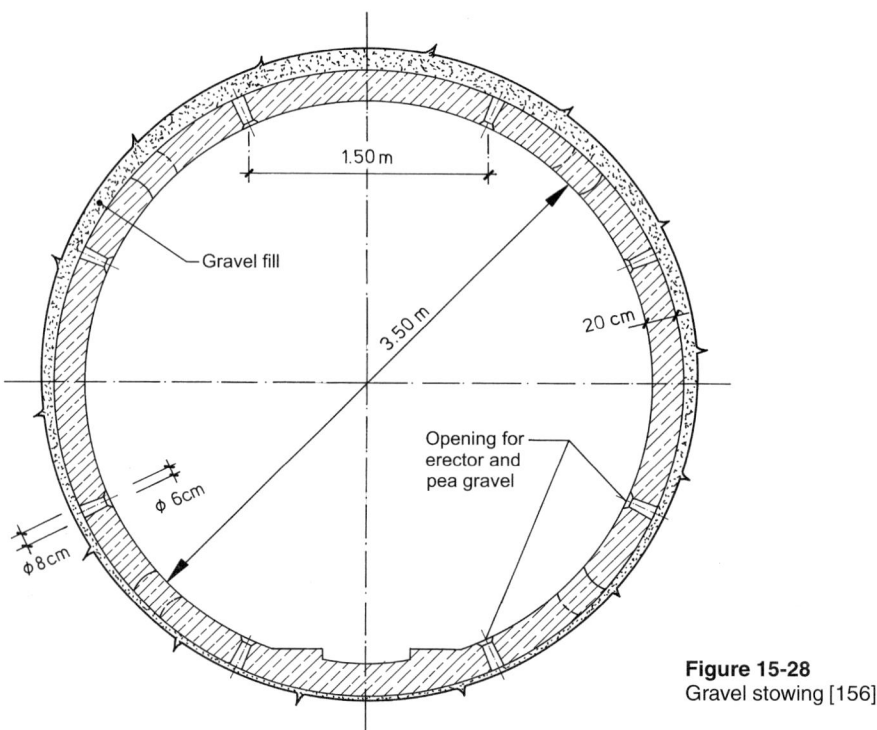

Figure 15-28
Gravel stowing [156]

as possible after the ring installation. Openings should be provided in the segments to attach the nozzles.

The voids in the gravel fill can also be injected with liquid mortar in a later step to prevent the draining effect of the permeable gap.

15.3.5.2 Mortar Grouting

When tunnelling through rock with low stability, the annular gap is injected with mortar. The necessary grouting pressure is coordinated with the existing ground and water pressure. Since the mortar is generally not taken into consideration for structural stability, no particular requirements are made regarding its strength in the final state. To justify the bedding coefficient value that has been assumed, it must at least reach the stiffness of the surrounding rock mass.

In order to be able to pump the material, the mortar must exhibit sufficient flowing characteristics as it is pressed into the gap, and depending on the rock conditions, part of the mixing water is lost into the surrounding rock, thus activating the granular consistency of the mortar as a supporting medium. At the same time, the filter water loss means a loss of volume for filling the gap, whose extent must be kept within bounds by low water content and high solid content.

A high compacted density is achieved by adding fine-grained material like fly ash. For rapid stabilization of the granular structure, cement is normally used as a bonding agent. The setting and stiffening behaviour must be adjusted so that the mortar can still be injected through the injecting lines even after longer standstills, to keep resulting flushing and cleaning measures to a minimum.

The grouting can be done either through openings in the segments or through the tail shield. For grouting through the segments, the segments are designed with closeable openings with threaded ends for the connection of injection lines. Another possibility is to use a plastic one-way valve integrated into the segments.

It is essential for tunnelling with low settlement specified that the annular gap should be injected as soon as possible and as near as possible behind the shield. In particular when the rock has low stand-up time or is not stable at all, there should be no room for rockfall behind the tail shield. Grouting of the annular gap directly through the tail shield has been made possible by the development of high capacity tail shield sealing systems like modern plastic seals or wire brushes seals. Grouting of the annular gap can be synchronized with the advance of the shield, which reduces the ground falling in.

15.3.6 Measures for Waterproofing Tunnels with Segment Linings

Single shell segmental linings are in most cases required to be sealed. Traffic tunnels have to be sealed against water inflow, while water tunnels must be sealed against the loss of water. Since the segmental lining has a high proportion of joints, the cost of guaranteeing the waterproofing is relatively high.

15.3.6.1 Sealing Bands

The great number of single-shell traffic tunnels with segmental lining in hard rock and loose ground are constructed to prevent the inflow of geological or groundwater by the use of seals in and along all ring joints. The sealing band is a prefabricated frame with vulcanized corners and is glued into the prepared edge grooves in the segments. The bracing of the segments to each other compresses the contacting seal profile sections and seals the joints (see Fig. 15-26). The contact pressure of the sealing frame must be higher than the water pressure acting on one side. The concrete surface of the groove must be free of defects to avoid water getting past.

The compression behaviour of the profile and the design of the groove must be coordinated so that concrete chipping behind the groove caused by splitting tension is prevented. The relationship between joint opening and compression force as well as control pressure and joint opening are shown as an example in Fig. 15-29 for profiles from the manufacturer Dätwyler.

The required contact pressure is applied to the ring joint by the reaction from the thrust cylinders and elastically stored. In the longitudinal joints, the sealing profile is com-

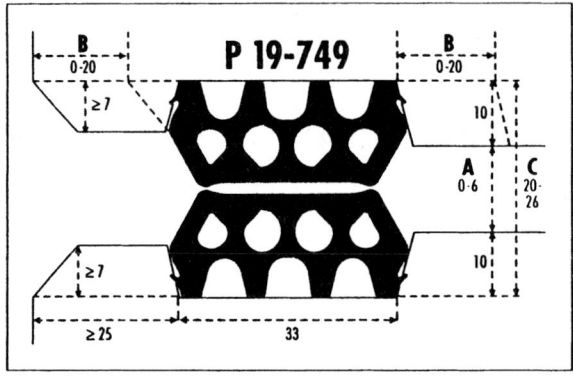

Figure 15-29
Elastomer sealing band

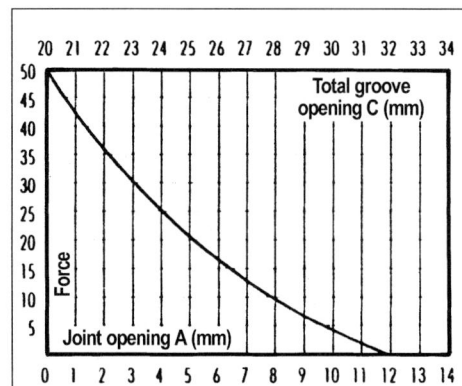
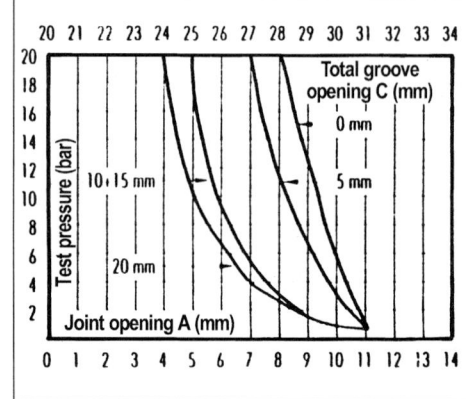

15.3 Lining with Concrete Segments

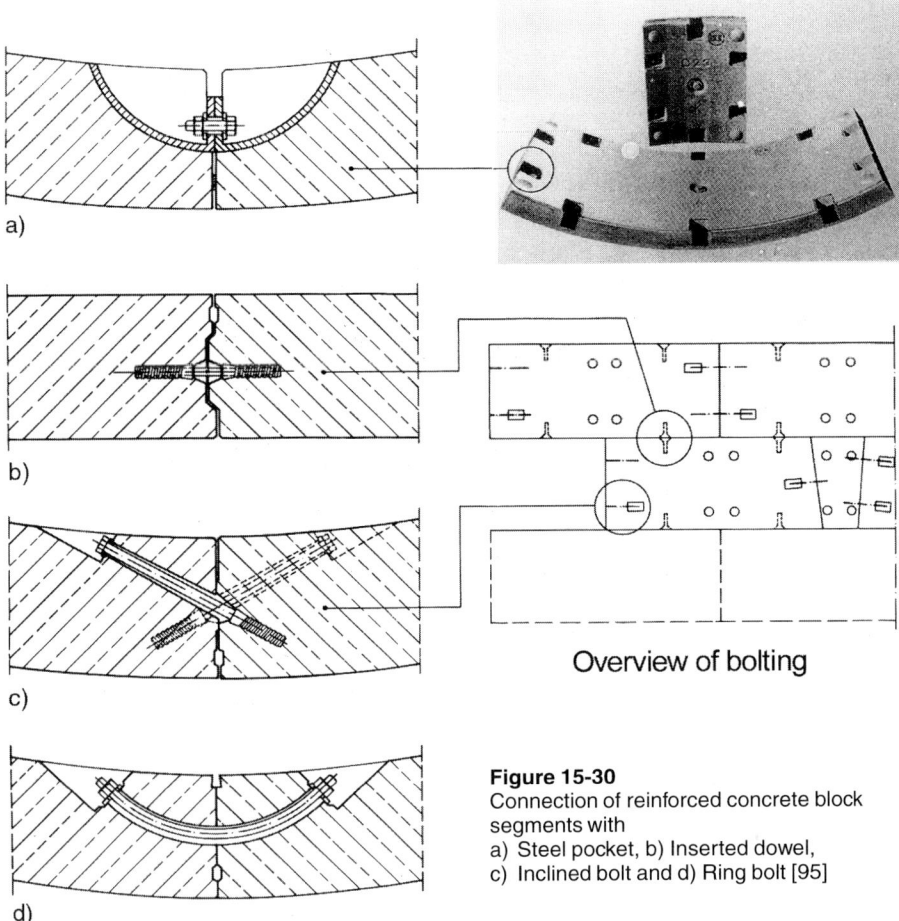

Figure 15-30
Connection of reinforced concrete block segments with
a) Steel pocket, b) Inserted dowel,
c) Inclined bolt and d) Ring bolt [95]

pressed by the axial compression load from ground and water pressure. "Breathing out" by the sealing gaskets should be temporarily prevented during the installation phase by a bolted connection. Figure 15-30 shows the possible variants. In portal and cross cut areas, the bolted connection should be made permanently.

Seal materials used at present are natural rubber types, plastics, elastomers, neoprene, silicone and swelling rubber types (Hydrotite, etc), which have to fulfil high requirements for durability. Considering the planned service life of the tunnel, the efficiency of the sealing gaskets, taking into account their relaxation as well as possible ageing effects, must still be preserved, even after a service life of over 100 years. As shown in Fig. 15-31, as seals relax, the compression is reduced, depending upon the composition of the seal material, to approximately 70% of the original value [56].

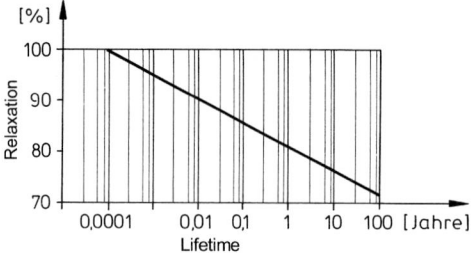

Figure 15-31
Relaxation of an elastomer profile at room temperature [56]

15.3.6.2 Injecting

Another possibility of ensuring the required waterproofing criteria lies in reducing the inflowing geological water by filling the joints in the surrounding rock by injecting.

For the construction of the Evinos tunnel, a water tunnel with a single-shell lining of hexagonal segments (Fig. 15-32), this method was used to limit the water losses to the contractually set quantity. A programme of injecting was developed for this purpose that intends contact injecting of the annular gap filled with pea gravel and subsequent systematic consolidation grouting of the surrounding ground.

The joints have to be so well sealed that the necessary grouting pressure can be applied. Mortar was applied to the longitudinal and ring joints for this purpose. During the backfill grouting, the filtering out of the excess water from the injection compound causes a stopping of the joint, so that the full injecting pressure can be applied and also the joint is thoroughly filled and the load transfer can be over the full surface [176].

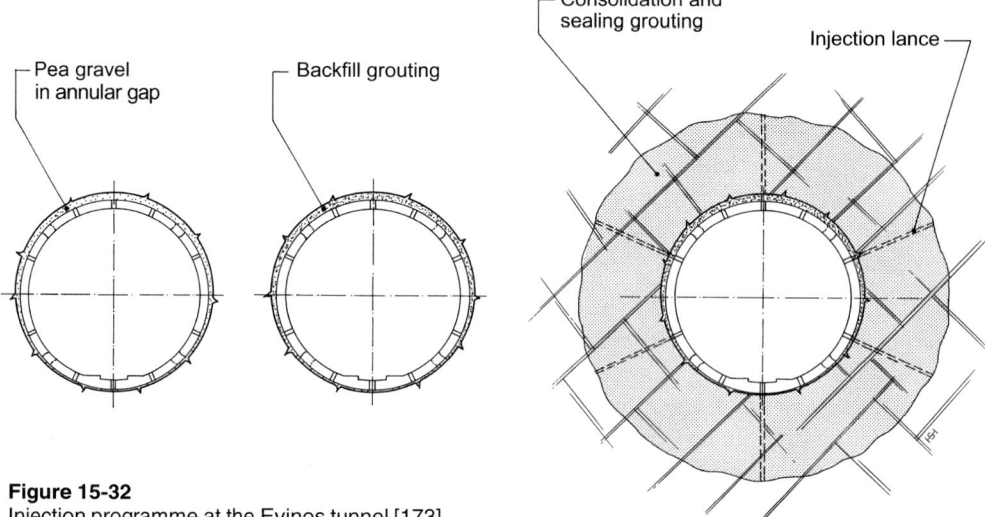

Figure 15-32
Injection programme at the Evinos tunnel [173]

15.3.7 Segment Production

Segments are normally manufactured in specially built precast concrete plants. For larger tunnelling projects, it can be more economical to set up a factory in the field specially for this, like for example for the Channel tunnel or the Belt tunnel. The areas required for this have to be considered in the construction site setup. Especially large projects need detailed logistical planning and development for a smooth supply of segments.

To observe the high requirements for the precision of segments, only steel formwork is practical. The precision requirements for manufacturing the segments are determined by the sealing function needed and the forcing stresses having to be minimised. Technological progress in moulds technology today enables tolerances, which are oriented on mechanical engineering standards. For railway tunnels in Germany, the strict dimensional tolerances from DS 853 [29] apply, as shown by excerpts in Table 15-2 and which have been the basis for specifications on many international projects.

Table 15-2
Dimensional tolerance for segments [29]

Segment width	± 0.5 mm
Segment thickness	± 2.0 mm
Segment curve length	± 0.6 mm
Flatness of longitudinal joints	± 0.3 mm
Flatness of ring joints	± 0.3 mm
Twist angle in longitudinal joints	± 0.04°
Taper angle of longitudinal joints	± 0.01°

The adherence to tolerances must be checked regularly and at short intervals in order to be able to recognize distortion of the moulds at an early stage of production. Infringement of tolerances may cause unplanned forcing of the segments during assembly or in operation. These can reach a magnitude, which is too much for the concrete or the reinforcement. If the tolerances cannot be kept in practice, then the effects on the segments should be investigated.

The longitudinal joint taper should be mentioned here as an example. With increasing taper angle of the longitudinal joint, the statically calculated uniform loading across the segment width can no longer be assumed. This results in eccentric loading of the joint in longitudinal direction, which may have to be considered with additional reinforcement.

The quality controlled precasting of the segments should be demonstrated in the construction phase with suitable quality control programs, such as QA manuals including appropriate work instructions.

15.3.8 Damages

Damage to segmental linings is mostly caused during construction. Types of damage can range from the formation of isolated cracks to large-scale spalling. The cause of damage in most cases is the creation of unacceptably large contact stresses as a result of high loading from advance the tunnelling machine combined with manufacturing and installation tolerances of the segment blocks with resulting geometrical lack of fit. Geometrical and kinematic studies should be performed in the design phase to investigate the segment design for any unwanted contact surfaces.

Figure 15-33 shows the example of the El-Salaam siphon, a water tunnel under the Suez canal, where spalling of the segments occurred at the longitudinal joint of the neighbouring segment to the key stone, which can be partially explained by the unfavourable detailing of the longitudinal joint with an interior tongue between the keystone and its neighbouring blocks.

If segments are solely used for temporary support behind the TBM as outer shell of a tunnel, damage can be regarded as uncritical, as long as the structural stability of the segment ring is not endangered and the resisting and transfer of the thrust forces is not impaired. In such cases, complete repair, even of serious damage, is not usually necessary.

With the use of single-shell and waterproof segmental linings, however, damage is critical where the structural safety and thus the water tightness is prejudiced. In such cases, the repair of the damage is inevitable, often requiring high expenditure and therefore representing a not to be underestimated cost factor.

Figure 15-33
Spalling of segment at the El-Salaam syphon under the Suez canal

Details are discussed below of possible causes of damage to single-shell, waterproof segment systems during segment installation, on applying the thrust forces, in the tail shield area and after leaving the shield, as well as repair methods

15.3.8.1 Damage During Ring Erection

The most common cause of damage to the concrete surfaces and seals during ring assembly is inexpert placement of the segments with the erector. In addition to this, damage occurs particularly with the insertion of the keystone due to the restricted space. The longitudinal joints of the new ring are not yet completely pushed together by the erector and the bolts, so the ring seems too large. The normally tapered keystone can only be pushed into place by applying a large amount of force. If the tolerances are insufficient or the erection of the rings is inaccurate, this may lead to wedging and unwanted contacts of the flanks and thus to spalling of the concrete. This needs to be taken into account in the design phase as well as in the construction phase.

15.3.8.2 Damage During Excavation

The necessary thrust force is exerted again by the thrust cylinders after the assembly of the ring. Since the last ring is still completely inside the shield, which means that it is not backfilled, each individual segment carries the longitudinal axial load exerted by the thrust cylinders independently. Each segment deflects until it reaches a stable situation. The ring only has limited bearing capacity. This situation is worsened by the one-sided thrust forces needed for steering round curves and correction radii.

Excessive thrust forces can in some cases cause longitudinal cracks in the middle third of the segment. Such cracks are often only noticed two to three rings behind the machine due to water occurrence or damp patches, but they usually form immediately when the jacking force is applied. Such damage can be observed more often with statically determinate systems than with statically indeterminately supported segments with more than two load transfer locations per segment. The reason for this is that even small assembly inaccuracies lead to the partial loss of a support and the segment is thus subjected to bending by the thrust cylinders without counter resistance. Even heavy bending tensile reinforcement cannot compensate for the high thrust forces if a support is not effective.

This effect can only be avoided by careful assembly of rings with a planned compression of the ring joints of the previously built ring. During the ring building process, the individual cylinders are retracted. At the same time, however, sufficient axial force must be exerted by bolting across the the ring joint to create enough friction and prevent shifting of the segments. Only exactly flush-mounted load distribution plates can enable the thrust forces to be transferred from from segment to segment without damage, like a virtual column (see Fig. 15-23).

15.3.8.3 Damage at the Shield Tail Seal

If the position of the segment ring is eccentric in the shield tail and the prestressed shield tail seal contacts the lining, it transfers concentrated loading onto the lining and attempts to push it back into the centre of the shield. If a ring is now assembled that is not centred on the shield tail, then it can be pushed into the centre of the shield by the shield tail seal brushing over it. If this deformation is resisted by an opposing concrete assembly, for example cam and pocket or tongue and groove, then damage is caused to the segment at the shield tail.

Damage like this can be minimised, for example, if tapered segment rings are used and these are positioned in such a way that the ring axis is centred as precisely as possible in the tail shield. In completed shield drives, experience shows that a possible source of errors is often that a ring is wrongly installed so as to increase the eccentricity in the tail shield.

It is essential to prevent harmful play adding up by constant monitoring of the the shield tail offset and control of the ring installation programme, or the rings to be installed. It should be pointed out that manual measuring of the tail shield offset is still the usual practice today. Local conditions and the need for construction progress often lead to insufficient checks, which is only human. The only solution is to use automated measuring systems. This is often not practical because of local conditions. The mechanisms, however, are being improved constantly (serviceability). Manual systems should belong to the past.

Continuous supervision and checking as well as careful analyses are the only possibility of avoiding damage.

15.3.8.4 Damage After Leaving the Shield

After leaving the tail shield, the segment ring is loaded by the mortar pressure from the grouting of the annular gap and ground and water pressure. This causes on account of the deformation that occurs a type of "trumpet effect". The loadings, which act on the ring under construction, are qualitatively the highest loads in the entire sequence of construction and completed conditions. What has an unfavourable effect here is the fact that the rear part of the ring is still mostly unbedded in the protection of the shield and is first gradually fully loaded with the advance.

An extreme twisting of the segment ring can occur on leaving the tail shield, which can cause corners to be broken off the contact surfaces.

Twisting like this can only be limited by careful installation of the rings with scheduled compaction of the longitudinal joints right next to the shield tail. In addition to the construction details, a coordinated, simultaneous grouting of the annular gap is a definite requirement for avoiding such damage. The data from the grouting of the annular gap should be recorded and evaluated continuously. Monitoring and analysis of the damage, which might our, makes feedback for optimisation possible.

It is also possible that the segment ring in the area where the mortar in the annular gap has not yet hardened becomes oval under buoyancy and this causes unacceptable deformation with concrete to concrete contact and spalling. The properties of the grouting mortar should be matched. Constant checking of the deformation of the segments will enable conclusions to be made about the cause of damage.

After leaving the shield tail, the segment rings are also loaded in the area where the grouting mortar is still soft by the wheel sets of the first backup. If the joints of the ring are unfavourably placed or if the grouting of the annular gap is incomplete, the load from the backup has to be carried over the joints, which can lead to deflection and serious damage to the joints. For this reason, the loading from the first backup car should be as spread and with as much distance from the shield tail as possible.

15.3.8.5 Repair of Damage

As long as the stability of the structure is not in question, damage to the interior surface of the lining can be repaired relatively easily. For concrete, local application of synthetic resin based repair mortar has proved to be effective. Larger areas of damage can be made up with layers of shotcrete after careful removal of the attacked or damaged concrete layers. Pressure jet sprayers are suitable for removal of the damaged concrete. If the damage affects the entire thickness of the segment and possibly the exterior sealing or its seating in the concrete, the lining must be completely opened.

Defective seals are the most frequent cause of damage, which can also involve expensive follow-up damage. When using flexible joint seals, injecting through the lining to the exterior surface has proved to work satisfactorily. If the damage is limited to the joint, as occurs more often at the keystone, then a hollow needle can be passed through between the sealing profiles. The choice of grouting material depends on local conditions. Cement paste, water glass or plastics have been used successfully.

Swelling rubber fabrications are also used for the repair of damage seals. Swelling rubbers increase to multiple of their volume on contact with water, which provides, when the deformation is continued, the necessary pressure for sealing the damaged areas. By providing a second, interior gasket groove, into which the swell rubber band is installed, the possibility of repair can be considered in advance in the segment design.

15.4 Cast in-situ Linings

15.4.1 General

Cast in-situ linings are used in TBM tunnels mostly as inner shell or as a two-shell construction in combination with an outer lining of shotcrete or segments. Regarding their production and function, they don't differ from the in-situ linings used in conventional rock tunnelling [87, 96].

15.4.2 Construction

The interior walls of a tunnel can be cast with or without reinforcement. The bending reinforcement required for structural reasons results from the relevant loading conditions. The necessity for a minimum reinforcement depends on the national reinforced concrete standards and/or the requirements of the client. Prefabricated mesh is generally used for reinforcement.

The thickness of cast in-situ concrete should be at least 30 cm, or 35 cm if reinforced.

The strength of the cast in-situ concrete is designed to meet structural requirements. It should not be overlooked here that with the increase of concrete strength, the heat of hydration increases while at the same time the energy of the concrete falls sharply. Considering the limitation of crack widths in service, unnecessarily high concrete strength should be avoided.

For railway and road tunnels in Germany, using waterproof concrete, the requirement is to demonstrate a crack width of $w_{k,cal} = 0.2$ mm [29] or $w_{k,cal} = 0.15$ mm [24]. A minimum penetration depth of water under 30 mm is also required.

A useful alternative to conventionally reinforced concrete is steel fibre concrete. The positive influence of the steel fibres shows especially in the load-bearing, deformation and cracking behaviour of concrete when used for typical forces and moments in tunnel linings with high axial compression loads and relatively low bending moments. But the engineering and construction advantages, which result from the use of steel fibre concrete, have also been proven. Applications in the underground and and rapid-transit railway tunnels in Dortmund showed the economical advantages. For the application and design of steel fibre concrete in tunnel construction see [34–36, 44, 92, 110, 112].

15.4.3 Manufacture

Concrete, reinforced concrete and steel fibre concrete are cast in sections in hydraulically folding mobile formwork, with the length per section usually between 8 and 12 m. For small cross-sections, an all-round formwork is used, which concretes the invert and the tunnel roof in one step without construction joints. For larger cross-sections, where the buoyancy force of the fresh concrete would be too high, the invert is concreted in advance and the arch is poured later with mobile arch formwork. This variant is also used with invert segments (see Section 9.3.3).

The minimum concrete strength required for striking can be calculated structurally. It is important to cure the concrete carefully.

The section joints serve as construction joints or also as expansion joints. Sealing is performed by the use of appropriate expansion joint strips, whose width depends on the prevalent water pressure but should not be less than 30 cm.

The fresh concrete is tipped into a concrete pump in the tunnel and is pumped through a concrete distributor ring into the formwork. The concrete distributor ring enables a

Figure 15-34
Formwork for concreting the in-situ inner shell with umbrella waterproofing over the traffic area, Murgenthal tunnel

consistent distribution of fresh concrete in the formwork. When concreting the arch, the permissible difference of concrete levels should be observed, which results from the construction and anchoring of the formwork. The compaction is done with external vibrators. In the design of the mobile formwork, it is important that the formwork can withstand the loading from concreting and compaction without allowing excessive deformations of the profile. A sufficient number of concreting windows should also be provided to be able to control the concreting.

When using steel fibre concrete, the negative effect of the steel fibres on the workability of the concrete should not be forgotten. It is recommended to introduce a quality assurance programme specially suited to concreting with steel fibre concrete in tunnels [90].

15.5 Shotcrete Layers as the Final Lining

Intensive research in the last few years has led to the situation that shotcrete can now be used as structural concrete under the relevant quality requirements [89]. Regarding the strength, this means that the strength class B 25 according to DIN 1045 can be achieved without problems. An important factor is the consistency of the material properties. An example for the problems facing the sprayed process is certainly the 5 N/mm^2 higher tolerance required in comparison with cast concrete: This means that it is assumed here that the range of variation of the compression strength for shotcrete is twice that of cast concrete. The consistency can be increased clearly by suitable processes.

When using shotcrete for the final lining with requirements for waterproofing, it should be noted that the complete spraying with steel arches and reinforcing mesh is very difficult, if not impossible and leads to leaks. Because of the possible spray shadow behind such built-in components, the water-tightness of waterproof shotcrete cannot be guaranteed [93]. The use of steel fibre concrete offers considerable advantages here [101].

15.6 Structural Investigations

The structural and and detailed design of tunnel structures has a special position in civil engineering. Starting with loading assumptions, through the formation of models and methods of calculation and up to selecting cross-sections and sizes under consideration of suitable factors of safety, the normal practice in concrete and reinforced concrete construction is coordinated with that of geotechnical engineering.

The special status of tunnel construction is mainly due to the geological considerations. The surrounding rock mass is a loading and load-bearing element at the same time. The loadings from the ground can only be determined with a certain statistical uncertainty, which is kept within limits by the experience of tunnel engineers. The load-bearing ability of the rock mass can, however, also only be evaluated with the same security, the difference here being that the load-bearing capability of the rock and the rock mass plays an important role. From scattered preliminary investigations, the engineer responsible for design must force the complex system of the rock mass into a model that he can process with the calculation methods available to him. This process has to make a multitude of assumptions, which have to be taken into account in the interpretation of the calculation results [96].

The possibilities for carrying out structural investigations for the evaluation of the structural stability of tunnel shells are varied. These can be differentiated into analytic and numerical procedures, with the latter strongly in the foreground.

Constantly improved material models and solution algorithms can be used to model the complex bearing behaviour of the construction material rock ever more realistically. The leap from the geological model, using a mechanical-mathematical model, to evaluating the practicality of a construction method or to design then tunnel shell has become clearer with improvements in recent years. The incompleteness of the geological model as a basis for preliminary design and the resulting encounters with local conditions during construction, which had not been foreseen, are for tunnel construction to be considered as empirical engineering, which demands constant attention and the employment of experienced specialists.

Because discussing the basics and methods of the consideration of tunnel structures would alone suffice to fill up the contents of a book, reference is made at this point to the literature. An overview of the calculation methods and processes is contained in [88] and [96]. These sources also refer to further reading.

16 Examples of Completed Tunnels

16.1 Tunnel Excavation with Gripper TBMs

16.1.1 Control and Drainage Tunnel, Ennepe Reservoir

Project

The construction of a control and drainage tunnel ($\varnothing$ 3.0 m) was required for the dam of the Ennepe reservoir, built at the start of the last century. This was to drain the dam and the underground rock and reduce the buoyancy. It was necessary to construct an access heading in order to excavate the tunnel (Fig. 16-1).

Figure 16-1
Ennepe dam with access heading [64]

Geological and Hydrogeological Conditions

14 boreholes were drilled around the dam down to 5m into the rock under the wall to investigate the geological conditions. According to the geological profile, the planned tunnel was to bore through intercalated clay and sandstone (Fig. 16-2).

Concept for the TBM and Tunnelling

The choice of a tunnelling method was to use a tunnel boring machine. An excavation by hand, which avoids vibration and is very adaptable to the geological conditions, is

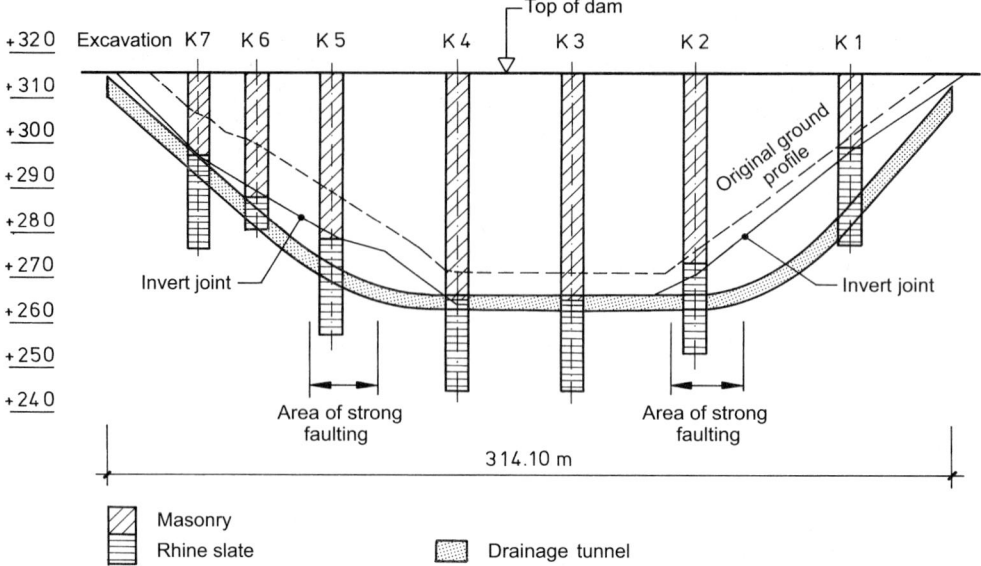

Figure 16-2
Geological profile of the drainage tunnel [64]

extremely expensive and scarcely practical today. This was ruled out, as was also the drill and blast method with the unavoidable vibration and risk associated with it.

The TBM was used to excavate the access heading as well as the control and drainage tunnel. Because of the tight curvature of the access heading, only a TBM of short construction and good turning capability could be used. A gripper TBM of an older type made by Robbins was available (Fig. 16-3).

The main beam of the machine has a length of 5.29 m. The backup with control mechanism, transformer and cable drum has a length of 9.40 m to the end of the conveyor belt. The tunnel diameter to be excavated was 3.00 m, with no lining system being provided in order to achieve the best possible drainage. In fractured rock sections, support was to be provided by anchors and reinforcing mesh. In order to make the tunnel and the work to be carried out there accessible, the control and drainage tunnel was built with an invert out of precast concrete elements with a water channel and steps in the inclined sections.

Project History

The tunnelling work for the access heading began in August 1997. The advancee had, however, to be abandoned after 35.50 m because, according to the opinion of the machine experts, the geological conditions being encountered here (diabase vein with an uniaxial compressive strength of 237 MN/m^2 and fault zones running parallel on one side and heavy sandstone in large blocks on the other) as well as the extremely tight

16.1 Tunnel Excavation with Gripper TBMs

Figure 16-3
Gripper TBM 81-113 (Robbins) with backup [64]

curve radius of 40 m made the use of a TBM technically and economically unsupportable. The next section was driven by drill and blast. After passing through the diabase vein, the TBM was installed again and the control and drainage tunnel, which lies exactly at the contact surface between the masonry dam and the undisturbed rock, was driven. The fault zones, which had been established in the preliminary investigation, caused considerable overbreak in some places and required a lining with liner plates in some places. Otherwise, satisfactory advance rates were reached. After extracting the TBM and clearing the tunnel, the accessible invert was installed at the end of 1998, which in the drainage tunnel consists of precast concrete elements, and at the junction and in the access heading consists of in-situ cast concrete.

16.1.2   Manapouri Underwater Tunnel, New Zealand

Project

The Manapouri project is an extension of the existing Manapouri hydro-power plant on the South Island of New Zealand. The power station uses the potential of the natural reservoir of the Te Anau and Manapouri lakes with a hydraulic head of about 180 m. For the extension of the existing power station, the construction of an underwater tunnel about 9.8 km long was necessary, which takes the processed water from the Manapouri underground power station to the sea. 9.6 km out of the total length were driven with a gripper TBM with a diameter of 10.05 m [113].

From the portal, which lies about 5 m under sea level, the tunnel has a fall of 12.5% for the first 370 m to 43 m under sea level. The remainder of the TBM tunnel lies between 43 and 15 m under sea level.

Geological and Hydrogeological Conditions

The project lies in the so-called Doubtful Sound Province, a complex sequence of metamorphic and plutonic rocks. The metamorphic rocks include gneiss, lime silicate, quartzite, amphibolite and marble. The plutonic rocks consist of diorite, gabbro, orthogneiss, granite, pegmatite and various ganggesteins.

The tectonics are dominated by the Great Alpine Fault, a subduction zone over 2000 km long, which runs parallel to the coast not far from the project. Altogether seven main faults cross the tunnel, mostly at a steep angle. Because of the tectonics, the rock mass is influenced by strong primary stresses, which lie between 16 and 30 MPa.

The jointing is very varied just like the strata, solid sections change in rapid sequence with blocky rock mass.

The measured uniaxial compressive strength of the rock is partially less than 300 MPa. The simultaneous presence of rock with various mechanical characteristics is significant for the excavation.

Water inflow of up to 1,100 l/s was expected and also experienced.

Concept for the TBM and Tunnelling

In contrast to the tendering of other projects, where the contractor is allowed the choice of tunnelling method and also of the suitable machine, in this case the tunnelling method and the machine are exactly specified. This covers details like the flat cutter head, the exclusion of any electro-hydraulic drive or also the cutter head revolution speed. Just as with many other projects, experts of a leading TBM manufacturer were involved with the preparation of the tenders.

The consortium under the leadership of STRABAG (ILBAU) decided to purchase a new gripper TBM (Table 16-1; Figs. 4-1a and 16-4).

The gripper TBM with single clamping is shown being assembled in front of the tunnel portal in Fig. 16-4. The steep downward descent at the start of the advance is recognisable. The TBM has only a relatively short shield in the stator area, to which the trailing fingers are fixed, through which rock anchors can be fixed.

For the transport of the excavation muck, railway operation or removal with tunnel dumpers had to be ruled out because of the steep gradient of the first section. For this reason a conveyor belt with a high-performance belt with 800 T capacity per hour and 900 mm belt width was chosen for the material transport (see Figs. 16-5 and 16-6).

The backup unit was specially equipped because of the high forecast quantity of formation water. One backup as supply unit, immediately behind the TBM, runs on rails (see Fig. 16-5). The remaining backup units run on rails, which are fixed to the crown with anchors.

16.1 Tunnel Excavation with Gripper TBMs

Table 16-1
Data for the Manapouri gripper TBM [128]

• Gripper TBM Atlas Copco Robbins	323–228
• Boring diameter with new cutting tools	10.05 m
• Main bearing	Three-row roller
• Cutter type	Robbins wedge lock, 17"
• Number of cutting rings	68
• Max. thrust force per cutter ring	267 kN
• Nominal total thrust force	18,156 kN
• Cutter head drive	Electric
• Motors	11×315 kW
• Main drive power	3,465 kW
• Cutter head revolution speed	5.07/2.53 rpm
• Torque at 5.07/2.53 rpm	6.344/9.516 kNm
• Total installed power	5.000 kVA
• Boring stroke	1.830 mm
• Machine belt, width	1.370 mm
• Machine belt, capacity	1.388 m^3/h
• Length retracted/extended	23.00/24.83 m
• Total weight	925 T
• Weight of heaviest single component	96 T

Figure 16-4
Side view of the Manapouri gripper TBM [163]

308   16 Examples of Completed Tunnels

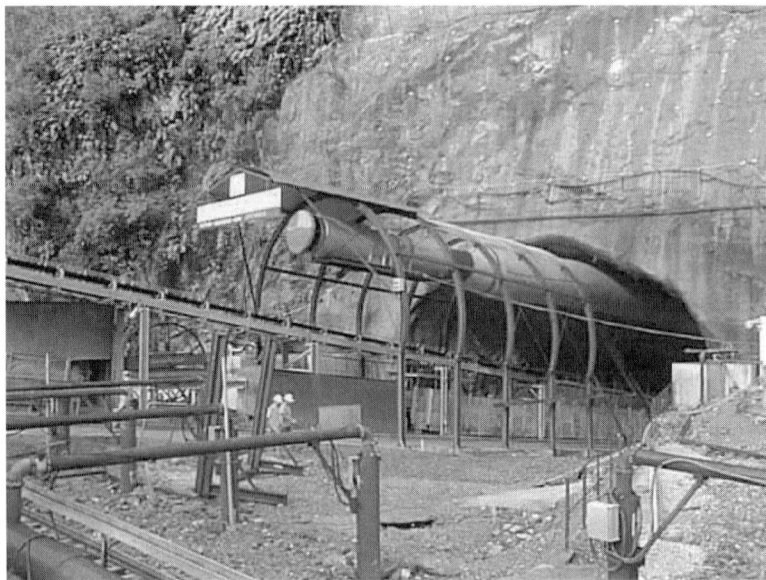

Figure 16-5
Steeply rising conveyor belt at the start of the underwater tunnel

Figure 16-6
Conveyor belt and 35 T locomotive tractor in the steep ramp

16.1 Tunnel Excavation with Gripper TBMs

Project History

The assembly of the TBM began in April 1998. The advance could be started only two months later.

From the start of work in June 1998 until the end of September 2000, the following performance was achieved in Manapouri [163]:

- Average daily advance   9.8 m
- Best day's advance   37.7 m
- Best month's advance   603.0 m
- Highest net penetration (through one stroke)   3.5 m/h
- lowest net penetration
  (over one stroke with nominal thrust pressure)   0.4 m/h

The large spread of values reflects the highly changeable geology. Except for short sections, the pattern of the tunnel face was characterised by two or more different rock types with different cuttability. Even in the first section of approx. 730 m with a 12.5% gradient, an average advance of 9.6 m/day was achieved.

The percentage division of activities on site is shown in Fig. 16-7.

What is remarkable is the high amount of time spent on cutter changes and cutter head repairs. This results essentially from the changeable geological conditions.

Changeable rock conditions affect mechanised tunnelling in hard rock even more than in soft rock. Mixed face conditions, meaning the simultaneous occurrence of rocks of different cuttability, leads to high dynamic loads, which become even more critical the closer the TBM operates to its capacity limit. The result is extreme vibrations of the cutter head as well as the entire machine body. In some cases, the dynamic loads occurring at individual cutters can reach ten times the nominal value under such conditions. Practical experience on site, in particular the analysis of the actual damage to cutter rings, confirms these values.

The cutters used on this machine, with a diameter of 432 mm (17"), are designed for for a nominal load of 267 kN. This corresponds to 231 bar of thrust pressure. After the start-up phase, the TBM was operated with a thrust pressure of 240 bar, but the nominal

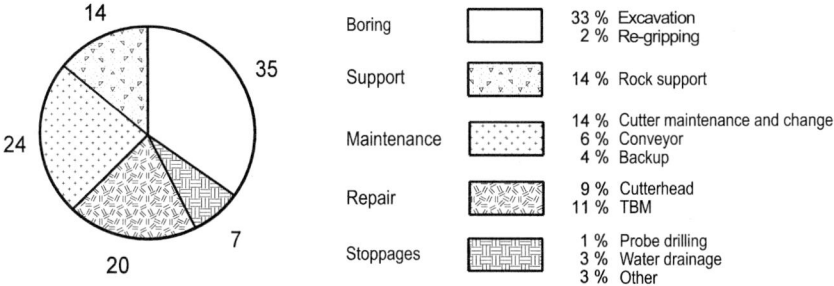

Figure 16-7
Percent proportion of activities during tunnelling [163]

contact pressure could rarely be used due to the mixed face conditions encountered. In most cases, the operator manually reduced the thrust pressure to limit vibration and damage to the cutter head and cutter rings. Nonetheless, after approx. 2 km cracks started to appear in the cutter head, particularly in the cutter housings. As a consequence, almost all housings were reinforced and repairs were made to the cutter head.

Cutter changes in hard rock excavation are a substantial factor causing downtime (see Fig. 16-7). Since the number of cutters increases approximately in proportion with the diameter, a larger portion of boring time is spent checking and changing cutters with large diameter TBMs. Although there is more space than with smaller machines, it is not possible to work in more than two places at the same time. While one quarter of the cutter head is being worked on, the other three quarters are not accessible.

As can be seen in Fig. 16-7, 14% of the entire construction period was spent on cutter checks and changing. With 55% utilization of the TBM and the same service life of the cutter, this value is slightly below 20%, which means that the time does not rise in complete proportional to the boring time. The reason for this is that with a higher degree of utilization of the TBM, the number of cutters having to be changed per stop increases. After more than 3.300 cutter changes, the downtime per cutter change, including checking, amounts to an average of 44 min. The gauge cutters are clearly overloaded, these are already operating on the borderline of maximum stress in the design of a TBM. The influencing factors rolling speed and cutting angle add together here in an unfavourable way.

The state of the technology in abrasive hard rock is cutters with highly alloyed steel such as H 13 (ASTM). Tests with cutters from all known manufacturers were performed on site. Various cutter profiles were also tested. In addition, tests were done with unconventional materials as well as rings from lesser known manufacturers. The latter attempts were catastrophic failures throughout. Generally, it can be said that the products of the well-known manufacturers provided good service lives, but are no match overall for the performance of a modern TBM. The cutter rings are stressed over the yield limit at nominal load or they splinter.

Rock support measures, as can be seen in Fig. 16-7, caused 14% of tunnelling interruptions. These interruptions are due to the short stand-up time of the rock. The entire support work required substantially more time, but most of the procedures, with exception of these 14%, could also be performed during the excavation. The originally planned four standard sections were found insufficient in the changing rock conditions. With increasing tunnel length, the number of standard sections increased as well; at the time of the report, this had grown to at least eight sections. In addition, there was also a substantial quantity of additional support measures in the contract, which were installed after the excavation.

The effect of formation water on the excavation produced extraordinary complications. Up to 1,100 l/s was anticipated based on experience with former tunnels and the expectations were found to be true. In the machine area, that is the first 30 m, up to 520 l/s was measured. A tunnel excavation under such conditions becomes increasingly difficult, even in large tunnels. The main problem is the fine material, which is sometimes washed out of the cutter head while boring. Further material is washed off the conveyors and settles in the invert metres high. Under such circumstances, continuous excavation with a conventional, rail-born backup becomes extremely difficult. In this case, the material had to be continuously cleared from the invert with wheeled loaders and a steep conveyor system. Although the TBM operation was obstructed, it never amounted to more than a few hours of downtime (Figs. 16-8 and 16-9).

The TBM advance was, however, stopped by a clause in the contract requiring two advance drainage drillings of 150 mm diameter in such cases. The intended purpose, to lower the water level and decrease the water inflow in the cutter head area, was not a success. Quite the opposite was the case, because the surrounding rock mass fed additional quantities of water into the tunnel.

Even measuring such heavy inflows of water is no simple task. The well-proven measuring weir is still the simplest solution. A circular aluminium segment with a right-angled weir, installed in the invert, records the quantity of water with sufficient accuracy. A measurement takes about half an hour.

Altogether, 34 complete working days were lost due to investigation and drainage drillings without a positive effect on the excavation. In order to achieve an improvement of the rock conditions for TBM operation and a reduction of risk, much more extensive measures, like injections, are necessary. A backup system like the one used in Manapouri, with sufficient space for clearing and water control in the invert, offers the best guarantee against downtime caused by water. Measures like the drainage drillings

Figure 16-8
Formation water in the backup area

Figure 16-9
Water in the tunnel

carried out here are not recommended because of the deleterious effect on continuous operation.

Summary

With 10 m diameter and 9.6 km in hard rock, the Manapouri underwater tunnel is far away from setting records in diameter and length. Nevertheless, the limits of technical and economical feasibility were certainly tested.

According to today's knowledge, the selection of a new Gripper TBM clearly exceeded the limits of cost effectiveness. The costs of the TBM operation would not withstand a comparison with modern drill and blast methods starting from both portals. Even considering a used TBM, which was not permitted in the contract, the expenditure of time and equipment would have been a lot less with drill and blast.

The second best solution from a purely economic point of view, in the opinion of the site manager, would have been a shield TBM with continuous segmental lining. Stand-up times for rock support would have been avoided and time would have been saved by omitting the subsequent partial lining, although the net boring performance would not have changed.

The analysis of the former tunnel excavation provides abundant knowledge, which has already been partially taken into consideration in the planning of new projects and machines. Apart from the selection of the system gripper TBM or shield TBM, there are two major issues facing future development of mechanised excavation in hard rock:

- The advance shows that the cutter head must be optimized for extreme rock conditions as found in Manapouri. Without being able to change cutter discs from behind, which has become the standard in hard rock tunnelling, an excavation like Manapouri would be inconceivable. On the other hand, this makes the structure of the cutter head much more complex, which naturally has an influence on the stiffness. With a 10 m diameter, a flat cutter head has clear structural disadvantages in comparison with a domed profile. A domed profile also enables a substantially more even load distribution onto the gauge cutters. The client had unfortunately specified the TBM to such an extent that such a solution was excluded from the start. A more rigid design of the cutter head could also be obtained with a flat profile.

- Cutters discs made of high alloy tool steel have been the state of the art for some time. With today's machine performance, the yield point of the steel is exceeded and a higher net penetration is negated by the extreme increase of cutter wear. Cutters with a larger diameter and higher contact pressure do not provide an advantage, because the net penetration is determined by the maximum surface pressure, which means by the yield point of the steel.

16.2 Tunnelling with Shield TBMs

16.2.1 San Pellegrino Tunnel, Italy

Project

The main part of the San Pellegrino Terme bypass is a tunnel of 2,300 m length, which was interrupted by a gorge 50 m wide. The two tubes measure 1,700 m and 600 m. To secure the neighbouring San Pellegrino spring, a pilot heading with a diameter of 3.90 m was driven first with the intention of enlarging it into the tunnel by blasting. After the header had been bored, it was decided to drive the tunnel with a shield TBM to preserve the spring.

Geological and Hydrogeological Conditions

The two tunnel sections pass through limestone called calcare di Castro, a hard lime, and the calcare di Zorzino, a less hard rock, with weather-susceptible and water-sensitive constituents. The transition between the limestone types is strongly disturbed (Fig. 16-10).

The fault zone has a horizontal thickness of about 10 m. It is characterised by strongly weathered rock with clay inclusions.

The water inflow in the fault zone was 10 l/s initially. In the course of the excavation this rose to 20 l/s.

314   16 Examples of Completed Tunnels

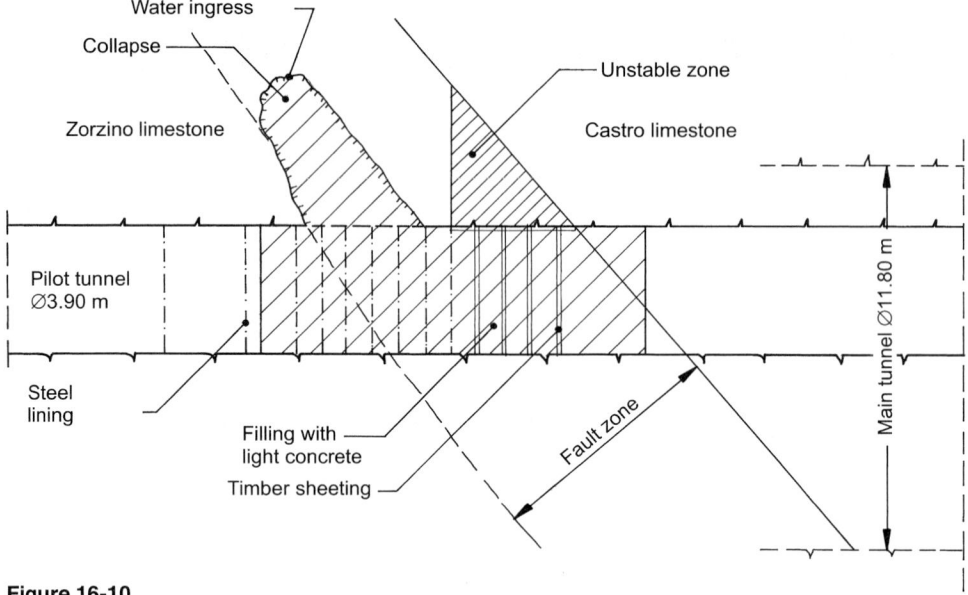

Figure 16-10
Fault zone at the changeover calcare di Castro to calcare di Zorzino [116]

Concept for the TBM and Tunnelling

Blasting was rejected out of consideration for the San Pellegrino spring, and the enlargement of the pilot heading as ruled out too. A Robbins gripper TBM with a diameter of 3.9 m was available to drive the pilot heading.

The full section was driven with a shield TBM from Herrenknecht AG with a cutter head diameter of 11.82 m (see Fig. 3-4a). The lining consisted of a six-part segment ring.

Project History

The drive of the pilot heading with a gripper TBM took place without considerable difficulties. Water penetrated, particularly in the first section of the drive. Various sections with an overall length of 120 m were supported with steel installation.

Before the shield TBM drove through, the steel installation in the pilot tunnel had to be dismantled. The rock, which had been exposed to air and dampness for approximately two years, reacted to the dismantling with spalling and ruptures.

For safety reasons, the steel installation was partially replaced tightenable plastic anchors, shotcrete and wood.

At the critical location in the transition area from calcare di Castro to calcare di Zorzino (Fig. 16-10), a collapse of approximately 30 m^3 occurred in the pilot heading due to the water courses in the faulted rock.

16.2 Tunnelling with Shield TBMs

To protect the San Pellegrino spring, neither synthetic resin injections nor plastic foam could be used. The backfilling of the collapse, together with a 12 m long tunnel section, was done with lightweight concrete, supplemented by cement injections. The remaining steel arches had been wrapped in polystyrene prior to backfilling and had to be cut into sections and transported laboriously through the top of the Shield TBM.

On starting to bore with the shield TBM it became apparent that the faulted rock in the transition zone was extremely wet. The backfill material in the pilot tunnel amounted to only 12% of the entire face area of the large TBM. Problems with the muck deposit slowed the excavation so much that within eleven hours the face became unstable and blocked the cutter head of the TBM.

After several unsuccessful attempts to free the cutter head, it was decided to investigate the zone of the collapse in more detail with probe drillings.

The results were:

- The shield was still in good rock (calcare di Castro).
- The tectonic fault situation is about 7–8 m thick in the calotte.
- The fault zone consists of a layer of clay with blocks and a layer of strongly weathered slaty lime with clay inclusions.
- Behind the fault zone is the calcare di Zorzino, which is indeed less stable than the calcare di Castro (because it is water-susceptible), but also well cuttable
- The water quantity is constant and is about 10 l/s.

Because of the difficult position, a first series of measures was undertaken:

- Implementation of drainage and injection drillings.
- Injections with Wilkit foam from the TBM, in order to remove the water flow out of the fault zone or out of the TBM and to improve the geomechanical properties.
- Implementation of a double pipe umbrella from the TBM as a support measure for the restarting of the advance.
- A rapid repair of the pilot heading with shotcrete and mesh reinforcement from the opposing portal, in order to be able to carry out any additional measures required from the pilot heading. This measure was necessary, because the steel arches had in the meantime been removed.

While the measures performed from the TBM took more time than intended, the refurbishment of the pilot tunnel went faster. At the end of September the miners were approx. 14 m in front of the shield TBM. The fault zone had been reached. Another collapse in the pilot tunnel forced a change in proceedings to the use of ground freezing. From an enlargement of the calotte, implemented from the pilot tunnel, freeze drillings were intended to form a connected frost body from the stable Zorzino limestone into the Castro limestone above the shield. This bridged over the collapse zone (Fig. 16-11).

Nine weeks after the blocking of the cutter head, the planned enlargement was performed from the pilot tunnel to the centre line of the tunnel cross section with a radius

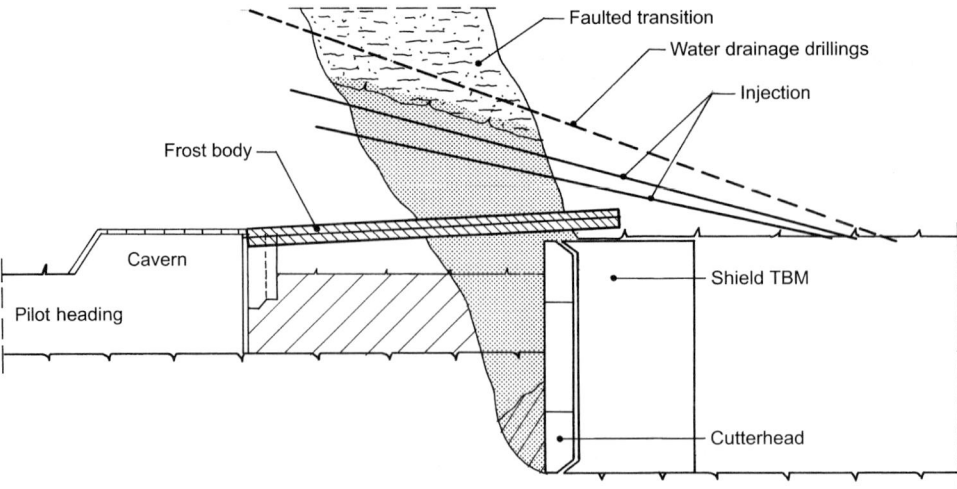

Figure 16-11
Planned frost body over cutter head and shield, starting from the cavern pilot heading [116]

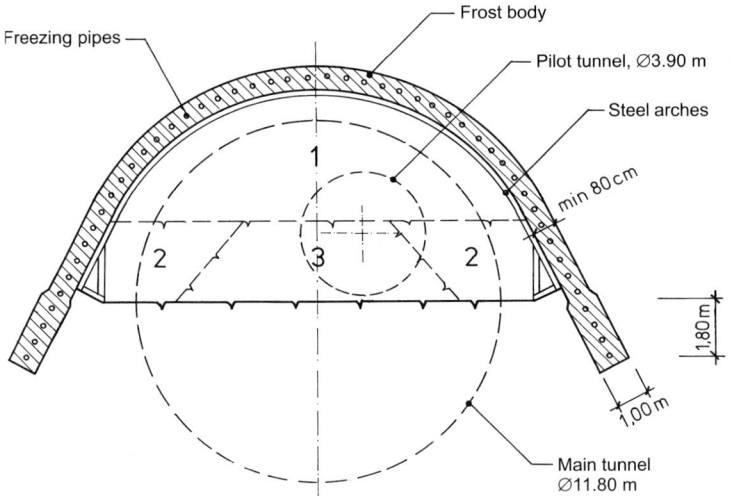

Figure 16-12
Frost body with two deeper shanks in the collapse area [116]

of 6.5–7.0 m and a length of approx. 8 m. The excavation was done in sections one metre long. The support consisted of steel installation (INP 160 double beams with widened supports) as well as mesh reinforcement and steel fibre shotcrete.

The aim of the project was to create a semicircular frost body with a length of approx. 22 m and a minimum thickness of 80 cm. To do this, 48 drillings were necessary in which steel tubes ($\varnothing$ 101.6 mm, thickness 12.5 mm) were inserted. The gap between rock and pipe was injected with cement and then freezing probes were inserted into the pipes and also injected. This work took up considerably more time than planned. Not until after two months was the equipment ready for the start of the actual freezing work.

Using liquid nitrogen, it was possible to create the frost body within 14 days, even if not completely with the intended dimensions (Figs. 16-11 and 16-12).

The construction of the frost body over the first 15–17 m, starting from the cavern, succeeded rapidly. In the collapse zone with flowing water, the frost body did not reach the target thickness despite the increasing the freezing capacity. The water drainage and sealing measures during TBM operation (Fig. 16-11) were not fully effective.

The excavation of the calotte under the protection of the frost body was started from the cavern two weeks after beginning the freezing work. After three weeks of careful excavation, the face was in front of the cutter head of the TBM. The collapses and injections had blocked the cutter head to such an extent that complete clearing of the area around the cutter head became unavoidable. 181 days after becoming blocked, the cutter head was again free to turn. Tunnelling could now be resumed through the supported zone until normal operation was again possible with high advance rates in the Zorzino limestone.

16.2.2 Zürich–Thalwil Twin-Track Tunnel, Section Brunau–Thalwil

Project

A completely new double railway track is being built to improve the existing railway line on the left bank of Lake Zurich for the additional trains of the Railway 2000 program. This is 10.7 km long, begins at the Langstrasse in Zurich, leads through the Seebahn cutting to Lochergut and further into a 9.4 km long tunnel under the Allmend Brunau and on to the southern tunnel portal in Thalwil. There, the new double track will merge back into the existing Seebahn line before the Thalwil railway station.

The 5.7 km long tunnel route up to the Nidelbad junction structure is designed as a double track tunnel (Fig. 16-13). The smallest radius is approx. 8,000 m and the climbing gradient is 5.3‰. The twin-track tunnel located in the molasse is in accordance with the typical construction of Swiss shield TBM tunnels (see Fig. 15-2 b).

From the connection structure at Nidelbad, the connection to the Seebahn line in Thalwil will at first be two single track tunnels and a connecting double track tunnel. The continuation of the twin-track tunnel from the Nidelbad junction structure as the Zimmerberg base tunnel is part of the plans for the AlpTransit project.

Geological and Hydrogeological Conditions

The new Zürich–Thalwil tunnel, coming from Zürich, first crosses the gravel of the Sihl, which consists of glacial lake deposits and moraines, before passing into the upper sweet water molasse (Fig. 16-14).

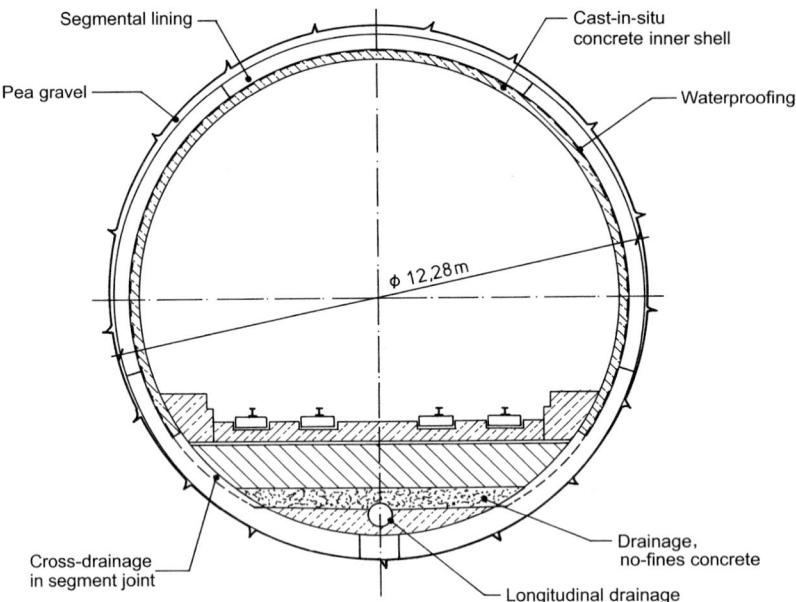

Figure 16-13
Standard section of the Brunau–Thalwil section

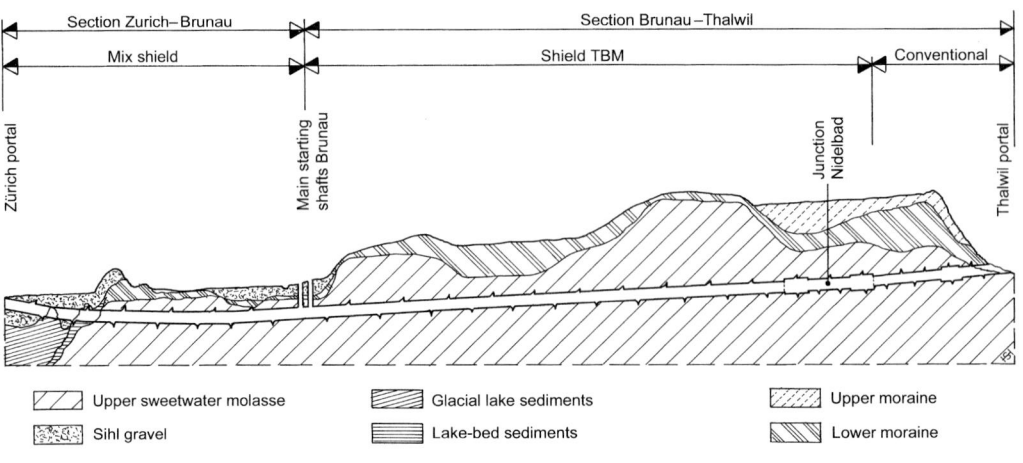

Figure 16-14
Longitudinal geological profile of the Zürich–Thalwil tunnel

16.2 Tunnelling with Shield TBMs

In this sweet water molasse as intercalation of marl, sandstone and siltstone, the tunnel runs very near to the surface until the shaft at Brunau. The first deep overburden is in the Brunau–Thalwil section.

The Sihl gravels represent an important groundwater reservoir. In the section with shallow rock overburden shortly before the shaft at Brunau, there is also a ground water lake. The molasse from the Brunau shaft to Thalwil turned out to be fully dry, with the exception of a few places with a meagre local inflow of water.

Concept for the TBM and Tunnelling

The section from Zürich to Brunau, with highly changeable geology, is predestined for the application of a Mixshield, which can be operated as a shield TBM or as a hydroshield. The section from Brunau to Thalwil, which lies fully in the molasse, was driven over 5.7 km with a shield TBM with diameter 12.35 m. The flying junction and branch tunnel as single-track bores to the existing tracks in Thalwil were constructed by blasting. The necessary enlargement of the shield tunnel (Nidelbad junction structure) for the Thalwil branch will be dealt with in chapter 11.

Both main tunnel works were performed through two shafts from the main installation site in Brunau. This installation site is connected for both road and rail access and all bulk goods are transported by train. The North shaft serves primarily for the Mixshield tunnelling. The South shaft serves the east drive towards Thalwil [81].

For the section from Brunau to Thalwil (contract section 3.01) a new Herrenknecht AG shield TBM was used, which was especially designed for these rock conditions. Table 16-2 points out the characteristics of the tunnelling equipment.

Table 16-2
Data for the shield TBM for the section from Brunau to Thalwil (contract section 3.01) [61]

• Herrenknecht shield TBM	S-139
• Overall length	170 m
• Installed electric power	3,300 kVA
• Drive type	electric
• Bored diameter	12.35 m
• Main drive power	1320 KW
• Cutter head revolution speed	1.95/3.9 rpm
• Max. thrust cutter head	16,625 kN
• Max. thrust force	63,700 kN
• Number of disc cutters	76
• Disc cutter diameter	17"
• Shield diameter	12.29 m
• Shield length	7.96 m
• Segment width	1.7 m
• Total weight	1,700 T

The shield TBM only has a short shield. The cutter head is equipped with adjustable front buckets. The good results obtained in earlier TBM drives with conveyor belts and the problems of vertical conveyor in the shaft persuaded the consortium to erect a high-performance conveyor belt system with a transport capacity of 1000 T/h and a belt width of 1000 mm for the muck transport.

Project History

The work began in September 1997 to sink both shafts with diameters of 22 m and 20 m and a depth of 30 m. The drive of the connecting tunnel between the shafts in partial excavation and the construction of the startup tube of 12 m length took place between February 1998 and May 1998.

The assembly on site of the TBM began in June 1998. It took approximately three months, mostly with two-shift operation (Fig. 1615).

Figure 16-15
Assembly of the central part with the drive unit into the prepared stator with shield body at top and shield with skin tail below

Tunnelling began at the end of September 1998, at first with limited advance. From the middle of November 1998, the advance had progressed so far that the backup extension and the conveyor system could by installed. The regular advance started in February 1999 and rapidly achieved very high and regular monthly advances (Fig. 16-16). The end of the double-track section was reached at the end of 1999. After the dismantling of the tunnelling system, the drill and blast work could start for the single track tunnel [17].

16.3 Inclined Shaft Tunnelling with Double Shield

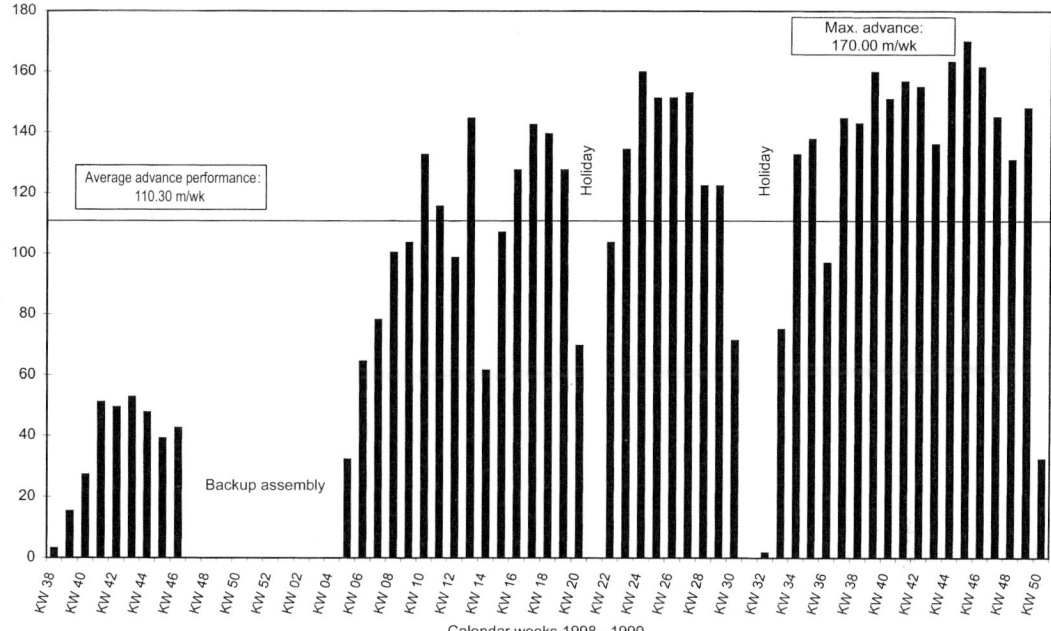

Figure 16-16
Advance rates of the shield TBM on the Brunau–Thalwil section (construction section 3.01) [116]

16.3 Inclined Shaft Tunnelling with Double Shield

16.3.1 Cleuson–Dixence Pressure Shaft

Project

To extend the Cleuson–Dixence hydro-electric power plant, an additional power plant with a pressure head of 1,883 m and a water volume of 75 m³ was constructed from 1993–1998. The construction of a pressure shaft was necessary as an inclined shaft with a gradient of up to 68% in some areas.

Geological and Hydrogeological Conditions

The inclined shaft lies in the Great St. Bernhard decke formation and in the Sion–Courmayeur zone. The formations from window F4 at the top to window F8 at the bottom are (Fig. 16-17):

- Verrucano: phyllitic slate and dolomite
- Upper Triassic: anhydrite, quartzite, dolomite
- Upper Carboniferous: mica schist with some anhydrite and dolomite
- Middle Triassic: intercalation of mica schists
- Lower carboniferous: mica schist with anhydrites, limes and dolomites
- "Du télégraphe" breccia: slightly tectonised breccia
- Sandstone and slate from "St. Christophe"

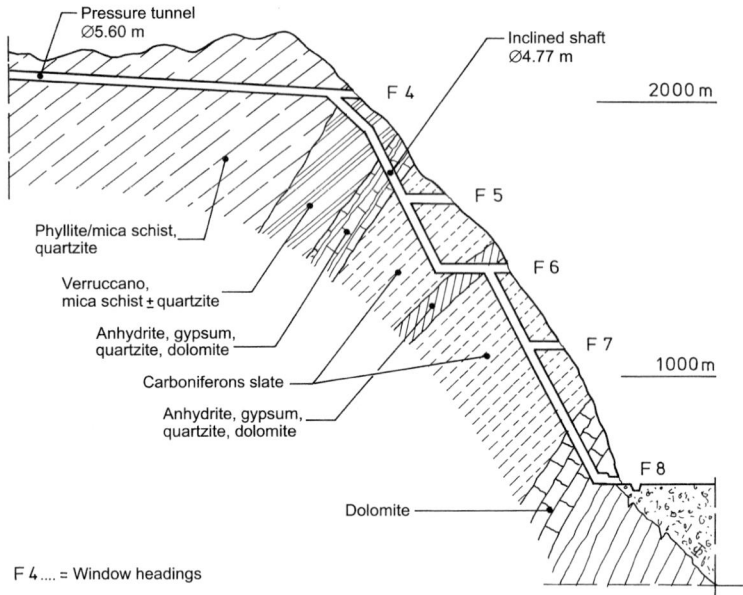

Figure 16-17
Simplified longitudinal geological profile through the pressure shaft with the windows F8–F4

In general, it can be said that these are difficult geological conditions for tunnelling. The classification according to SIA 198 was, for submission purposes, classes III and V.

Concept for the TBM and Tunnelling

The shaft section in Verrucano had already been evaluated as difficult to drive at the project planning stage. The drill and blast method was therefore tendered and implemented from window F4 downwards.

A Double Shield TBM was to the shaft in all formations from window F8 to the breakthrough at the border of the upper Triassic to Verrucano [77]. A Robbins double shield TBM was purchased to do the work (Table 16-3; see also Figs. 16-4 and 3-3 b).

Table 16-3
Data for the Cleuson–Dixence double shield TBM [128]

• Robbins double shield TBM	154–273
• Excavation diameter (1st cutter head)	4.77 m
• Main drive power	1,260 KW
• Cutter head thrust	6,800 kN
• Weight	40,000 kN
• Reinforced concrete segments	6 and 1 part
• Internal diameter	4.0 m
• Performance of the shaft winding gear	1,000 KW

16.3 Inclined Shaft Tunnelling with Double Shield

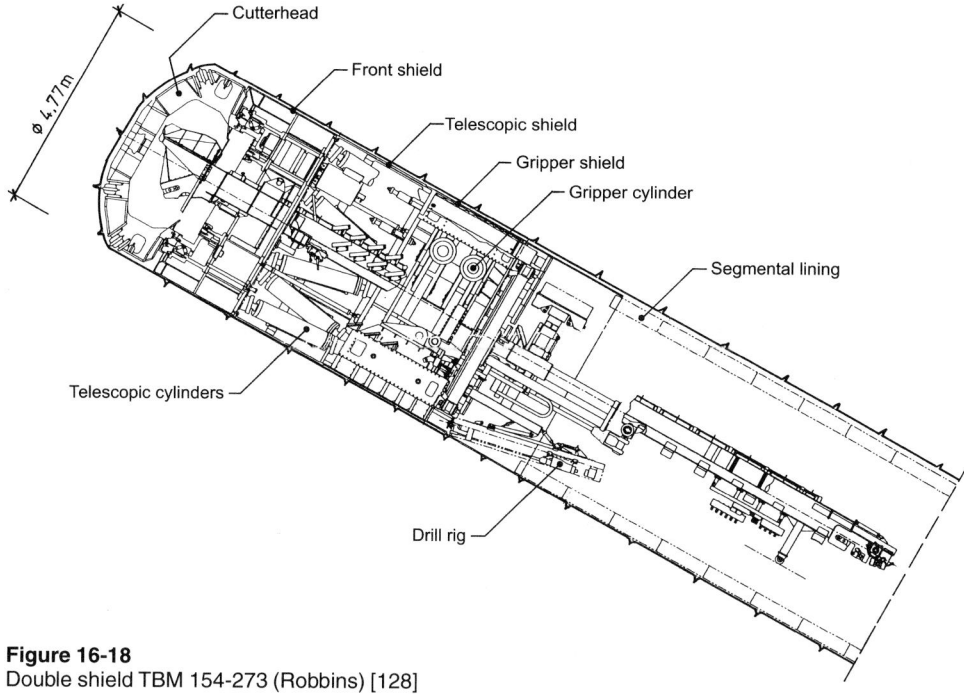

Figure 16-18
Double shield TBM 154-273 (Robbins) [128]

The changing gradients of the pressure shaft made hydraulic transport of the muck seem unsuitable. The decision was made to use rail transport with shaft winding gear.

Reinforced concrete segments with an internal diameter of 4.0 m and a wall thickness of 30 cm were intended for the support. The 1.2 m long segment rings consists of seven elements, where the elements 6a and 6b together with the keystone occupy the same angle from the centre of 60° as each single standard segment.

Project History

For the assembly of the TBM and backups, a starting shaft in the intended position was excavated by drill and blast from window F8 with 68% incline. The chamber at the bottom was constructed with sufficient space for the assembly of the TBM and also with an assembly platform as a launch pad for the TBM. The assembly platform could be tilted upwards for TBM assembly into an inclined position at the shaft gradient.

In June 1994, the tunnelling system was ready to start the shaft drive. Tunnelling began with average daily advances slightly above 7 m. At TM 630, several teeth broke off the crown wheel of the main drive due to a material defect. The interruption of tunnelling for the complex crown wheel change, which had to be brought in divided into ten segments, lasted approximately three months.

The subsequent tunnel advance in Carboniferous and Triassic proved to be difficult. Water inflows and collapses in front of the cutter head and above the shield forced the use of stabilizing measures in front of the cutter head. Drainage drillings and plastic injections, non-foaming and foaming, led to tunnelling interruptions, which resulted in a decrease of average advance to approximately 3 m per working day.

The spherical cutter head, standing out 1.9 m in front of the shield, proved to be unsuitable for the slate. It turned out to be unsuitable for the stabilising measures and because of the high friction component during rotation. The wear on this cutter head reached exceptional levels. The joint venture therefore decided to change the cutter head at window F6 for a new, flat construction protruding only 70 cm in front of the shield.

Considerably less collapses occurred in the mica schists of the upper Carboniferous, not least because of the flat cutter head. The water inflow also decreased to small quantities. The average daily advance rose again to a respectable 8 m. The double shield proved to be unfavourable in the unstable rock. The withdrawal of the rear gripper shield was often hindered by material falling down between the shields. The telescopic shield had to be taken apart several times in order to remove the trapped material. Except for a few hundred meters, the double shield was operated as a single shield.

The difficulties driving through the altogether worse than forecast rock conditions led to an reconsideration of the geology and tunnelling concept between window F5 and the border Triassic–Verrucano. Probe drilling ahead from window F5, through the Triassic towards the Verrucano, showed destructional dolomitic rocks and quartzite, which had dissolved to sand. There was little hope of being able to drive the double shield TBM through such rock conditions. The consortium therefore decided on a combined shield TBM from Lovat Inc. based on a shield TBM and an EPB Shield for these questionable 400 m (Table 16-4).

Table 16-4
Data of the Cleuson–Dixence combined shield

• Lovat Single shield TBM	TBM/EPB
• Excavation diameter	4.4 m
• Disc cutters	23 × 15″
• Cutter head drive	electric
• Length of shield including cutter head	7.0 m
• Overall weight	145 T
• Segments	steel segments

For approx. 2/3 of the 400 m section starting from window F5, open TBM excavation was possible. The remaining approx. 100 m distance consisted of fine white quartz sand. In this section, the Lovat machine proved to operate very well as an earth pressure shield. Altogether over this 400 m distance an average daily advance rate of 5.8 m was achieved.

The rock support of this section consisted of steel segments made of rolled profiles HEA 160 and a steel plate outer skin 4 mm thick. A ring consists of seven elements and

16.3 Inclined Shaft Tunnelling with Double Shield

Figure 16-19
Steel lining of the section driven with the Lovat combined shield

a key. The ring width amounted to 1 m. To support the TBM in the shaft, the profiles are connected with 24 longitudinal supports of the type HEA 160. They resist the supporting forces of the 24 thrust cylinders of altogether 2280 t (Fig. 16-19).

Extraordinary efforts by the contractor and the parallel application of the second shaft TBM led to the completion of the tunnelling operation on time in November 1996.

Summary

The application of tunnel boring machines in the pressure shaft exceeded several borders – borders of technology, but also borders of cost effectiveness and therefore the normal contract framework.

During the construction of the original plant, substantial sections, which were to be driven with the present shaft, were considered as not practical. The joint venture took a chance taking the work. During the construction, the shape of the cutter head proved to be unsuitable and also the application of a double shield machine, despite its oft praised advantages, turned out to be incorrect.

Tunnelling by drilling and blasting would not have been as efficient regarding schedule, cost or safety as the application of a TBM in this borderline area. Correspondingly, the contract parties must also create contract provisions, which consider the interests of both sides in such borderline conditions.

References

[1] Alber, M.: Geotechnische Aspekte einer TBM-Vertragsklassifikation. Diss. TU Berlin. Berlin: Mensch und Buch 1999.
[2] Anheuser, L.: Neuzeitlicher Tunnelbau mit Stahlbetonfertigteilen. Beton- und Stahlbetonbau 1981.
[3] ARGE ADA (Atlas Druckstollen Amsteg): Company information.
[4] Asea Brown Boveri (ABB): Company informationen.
[5] Atlas Copco: Company information.
[6] Ayaydin, N.: Entwicklung und neuester Stand der Gebirgsklassifizierung in Österreich. Felsbau 12 (1994), No. 6, 413–417.
[7] Barton, N.: TBM Tunnelling in jointed and faulted rock. Rotterdam: Balkema 2000.
[8] Barton, N.; Lien, R.; Lunde, J.: Engineering classification of rock masses for the design of tunnel support. Rock Mechanics 6 (1974), No. 6, 189–236.
[9] Baumann, L.; Zischinski, U.: Neue Lösungen und Ausbautechniken zur maschinellen „Fertigung" von Tunneln in druckhaftem Fels. Felsbau 12 (1994), No. 1, 25–29.
[10] Baumann, Th.: Tunnelauskleidungen mit Stahlbetontübbingen. Bautechnik 69 (1992), No. 1, 33–35.
[11] Beckmann, U.: Einflussgrößen für den Einsatz von Tunnelbohrmaschinen. Diss. TU Braunschweig. Mitteilungen des Instituts für Grundbau und Bodenmechanik, No. 10. 1982.
[12] Bessolow, W.; Makarow, O.: Erfahrungen beim Vortrieb des Severomuisk-Tunnels an der Baikal-Amur-Magistrale. In: Probleme bei maschinellen Tunnelvortrieben? Tagungsband Symposium TU München (1992), 163–174.
[13] Bielecki, A.; Schreyer, J.: Eignungsprüfungen für den Tübbingausbau der 4. Röhre des Elbtunnels, Hamburg. Tunnel 15 (1996), No. 3, 32–39.
[14] Bieniawski, Z.T.: Engineering rock mass classifications. New York: Wiley 1989.
[15] Bieniawski, Z.T.: Geomechanics classification of rock masses and its application in tunneling. 3. Kongreß der IGFM. Advances in Rock Mechanics, Vol. IIa, 27–32. Denver: 1974.
[16] BLS AlpTransit AG: Lötschberg-Basislinie. Project information.
[17] Bolliger, J.: TBM-Vortriebe beim SBB-Tunnel Zürich–Thalwil. Felsbau 17 (1999), No. 5, 460–465.
[18] Brand, W.: Entwicklung des maschinellen Vortriebs: Ein Vergleich von Steinkohlenbergbau und Tunnelbau. In: Forschung + Praxis 30, Düsseldorf: Alba 1986, 76–82.
[19] Brunar, G.; Powondra, F.: Nachgiebiger Tübbingausbau mit Meypo-Stauchelementen. Felsbau 3 (1985), No. 4, 225–229.
[20] Brux, G.: Tübbinge aus Stahlfaserbeton zur Erprobung. Eisenbahningenieur 49 (1998), No. 11, 80–81.
[21] Büchi, E.: Einfluss geologischer Parameter auf die Vortriebsleistung einer Tunnelbohrmaschine. Diss. Universität Bern. Self-published 1984.

[22] Büchi, E.; Mathier, J.-F.; Wyss, Ch.: Gesteinsabrasivität – ein bedeutender Kostenfaktor beim mechanischen Abbau von Fest- und Lockergestein. Tunnel 14 (1995), No. 5, 38–44.
[23] Büchi, E.; Thalmann, C.: Wiederverwendung von TBM-Ausbruchsmaterial – Einfluss des Schneidrollenabstandes. In: TBM-Know How zum Projekt NEAT, AC-Robbins Symposium Luzern 1995.
[24] Bundesministerium für Verkehr: Zusätzliche technische Vertragsbedingungen und Richtlinien für den Bau von Straßentunneln, Teil 1, Geschlossene Bauweise. 1995.
[25] Buvelot, R.; Jonker, J; Maidl, B.: Die Schildtunnel der Betuwelinie: Erfahrungen mit einem neuartigen Ausschreibungs- und Vergabeverfahren. In: Forschung + Praxis 37. Düsseldorf: Alba 1998, 26–30.
[26] Cording, E.J.; Deere, D.U.; Hendron, A.J.: Rock engineering for underground caverns. Symposium on Underground Rock Chambers. Phoenix: 1972.
[27] Dahl, J.: Anwendung des Stahlfaserbetons im Tunnelbau. In: Fachseminar Stahlfaserbeton, 04.03.1993. Institut für Baustoffe, Massivbau und Brandschutz, TU Braunschweig, No. 100, 1993.
[28] Dahl, J.; Nußbaum, G.: Neue Erkenntnisse zur Ermittlung der Grenztragfähigkeit im Bereich der Koppelfugen. In: Taschenbuch für den Tunnelbau 1997. Essen: Glückauf 1996, 291–319.
[29] Deutsche Bahn AG: Richtlinie 853. Eisenbahntunnel planen, bauen und instandhalten. 1998.
[30] Deutsche Gesellschaft für Geotechnik (Hrsg.): Empfehlungen des Arbeitskreises Tunnelbau (ETB). Berlin: Ernst & Sohn 1995.
[31] Deutscher Ausschuss für unterirdisches Bauen (DAUB): Empfehlung für Konstruktion und Betrieb von Schildmaschinen. In: Taschenbuch für den Tunnelbau 2001. Essen: Glückauf 2000, 256–288.
[32] Deutscher Ausschuss für unterirdisches Bauen (DAUB): Empfehlung zur Planung und Ausschreibung und Vergabe von Schildvortrieben. 1993.
[33] Deutscher Ausschuss für unterirdisches Bauen (DAUB): Empfehlungen zur Auswahl und Bewertung von Tunnelvortriebsmaschinen. Tunnel 16 (1997), No. 5, 20–35.
[34] Deutscher Beton-Verein e.V.: Merkblatt Bemessungsgrundlagen von Stahlfaserbeton im Tunnelbau. Wiesbaden: Self-published 1996.
[35] Deutscher Beton-Verein e.V.: Merkblatt Stahlfaserbeton. Wiesbaden: Self-published 2001.
[36] Dietrich, J.: Zur Qualitätsprüfung von Stahlfaserbeton für Tunnelschalen mit Biegezugbeanspruchung. Diss. Ruhr-Universität Bochum. Technisch-wissenschaftliche Mitteilungen des Instituts für Konstruktiven Ingenieurbau der Ruhr-Universität Bochum, No. 92–4. 1992.
[37] DIN 4020 Geotechnische Untersuchungen für bautechnische Zwecke. 1990.
[38] DIN 18312 Allgemeine Technische Vertragsbedingungen für Bauleistungen; Untertagebauarbeiten. 1992.
[39] Duddeck, H.: Tunnelauskleidungen aus Stahl. In: Taschenbuch für den Tunnelbau 1978, 159. Essen: Glückauf 1977, 159–194.

[40] Dyckerhoff und Widmann AG: Company information.
[41] EN 815 Sicherheit von Tunnelbohrmaschinen ohne Schild und gestängelosen Schachtbohrmaschinen zum Einsatz im Fels. 1996.
[42] Eves, R.C.W.; Curtis, D.J.: Tunnel lining design and procurement. In: The Channel Tunnel. Part 1: Tunnels. Civil Engineering Special Issue 1992, 127–143.
[43] Ewendt, G.: Erfassung der Gesteinsabrasivität und Prognose des Werkzeugverschleißes beim maschinellen Tunnelvortrieb mit Diskenmeißel. Diss. Ruhr-Universität Bochum. Bochumer Geologische und Geotechnische Arbeiten, No. 33. 1989.
[44] Feyerabend, B.: Zum Einfluss unterschiedlicher Stahlfasern auf das Verformungs- und Rissverhalten von Stahlfaserbeton unter den Belastungsbedingungen einer Tunnelschale. Diss. Ruhr-Universität Bochum. Technisch-wissenschaftliche Mitteilungen des Instituts für Konstruktiven Ingenieurbau der Ruhr-Universität Bochum, No. 95–8. 1995.
[45] Fliegner, E.: Der Seikan-Unterwassertunnel, eine wagemutige Pionierleistung der Japaner. Strassen- und Tiefbau 35 (1981), No. 3, 31–49.
[46] Flury, S.; Rehbock-Sander, M.: Gotthard-Basistunnel: Stand der Planungs- und Bauarbeiten. Tunnel 17 (1998), No. 4, 26–30.
[47] Gehring, K.: Classification of Drillability, Cuttability, Borability and Abrasivity in Tunnelling. Felsbau 15 (1997), No. 3, 183–191.
[48] Gehring, K.: Leistungs- und Verschleißprognosen im maschinellen Tunnelbau. Felsbau 13 (1995), No. 6, 439–448.
[49] GeoExpert AG: Company informationen.
[50] Girmscheid, G.: Baubetrieb und Bauverfahren im Tunnelbau. Berlin: Ernst & Sohn 2000.
[51] Grandori, C.: Development and current experience with double shield TBM. In: Proc. of the 8th RETC (1987), 509–514.
[52] Grandori, C.: Fully mechanized tunnelling machine and method to cope with the widest range of ground conditions – Experiences with a hard rock prototype machine. In: Proc. of the 3rd RETC (1976), 355–376.
[53] Grandori, C.; Dolcini, G.; Antonini, F.: The Los Rosales water tunnel in Bogota. In: Proc. of the 10th RETC (1991), 561–581.
[54] Grandori, R.; Jäger, M.; Antonini, F.; Vigl, A.: Evinos-Mornos-Tunnel – Greece. In: Proc. of the 12th RETC (1995), 747–767.
[55] Grimstad, E.; Barton, N.: Updating of the Q-System for NMT. In: Proc. of the International Symposium on Sprayed Concrete – Modern Use of Wet Mix Sprayed Concrete for Underground Support, Fagernes. Oslo: Norwegian Concrete Association. 1993.
[56] Haack, A.: Vergleich zwischen der einschaligen und zweischaligen Bauweise mit Tübbingen bei Bahntunneln für den Hochgeschwindigkeitsverkehr. Forschung + Praxis 36. Düsseldorf: Alba 1996, 251–256.
[57] Hamburger, H.; Weber, W.: Tunnelvortrieb mit Vollschnitt- und Erweiterungsmaschinen für große Durchmesser im Felsgestein. In: Taschenbuch für den Tunnelbau 1993. Essen: Glückauf 1992, 139–197.

[58] Henneke, J.; Lange, R.; Setzepfandt, W.: Hydraulische Bergeförderung beim maschinellen Vortrieb des Radau-Stollens. Glückauf 116 (1980), No. 9, 426–431.

[59] Henneke, J.; Weber, W.: Entwicklungsstand des Bohrens von Tages- und von Blindschächten. Glückauf 127 (1991), No. 21/22, 978–988.

[60] Hentschel, H.: Erkundungsstollen unter dem Mont Raimeux. Tunnel 16 (1997), No. 6, 8–14.

[61] Herrenknecht AG: Company information.

[62] Hochtief AG: Freudensteintunnel – Pilottunnel Los 2. Site brochure.

[63] Hochtief AG: Freudensteintunnel – Pilottunnel Los 1. Site brochure.

[64] Holzmann AG: Kontroll- und Drainagestollen für die Ennepestaumauer. Site brochure. 1998.

[65] Holzmann AG: Tunnelbau im Raum Rhein-Ruhr. Site brochure. 1996.

[66] Holzmann AG: Untertagebauten des LEP für die Forschungsorganisation CERN in Genf. Site brochure. 1987.

[67] Horst, H.: Neue Konzepte für Vollschnitt-Vortriebsmaschinen. Glückauf 113 (1977), No. 2, 70.

[68] Isendahl, H.; Wild, M.: Gusseisen im Tunnelbau. In: Taschenbuch für den Tunnelbau 1979. Essen: Glückauf 1978, 213–239.

[69] Jäger Baugesellschaft mbH: Company information.

[70] Janßen, P.: Tragverhalten von Tunnelausbauten mit Gelenktübbings. Diss. TU Braunschweig. Berichte aus dem Institut für Statik der TU Braunschweig, No. 83/41. 1983.

[71] John, M.; Purrer, W.; Myers, A.; Fugeman, C.D.; Lafford, G.M.: Planung und Ausführung der britischen Überleitstelle im Kanaltunnel. Felsbau 9 (1991), No. 1, 37–47.

[72] Jones, B.: Tunnelling in a bee line! Tunnels & Tunnelling 16 (1984), No. 12, 41.

[73] Kirkland, C.J.; Craig, R.N.: Precast Linings for High Speed Mechanised Tunnelling. In: Forschung und Praxis 36. Düsseldorf: Alba 1996, 119–123.

[74] Kirschke, D.; Pommersberger, G.: Der Freudensteintunnel – Ein neuer Maßstab für den Stand der Technik. AET – Archiv für Eisenbahntechnik 44 (1992), 131–156.

[75] Klawa, N.; Haack, A.: Tiefbaufugen. Berlin: Ernst & Sohn 1989.

[76] Knickmeyer, W.: Neue Vortriebssysteme – Wege zu höherer Wirtschaftlichkeit. Glückauf 122 (1986), No. 3, 230–235.

[77] Kobel, R.: Schrägschacht für das Kraftwerk Cleuson–Dixence in den Walliser Alpen. Tunnels & Tunnelling 27 (1995), Bauma Special-Issue, 26–32

[78] Könings, H.-D.: Bau der längsten Eisenbahntunnel in Deutschland für die NBS zwischen Erfurt und München. In: Tunneltechnologie für die Zukunftsaufgaben in Europa. Rotterdam: Balkema 1999, 83–95.

[79] Krenkler, K.: Chemie des Bauwesens. Band 1. Berlin: Springer-Verlag 1980.

[80] Kühl, G.: Dywidag-Stahlrohrgelenkschild – Langstreckenvortriebe in hindernisreichen, stark wechselnden Böden und im Felsgestein – Baustellenerfahrungen. In: Microtunnelbau. Rotterdam: Balkema 1998, 75–78.

[81] Künzli, G.: Zimmerberg-Basistunnel: Die Bauausführung. In: Städtischer Tunnelbau. Int. Symposium ETH Zürich. 1999.

[82] Kuhnhenn, K.; Prommersberger, G.: Der Freudensteintunnel – Tunnelbau in schwellfähigem Gebirge. In: Forschung + Praxis 33. Düsseldorf: Alba 1990, 137–143.

[83] Kutter, H.K.: Voruntersuchungen für einen Abwasserstollen: Ermittlung der für einen mechanischen Ausbruch maßgeblichen Gesteinsparameter. Berichte 4. Nat. Tag. der Ing.-Geol. Goslar (1983), 75–86.

[84] Kuwahara, H.; Fukushima, K.: The first application for the hexagonal tunnel segment in Japan. In: Tunnels for People. Rotterdam: Balkema 1997, 405–410.

[85] Lombardi, G.: Entwicklung der Berechnungsverfahren im Tunnelbau. Bauingenieur 75 (2000), No. 7/8, 372–381.

[86] Lovat Inc.: Company information.

[87] Maidl, B.: Handbuch des Tunnel- und Stollenbaus, Band I: Konstruktion und Verfahren. Essen: Glückauf 1988.

[88] Maidl, B.: Handbuch des Tunnel- und Stollenbaus, Band II: Grundlagen und Zusatzleistungen für Planung und Ausführung, 2. Auflage. Essen: Glückauf 1994.

[89] Maidl, B.: Handbuch für Spritzbeton. Berlin: Ernst & Sohn 1992.

[90] Maidl, B.: Konstruktive und wirtschaftliche Möglichkeiten zur Herstellung von Tunneln in einschaliger Bauweise. Schlussbericht des Forschungsvorhabens im Auftrag des Bundesministeriums für Verkehr der Bundesrepublik Deutschland. 1993.

[91] Maidl, B.: Planerische und geotechnische Voraussetzungen in der Ausschreibung für einen Maschinenvortrieb als Nebenangebot. In: Forschung + Praxis 38. Düsseldorf: Alba 2000, 171–178.

[92] Maidl, B.: Stahlfaserbeton. Berlin: Ernst & Sohn 1992.

[93] Maidl, B.; Berger, Th.: Empfehlungen für den Spritzbetoneinsatz im Tunnelbau. Bauingenieur 70 (1995), No. 1, 11–19.

[94] Maidl, B.; Handke, D.; Maidl, U.: Developments in mechanised tunnelling for future purposes – Entwicklungen im maschinellen Tunnelbau für die Zukunftsaufgaben. In: Tunnelbau. Rotterdam: Balkema 1998, 1–17.

[95] Maidl, B.; Herrenknecht, M.; Anheuser, L.: Maschineller Tunnelbau im Schildvortrieb. Berlin: Ernst & Sohn 1994.

[96] Maidl, B.; Jodl, H.G.; Schmid, L.; Petri, P.: Tunnelbau im Sprengvortrieb. Berlin: Springer-Verlag 1997.

[97] Maidl, B.; Kirschke, D.; Heimbecher, F.; Schockemöhle, B.: Abdichtungs- und Entwässerungssysteme bei Verkehrstunnelbauwerken. Forschung Straßenbau und Straßenverkehrstechnik, Heft 773. Bonn: Bundesdruckerei 1999.

[98] Maidl, B.; Maidl, U.: Maschineller Tunnelbau mit Tunnelvortriebsmaschinen. In: Zilch, K.; Diederichs, C.J.; Katzenbach, R. (Hrsg.): Handbuch für Bauingenieure. Berlin: Springer-Verlag 2001.

[99] Maidl, B., Maidl, U.: Planerische und geotechnische Voraussetzungen in der Ausschreibung für einen Maschinenvortrieb als Nebenangebot. In: Taschenbuch für den Tunnelbau 2001. Essen: Glückauf 2000, 231–255.

[100] Maidl, B.; Maidl, U.; Einck, H.B.: Der Eurotunnel. Essen: Glückauf 1994.

[101] Maidl, B.; Ortu, M.: Einschalige Bauweise. In: Festschrift Prof. Falkner, Institut für Baustoffe, Massivbau und Brandschutz der TU Braunschweig, No. 142. 1999.

[102] Maidl, U.: Aktive Stützdrucksteuerung bei Erddruckschilden. Bautechnik 74 (1997), No. 6, 376–380.

[103] Maidl, U.: Einsatz von Schaum für Erddruckschilde – Theoretische Grundlagen der Verfahrenstechnik. Bauingenieur 70 (1995), 487–495.

[104] Maidl, U.: Erweiterung des Einsatzbereiches von Erddruckschilden durch Konditionierung mit Schaum. Diss. Ruhr-Universität Bochum 1994. Technisch-Wissenschaftliche-Mitteilungen des Instituts für konstruktiven Ingenieurbau, No. 95–4. 1995.

[105] Marty, T.; Wüst, M.; Loser, P.: Tunnel Uznaberg: Erschütterungsarmer Vortrieb. Tunnel 19 (2000), No. 4, 71–79.

[106] Mülheim a.d. Ruhr/ARGE: Stadtbahn Mülheim an der Ruhr – Bauabschnitt 8/ Ruhrunterquerung. Site brochure.

[107] Murer, T.: Straßentunnel Umfahrung Locarno – Einsatz einer einstufigen Erweiterungsmaschine. In: Probleme bei maschinellen Tunnelvortrieben. Symposium TU München 1992, 65–78.

[108] Murer AG: Company information.

[109] Myrvang, A.: Tunnelvortrieb mit Tunnelbohrmaschinen in gebirgsschlaggefährdetem Hartgestein. In: TBM-Know How zum Projekt NEAT. AC-Robbins Symposium Luzern 1995.

[110] Nitschke, A.: Tragverhalten für Stahlfaserbeton für den Tunnelbau. Diss. Ruhr-Universität Bochum. Technisch-wissenschaftliche Mitteilungen des Instituts für Konstruktiven Ingenieurbau der Ruhr-Universität Bochum, No. 98–5. 1998.

[111] ÖNORM B 2203: Untertagebauarbeiten. Werkvertragsnorm. 1994.

[112] Ortu, M.: Rissverhalten und Rotationsvermögen von Stahlfaserbeton für Standsicherheitsuntersuchungen im Tunnelbau. Diss. Ruhr-Universität Bochum. Fortschritt-Bericht VDI, Reihe 4, No. 164. Düsseldorf: VDI 2000.

[113] Papke M., Heer B.: Building the second Manapouri tailrace tunnel. Tunnels & Tunneling 31 (1999), No. 5, 61.

[114] Pelzer, A.: Die Entwicklung der Streckenvortriebsmaschinen im In- und Ausland. Glückauf 90 (1954), 1648–1658.

[115] Philipp, G.: Schildvortrieb im Tunnel- und Stollenbau; Tunnelauskleidung hinter Vortriebsschilden. In: Taschenbuch für den Tunnelbau 1987. Essen: Glückauf 1986, 211–274.

[116] Prader AG: Company information.

[117] Rahmenrichtlinie 89/391 des Rates über die Durchführung von Maßnahmen zur Verbesserung der Sicherheit und des Gesundheitsschutzes der Arbeitnehmer bei der Arbeit. 8. Einzelrichtlinie im Sinne des Art. 16 Abs. 1 der Richtlinie 89/391/ EWG. Brüssel 1989.

[118] Rauscher, W.: Untersuchungen zum Bohrverhalten von Tunnelbohrmaschinen im Fels bei wechselnden Gebirgseigenschaften. Diss. TU München. Self-published 1984.

[119] Reder, K.-D.: Baugeologische, geotechnische und baubetriebliche Abhängigkeiten zwischen Gebirge–Maschine–Mensch, dargestellt anhand drei ausgewählter TBM-Vortriebe im alpinen Raum. Diss. Universität Salzburg 1992.

[120] Reinhardt, M.; Weber, P.: Gebirgsbedingte Erschwernisse beim maschinellen Stollenvortrieb im verkarsteten devonischen Massenkalk. Bautechnik 24 (1977), No. 5, 169–174.

[121] Richtlinie 92/57/EWG des Rates über die auf zeitlich begrenzte oder ortsveränderliche Baustellen anzuwendenden Mindestvorschriften für die Sicherheit und den Gesundheitsschutz. Brüssel 1992.

[122] Robbins, R.J.: Economic Factors in Tunnel Boring. South African Tunnelling Conference, Johannesburg 1970.

[123] Robbins, R.J.: Hard rock tunnelling machines for squeezing rock conditions: Three machine concepts. In: Tunnels for people. Rotterdam: Balkema 1997, 633–638.

[124] Robbins, R.J.: Large diameter hard rock boring machines: State of the art and development in view of alpine base tunnels. Felsbau 10 (1992), No. 2, 56–62.

[125] Robbins, R.J.: Mechanized tunnelling – progress and expectations. In: Tunnelling × 76. London: Institution of mining and metallurgy 1976.

[126] Robbins, R.J.: Tunnel boring machines for use with NATM. Felsbau 6 (1988), No. 2, 57–63.

[127] Robbins, R.J.: Tunnel machines in hard rock. In: Civil engineering for underground rail transport. London: Butterworth 1990, 365–386.

[128] The Robbins Company: Company information.

[129] Robinson, R.A.; Cording, E.J.; Roberts, D.A.; Phienweja, N.; Mahar, J.W.; Parker, H.W.: Ground deformations ahead of and adjacent to a TBM in sheared shales at Stillwater tunnel, Utah. In: Proc. of the 7th RETC (1985), 34–53.

[130] Rottenfußer, F.: Trinkwasserstollen Schäftlarn–Baierbrunn – Neue Methoden und Erfahrungen bei der Naßförderung. In: Forschung + Praxis 23. Düsseldorf: Alba 1980, 46–51.

[131] Rutschmann, W.: Mechanischer Tunnelvortrieb im Festgestein. Düsseldorf: VDI 1974.

[132] Sanio, H.-P.: Nettovortriebsprognose für Einsätze von Vollschnittmaschinen in anisotropen Gesteinen. Diss. Ruhr-Universität Bochum. Bochumer Geologische und Geotechnische Arbeiten, No. 11. 1983.

[133] Scheifele, C.; Gugger, B.; Wildberger, A.; Weber, R.: Erfahrungen beim Vortrieb und Ausbau des Tunnels Sachseln. Tunnel 13 (1994), No. 6, 15–24.

[134] Schimazek, J.; Knatz, H.: Die Beurteilung der Bearbeitbarkeit von Gesteinen durch Schneid- und Rollenbohrwerkzeuge. Erzmetall 29 (1976), No. 3, 113–119.

[135] Schmid, L.: Einsatz großer Tunnelbohrmaschinen verschiedener Bauart in der Schweiz – Leistungen und Wirtschaftlichkeit. In: Forschung + Praxis 29. Düsseldorf: Alba 1980, 70–75.

[136] Schmid, L.: Felsanker im Untertagebau – ein Segen? Schweizer Ingenieur- und Architekt 105 (1987), No. 25, 780–782.

[137] Schmid, L.: Ingenieurgemeinschaft Wisenberg-Tunnel. Versuche zur Sulfatbeständigkeit des Betons. 1993 (unpublished).
[138] Schmid, L.: Methangasführung Tunnel Murgenthal und Werkvertrag. 1995 (unpublished).
[139] Schmid, L.: Prüfung und Bewertung der Angebote Tunnel Murgenthal, Oenzberg und Flüelen. 2000 (unpublished).
[140] Schmid, L. Tunnelbauten im Angriff aggressiver Bergwässer. Schweizer Ingenieur- und Architekt 113 (1995), No. 44, 1012–1018.
[141] Schmid, L.: Ventilationsschema für Tunnel Oenzberg und Flüelen. Vorgaben der Submission. 1999 (unpublished).
[142] Schmid, L.: Versuche Blasversatz Gubristtunnel. 1979 (unpublished).
[143] Schmid, L.: Wie begründet sich der hohe Anteil von Maschinenvortrieben in der Schweiz. In: Tunneltechnologie für die Zukunftsaufgaben in Europa. Rotterdam: Balkema 1999.
[144] Schmid, L.; Ritz, W.: Bauausführung des Gubristtunnels. Schweizer Ingenieur und Architekt 103 (1985), No. 23, 544–547.
[145] Schmidt, Kranz & Co.: Company information.
[146] Schneider, J.: Sicherheit und Zuverlässigkeit im Bauwesen. Grundlagen für Ingenieure. Stuttgart: Teubner 1996.
[147] Schockemöhle, B; Heimbecher, F.: Stand der Erfahrungen mit druckwasserhaltenden Tunnelabdichtungen in Deutschland. Bauingenieur 74 (1999), No. 2, 67–72.
[148] Schoeman, K.D.: Buckskin mountains and Stillwater tunnel: Developments in technology. In: Tunnelling and Underground transport: Future developments in technology, economics and policy. Amsterdam: Elsevier 1987, 92–108.
[149] Schreyer, J.; Winselmann, D.: Eignungsprüfungen für die Tübbingauskleidung der 4. Röhre Elbtunnel – Ergebnisse der Großversuche. In: Forschung + Praxis 38. Düsseldorf: Alba 1995, 102–107.
[150] Schreyer, J.; Winselmann, D.: Eignungsprüfungen für die Tübbingauskleidung der 4. Röhre Elbtunnel – Ergebnisse der Scher-, Abplatz-, Verdrehsteifigkeits- und Lastübertragungsversuche. In: Taschenbuch für den Tunnelbau 1999. Essen: Glückauf 1998, 337–352.
[151] Schweizerische Unfallversicherungsanstalt (SUVA): Rettungskonzept für den Untertagebau, No. 88112. 1996.
[152] Schweizerischer Baumeisterverband (SBV): Richtpreise und Leistungswerte für den Untertagbau.
[153] Schweizerischer Ingenieur- und Architekten-Verein (SIA): Norm 196: Baulüftung im Untertagbau. 1998.
[154] Schweizerischer Ingenieur- und Architekten-Verein (SIA): Norm 198: Untertagbau. 1993.
[155] Schweizerischer Ingenieur- und Architekten-Verein (SIA): Norm 199: Erfassen des Gebirges im Untertagbau. 1998.
[156] S.E.L.I. S.p.A.: Company information.

[157] Simons, H.; Beckmann, U.: Vergütung für TBM-Vortriebe auf der Grundlage angetroffener Gebirgseinflüsse. Tunnel 1 (1982), No. 4, 224–234.
[158] Spang, K.: Felsbolzen im Tunnelbau. Communication des Laboratoires de Mécanique des Sold et de Roches. Ecole Polytechnique Fédérale de Lausanne, No. 113. 1986.
[159] Stack, B.: Handbook of Mining and Tunnelling Machinery. Chichester: Wiley 1982.
[160] Stein, D.; Möllers, K.; Bielecki, R.: Leitungstunnelbau. Berlin: Ernst & Sohn 1988.
[161] Steiner, W.: Tunnelling in squeezing rocks: Case histories. Rock Mechanics and Rock Engineering 29 (1996), No. 4, 211–246.
[162] Steinheuser, G.; Maidl, B.: Projektierung und Wahl des Bauverfahrens aufgrund der geologischen Voruntersuchungen dargestellt am Beispiel des Straßentunnels Westtangente Bochum. Tiefbau 96 (1984), No. 2, 74–86.
[163] Strabag/E. Gschnitzer: Baustellenreport Manapouri, Company information.
[164] Strohhäusl, S.: Eureka Contun: TBM tunnelling with high overburden. Tunnels & Tunnelling 28 (1996), No. 5, 41–43.
[165] Teuscher, P.: Sondierstollen Kandertal. In: AlpTransit: Das Bauprojekt – Schlüsselfragen und erste Erfahrungen. Dokumentation SIA D 0143 (1997), 95–101.
[166] Teuscher, P.; Keller, M.; Ziegler, H.-J.: Lötschberg-Basistunnel: Erkenntnisse aus der Vorerkundung. Tunnel 17 (1998), No. 4, 32–38.
[167] Thompson, J.F.K.: Flexible All-Purpose Segmental Tunnelling by Tunnel Boring Machine. Forschungsantrag: FAST by TBM. Ruhr-Universität Bochum. 1996 (unpublished).
[168] Tiefbau Berufsgenossenschaft: Tunnelbau – Sicher arbeiten. No. 484. 1990.
[169] Tonscheidt, H.W.: Entwicklungsstand des Schachtabteufens mit gestängelosen Schachtbohrmaschinen. Glückauf 125 (1989), No. 13/14, 760–768.
[170] Trösken, K.: Erfahrungen mit Streckenvortriebsmaschinen im Ruhrbergbau und in anderen Bergbauländern. Glückauf 99 (1963), 1327–1341.
[171] Turbofilter GmbH Entstaubungstechnik, Company information.
[172] University of Trondheim – The Norwegian Institute of Technology: Hard Rock Tunnel Boring. Project Report 1–83, 1983/88.
[173] Vigl, A.: Planung Evinos-Tunnel. Felsbau 12 (1994), No. 6, 495–499.
[174] Vigl, L.; Gütter, W.; Jäger, M.: Doppelschild-TBM – Stand der Technik und Perspektiven. Felsbau 17 (1999), No. 5, 475–485.
[175] Vigl, L.; Jäger, M.: Tunnel Plave–Doblar, Slowenien, 7 m Doppelschild-TBM mit einschaligem, hexagonalem Volltübbing-Auskleidungssystem. Vortrag auf dem Österreichischen Betontag 2000, Wien.
[176] Vigl, L.; Pürer, E.: Mono-shell segmental lining for pressure tunnels. In: Tunnels for People. Rotterdam: Balkema 1997, 361–366.
[177] Wagner, H.; Schulter, A.; Strohhäusl, S.: Gleitsegmente für flache und tiefe Tunnel. In: Forschung + Praxis 36. Düsseldorf: Alba 1996, 257–266.
[178] Wallis, S.: Knoxville Smallbore. World Tunnelling 10 (1997), No. 3, 67–71.
[179] Wanner, H.: Klüftigkeit und Gesteins-Anisotropie beim mechanischen Tunnelvortrieb. Rock Mechanics, Suppl. 10 (1980), 155–169.

[180] Wanner, H.; Aeberli, U.: Bestimmung der Vortriebsgeschwindigkeit beim mechanischen Tunnelvortrieb. Forschungsvorhaben Atlas Copco Jarva AG – ETH Zürich. Zwischenbericht 1978 – unpublished.

[181] Weber, W.: Standortbestimmung des TBM-Vortriebes unter besonderer Berücksichtigung der geplanten Alpenbasistunnel. Felsbau 12 (1994), No. 1, 12–18.

[182] Wirth Maschinen- und Bohrgerätefabrik GmbH, Company information.

[183] Zbinden, P.: AlpTransit Gotthard-Basistunnel: Geologie, Vortriebsmethoden, Bauzeiten und Baukosten, Stand der Arbeiten und Ausblick. In: Forschung + Praxis 37. Düsseldorf: Alba 1998, 11–15.

[184] Zbinden, P.: AlpTransit Gotthard: Eine neue Eisenbahnlinie durch die Schweizer Alpen. Bauingenieur 73 (1998), No. 12, 542–546.

[185] Zwicky, P.: Erfahrungen mit Tunnelabdichtungen in der Schweiz von 1960 bis 1995. In: Forschung + Praxis 37. Düsseldorf Alba 1998, 130–132.

Index

Abrasiveness   41, 220
– ABR index   43 ff.
– classification scheme   45
Action plan   109
Advance formula   46
Advance rate   53 f., 130 ff., 219, 223, 321
Advance support   19
Alternative proposal   253 ff.
– examples   253 ff.
Anchor drills   117
Anchors   144 f.
– permanent   145
– temporary   145
Annular gap   269
– blowing or filling   19
– grouting annular gap   290 f.
Annular gap, filling   19, 278, 290 f.
– with gravel   290
– with mortar   291
Articulated steel pipe shield   175 f.
Award criteria   244 f.
Award law   243
Awarding of contracts   243 ff.

Back-cutting technology   58 ff.
Backup   89 ff.
Backup area, equipment   89
Backup designs   93
Backup running gear   97
Ball bearing   53 f.
Bayonet system   34
Bell section   133 f.
Belt cassette   84 ff.
Bentonite lubrication   150
Blind drilling   64
Bore stroke, maximum   17
Boreholes   26
– drillhole photography   198 f.
Boring advance, *see* tunnelling advance
Boring operation   25
Boring principle   1
Boring process   25, 199 ff.
– factors influencing   25
– performance of the boring process   25
Boring system   16
Bouygues tunnelling machine   57
Breakdown torque   51

Bucket edge   30
Bucket system   29 f.
Buckets   28 ff.
Building site regulations   104
Building site setup   108
Building workers protection regulations   106

CAI test   42, 44, 47
Cast iron segments, tubbings   269
Cavings   108, 111
Cerchar abrasiveness index test (CAI test)
   42, 44, 47
Chips   17, 35 ff.
Cladding friction, shield cladding friction   123
Clamping   17, 67 ff., 71 ff., 121 f., 202 f.
Clamping force   67 ff., 202
– limits   17 f.
Clamping pressure   70
Clamping systems   20, 67 ff.
– double clamping systems   68
– single   69
Classification   207 ff.
– according to cuttability and abrasiveness
   220
– according to rock properties   208
– according to support measures required
   222 ff.
– cuttability   207, 220, 222
– excavation classes   241
– in Austria   228 ff.
– in Germany   223
– in Switzerland   234 ff.
– rock classification   207
– tunnelling classification   207
Classification parameters   209 f.
Classification systems   208
Clearing   28
Cleuson-Dixence pressure shaft   321 ff.
CLI   216, 218 f.
Compressive strength   47, 242
– uniaxial   38
Concrete segments   137, 269, 271, 273, 275,
   277, 279, 281, 283, 285, 287, 295,
   see also Segments
Concrete shells, cast in-situ   299 ff.
– casting   300
Concreting distributor ring   300

Conex lining system   275 f.
Construction phase   263
Construction process, combined   177
Construction safety   99, 103 ff.,
   see also integrated safety plan
– in Austria   105
– in Germany   104
– in Switzerland   105
Construction ventilation   99 ff.
Continuous Miner   60
Contract   243 ff.
Control and drainage heading, Ennepe
   reservoir   303
Conventional tunnelling, see Drilling and
   blasting
Conveyor belt   29, 60, 75, 84 ff., 171, 308
– limit of performance   29
Conveyor transport   84 ff.
– advantages   87
Conveyor transport system   18
Correction segment rings   125
Cost control   262
Costs   262
Conex segments   275
Cross-clamping, X-type   67 ff.
Cross-section enlargement   178, 181
Crossed roller bearing   53 f.
Crown arches   134 f.
Curves, driving round   125, 272
Cuttability   239, 242
Cuttability class   207, 220, 222, 235
Cutter changing   32, 34, 42 f.
Cutter development   32
Cutter disc wear   41–49
Cutter head   26 ff.
– functions   26
– revolution speed   25
– rotation   125
Cutter head construction   31
Cutter head perimeter   18
Cutter head shapes   27 ff.
Cutter head shield   22, 152
Cutter head shovels, see buckets   17
Cutter life   32, 43 ff.
Cutter life index (CLI)   216, 218 f.
Cutter loading
– average   32
– critical   38
Cutter pressure, limit   17
Cutter rings, life   41 f.
Cutter spacing   36

Cutter track spacing   17, 36
Cutting angle, influence   33
Cutting tools   32 ff.
– diameter   32

Danger catalogue   105
Danger patterns   107
Dedusting   100 ff.
Design phase   262
Development history   1 ff.
Development tendencies   12
Disc cutter   17, 32 ff.
– method of working   34
Disc cutter   housing   32 f., 49 f., 200
Disc cutter mounting   32 f.
Discs   17
– diameter   32
Documentation   195
Double shield TBM   22
– advantages   23
– disadvantages   23
– main components   22 f.
– working principle   159 f.
Dowelled (Conex) segments   275
Drifting off   124
Drill hole photography   26, 115, 198 f.
Drilling and blasting   14, 99, 144, 191, 223,
   236, 243
Drilling and injection channels   31 f.
Drilling equipment   117 f.
Dust   99 f.
Dust removal   100
– dry dust remover   100, 101
– wet dust remover   100

Earth pressure shield   169
Effective capacity   30
Electric main drive
– with frequency control   51
– with totally enclosed water-cooled motors   51
Electro-hydraulic main drive   51
Employees, protection of   104
Enlargements   181 ff.
– Muelheim underground railway   181
– Nidelbad   junction structure   190
– Sachseln road tunnel   181
– scheme   191
– service tunnels for the Channel tunnel   182
– Vereina tunnel   181
– Zürich-Thalwil railway tunnel   181
– Zürich-Thalwil tunnel   190

Index

EPB shields 168
– scope of application 168
Escape plan 108
ESR 215 f., 218
Evaluated numbers 209
Excavation 108
Excavation class 235 ff., 241
– determination of 240
– in national standards 130
Expanding segments 276 f.

Face 15, 25, 30, 46, 55
Face support 23
Fault zone 18, 198, 203, 314
Filters 101
Fire in the tunnel 108
Force-wear relationship 47
Formula for forecasting tool wear 46
Front bucket slots 30
Front buckets 29 ff.
Front shield 22
Full shields 155
Full-face excavation 235

Gas 197
Gas emissions 109
Gauge buckets 31
Gauge cutter 35
Geological engineering investigations 195 ff.
Geology 196
Geomechanical classification 208
Grill bars 28
Gripper 18
Gripper TBM 20 ff.
– scope of application 19
Grippers 17, 60
Ground conditioning 167
Ground freezing 147, 316
Groundwater displacement 23
Grouting mortar 291
Grouting pressure 291
Guidelines 104, 110, 223

Hexagonal or honeycomb segments 274
High-pressure water jet 35
History of the TBM 1 ff.
Honeycomb segments, see hexagonal segments
Honigmann shaft boring 64
Hydraulic clutches 51
Hydro-geology 197

In-hole hammer drill 115
Inclined shafts 321, 323, 325
– tunnelling 321
Injection channels 31 f.
Injection lances 115
Injections 115, 294
Inner Kelly 67
Inner shell 265 f., 301
Innovation route 12
Integrated safety plan 106 ff.
Intermediate starting points
– Euerwang tunnel 178
– Irlahüll tunnel 178
– NEAT base tunnel, Gotthard 178
– NEAT base tunnel, Lötschberg 178
Intermediate store 82
Invert segments 137
Investigation
– seismic 115
– with boreholes 115
Investigation and improvement of the ground 115
Investigation drillings 198
Investigation headings 179 ff.
– Amberg motorway tunnel 177
– construction 179
– fault zone 184
– Freudenstein tunnel 179
– Gotthard base tunnel 184
– investigation scheme 184
– Kandertal 186, 189
– Lötschberg base tunnel 186
– outside the cross-section 179
– Pfaender tunnel 177
– Piora-Mulde 184
– probe 180
– results 187
– Seikan tunnel 178
– tunnelling scheme 187
– working time 180
Inward displacement 34

Jamming 125
Joint alteration number (Q system) 212, 215
Joint construction 286
Joint detailing 282 ff.
Joint roughness number (Q system) 212, 214
Joint set number (Q system) 212, 214
Joint spacing 41
Joint systems 212, 214

Joint water reduction number (Q system) 212, 216
Jointed rock mass 199 f.
Joints 40, 197

Keystone 271
Keystone installation 272
Kilometres travelled (cutter) 41

Lattice girder 136
Lighting 108
Liner plates 136
Lining 265 ff.
– construction principles 265 ff.
– construction types 267
– double-shell 265 ff.
– shotcrete shells 301
– single-shell 265 ff.
– sizing 265
– with concrete segments 269
Lining systems, yieldable 281 ff.
Longitudinal joints 269, 282 ff.
– with convex-concave contact surfaces 285
– with flat contact surfaces 283
– with tongue and groove 286
– with two convex contact surfaces 284

Machine clamping 203 ff.
Machine support, front 22 ff.
Machine types 20 ff.
Machine, relocation of 17
Main beam 68, 72
Main bearing 53
Main drive 50
– types of 51
Main duct 99
Main dust-removal equipment 92
Main muck transport 64
Management plans 106
Manapouri underwater tunnel 305 ff.
Material flow 29
Material transport 75
Material transport
– at the machine 75
– diesel or electric operation 79
– rail transport 77
– trackless operation 81
Matrix 239
Measures
– material 109
– personnel 109

– structural 109
Mesh erector 118
Micro cracks 200
Micro machines 23, 172
Milling 15
Mini Fullfacer 58, 203
Mini TBM 172
Mobile formwork
– full-section mobile formwork 300
– mobile arch formwork 300
Mobile Miner 58
Mont-Cenis tunnel 1
Muck cars 79
Muck, chip-shaped 36
Muck removal system 18, 163
– rail 18
Muckring conveyor belt arrangement 171

Natural gas 99
Natural gas danger 110 ff.
Negotiation process 249 ff.
Net boring rate 238
Net penetration 17

Occupational safety 103
ÖNORM B 2203 (Austrian standard) 228 f.
Outer Kelly 67
Outer shell 265 f.

Penetration 38 ff.
– limit area 29
Penetration rate 40 f., 216
Penetration tests 40
Permissible workplace concentration (MAK) 100
Pilot headings 178, 180
– advantages 180
– blasting pattern 191
– conventional tunnelling 180
– Milchbuck tunnel 180
– tunnelling concept 190
– Uznaberg tunnel 180, 188 ff.
Pipe jacking 174
– press-boring pipe jacking 174
– shield pipe jacking 174
Preliminary transport 65
Price matrix 242
Probe heading 178
– Tunnel de Raimeux 179
Propulsion cylinders, twisting 74

Index

Q control plan, procedure 245 f.
Quality management (QM) 245, 247
– project-oriented 245

Raise boring 62
Reamer machines 56 ff., *see* Reamer TBM
Reamer TBM (enlargement TBM) 19, 56
Reaming machines 23
Recommendations 196
Reflection seismology 26, 196, 198
Reinforced concrete segment ring 225
Removal of muck
– hydraulic 163
– in the excavation area 29
Ring building 271 ff.
Ring joint level 286
Ring joints 270, 286 ff.
– cam and pocket system 288 f.
– flat ring joint detail 287
– tongue and groove system 287
Risk 107
Risk analysis 107
Risk assignment 246, 248
RMR system 208 ff.
– classification parameters 210
– connection between Q and RMR 220
– correction 211
– for mechanised tunnelling 209
– linkage to ÖNORM B 2203 211
Rock bursts 205
Rock classes
– determination 211 - evaluation 211
Rock conditions, description 197
Rock Mass Rating system, *see* RMR system
Rock quality 209 f.
Rock Quality Designation (Q system) 209
Rock support 117, 129 ff., 239
– in the machine area 133
– influences 205
Rock types 229
– for cyclical and continuous advance 230 ff.
Rockbolts, *see* anchors
Roller assembly 32 f.
Roller bearing 32
Rolling 125 f.
Roof shields 22, 111, 149
RQD, *see* Rock Quality Designation

Safety 103 ff.
– coordinators 104
– excavation classes 205

– integrated 103
– integrated safety plan 106
– occupational 103
– safety aims 107
– safety plan 104, 107
– safety planning 103 ff.
Safety and health plan *(SiGe plan)* 105
San Pellegrino tunnel 320 ff.
Screw conveyor muck removal 165 ff.
Sealing bands 292
Section joints 300
Segment systems
– rhomboidal 275
– trapezoidal 275
Segmental lining 137
– pre-loading 271
Segments 18 f., 111, 137, 236, 269 ff.
– block segments with right-angled plan 271
– cam and pocket segments 288
– cast iron segments 269
– concrete segments 137
– construction types 271
– construction variants 271
– damage 296 ff.
– dimensional tolerances 295
– expanding segments 276
– hexagonal or honeycomb segments 274
– invert segments 18, 137
– production 295
– segment reinforcement 137
– spiral segment 274
– steel fibre concrete 289
– steel segments 269
Shaft boring 63 ff.
– BorPak 66
– drill stem-led 63 ff.
– without drill stem 63 ff.
– without pilot hole 63 ff.
Shaft boring process 62 ff.
Shaft raising equipment 62
Shearing off theory 33
Shield 22
– contra-rotation 125
Shield bearing force 123
Shield cladding friction 123
Shield friction 74 Shield pipe jacking 172
Shield tail 19
Shielded TBM 22
Shields with screw conveyor muck removal 165 f.
– closed mode 167

– EPB mode   169
– machine types   167
– open mode   167
– working principle   165
Shotcrete construction method   177 ff.
Shotcrete in the machine area   19
Shotcrete support   129, 142
– in the backup area   143
– in the machine area   142
SIGMA   218
Single clamping   20
Single shield TBM   22, 155
Sink (downward) boring   61
Site electricity supply   108
Sliding shoe   67
Slurry shield machines   163
– working principle   163
Soil consolidation   31
Soil investigations   26
Special machines for non-circular sections   23
Special processes   177
Special types of TBM   55 ff.
Spiral segment   274
Splitting process of rock   33
Stability index   218
Stackers   82 f.
Stalling moment   52
Standards   223
Steel arch support   133 ff.
Steel fibres   140
Steel ring equipment   118
Steel segments   269
Steering   119 ff.
– gripper TBM with single clamping   119
– gripper TBM with X-type bracing   122
– single shield TBM   122
Stress reduction factor SRF (Q system)   217
Stresses, primary   200
Structural investigations   302
Structural stability   221 f.
Supply trailer   92 f.
Support   22, 108
– final   22, 265
– localised   144
– temporary   19, 22
Support classes   210
Support measures   18, 222, 235 ff.
– evaluation   233
Support of the crown   19
Support system   18, 130

Surveying   126 f.
Surveying errors   125
Surveying systems   126
Synthetic resin injections   147

Tail shield   19
Target chamber   62
TBM
– closed shield   23
– open   19 ff.
– with cutter head shield   22
– with roof shield   22
– with roof shield and side steering shoes   20 f.
– with shield   19, 22
TBM tunnelling
– advantages   15
– disadvantages   15
– system groups   16
TBM with roof shield and side steering shoes   20, 150
TCI cutters   32, 38
Telescopic jacks   22, 159
Telescopic joint   22, 159
Telescopic shield   159
Telescopic shield TBM   22
Tender evaluation   244
Tendering   243 ff., 263
Tendering phase   266
Third band spectrum   113
Thrust   67
– with gripper clamping   67
– with shield TBM   74
Thrust force   67
Thrust jacks, twisting   74
Thrust ring   157
Thrust system   17
Torque   51 ff.
Torque, loss of   29
Trackless operation   81
Trailing buckets   30 f.
Trailing finger roof   111, 150
Train timetable   80
Transfer chutes   18
Transport chutes   31, 75
Transport equipment   108
Tunnel boring machines (TBM)   15
– application possibilities   15
– basic principles   16
– development history   1 ff.
– development tendencies   12 f.

– innovation route   12
– machine systems   20
– shielded   157
– with full-face excavation   20 ff.
– with partial-face excavation   23
Tunnel systems, water draining   271
Tunnel waterproofing
– drained   267 f.
– for tunnel boring machines   222 ff.
– pressure water-resistant   267 f.
Tunnelling classes   227, 235 f., 239
– according to ÖNORM   237
– for shield machines   225 ff.
– matrix   230, 235
– with full-face excavation   226
Tunnelling cost   131 f.
Tunnelling machines   20
Tunnelling works   237
Turbo couplings, hydraulic clutches   51
Two-phase boring   175 f.

Underground ventilation   99 ff., 108

Ventilation   99 ff.
Ventilation schemes   99
Ventilation systems   99
Vertical shafts, Swiss classification   237
Vertical steering   123 f.
Vibration   111 ff.

Walking blade gripper TBM, expanding   153 f.
Waterproofing measures   291
Wear   41 ff., 242
Wear forecast formula   48
Wear index   219
Wear tests   42
Wedge lock system   34
Work accidents   103
Working areas   236 f.

X-type gripping   67 ff.

Yieldable lining systems   281 ff.

Zürich–Thalwil two-track tunnel   317 ff.

BOOK RECOMMENDATION

Deutsche Gesellschaft für Geotechnik e.V. (ed.)
Recommendations on Excavations – EAB
2nd revised edition
2008. Approx 300 pages. Hardcover.
Approx EUR 69.–

ISBN: 978-3-433-01855-2

€ Prices are valid in Germany, exclusively, and subject to alterations. Prices incl. VAT. Books excl. shipping. Journals incl. shipping. 008358016_my

Recommendations on Excavations EAB

The aim of these recommendations is to harmonize and further develop the methods, according to which excavations are prepared, calculated and carried out.
Since 1968, these have been worked out by the TC "Excavations" at the German Geotechnical Society (DGGT) and published since 1980 in four German editions under the name EAB. The recommendations are similar to a set of standards.
They help to simplify analysis of excavation enclosures, to unify load approaches and analysis procedures, to guarantee the stability and serviceability of the excavation structure and its individual components, and to find out an economic design of the excavation structure.
For this new edition, all recommendations have been reworked in accordance with EN 1997-1 (Eurocode 7) and DIN 1054-1. In addition, new recommendations on the use of the modulus of subgrade reaction method and the finite element method (FEM), as well as a new chapter on excavations in soft soils, have been added.

www.ernst-und-sohn.de

Ernst & Sohn Verlag für Architektur und technische Wissenschaften GmbH & Co. KG
Für Bestellungen und Kundenservice: Verlag Wiley-VCH, Boschstraße 12, D-69469 Weinheim
Tel.: +49(0)6201 606-400, Fax: +49(0)6201 606-184, E-Mail: service@wiley-vch.de

Stabilizing, Sealing, Filling – providing optimum safety.

Consolidation Line

WEBAC₀ Chemie GmbH
Fahrenberg 22
22885 Barsbüttel/Hamburg • Germany
Tel.: +49 (0)40 670 57-0
Fax: +49 (0)40 670 32 27
info@webac.de • www.webac.de